AF545209

# Aquatic Biodiversity in India

## The Present Scenario

**The Editors**

**Dr . D.R. Khanna**, Ph.D., FIAES, FASEA, FNC, FZSI, FSES is reader in Department of Zoology and Environmental Science, Gurukula Kangri University, Hardwar (U.A.), India. He is a well known Limnologist and has published about 65 research papers and articles on limnology and aquatic microbiology. He has edited 6 books in the field of Environment and written 5 books in limnology and zoology, and a book of Yajya *"Yajya evam Vayu Pradushan."* Dr. Khanna is an editor of two journals "Environment Conservation Journal" and "Journal of Natural Conservators" and member of editorial board of three journals. He has also organized several National Seminars as organizing secretary/joint organizing secretary. He is awarded by ASEA Excellence Award–2000, and Nature Conservators Impetus Gold Medal–2003.

**Prof. A.K. Chopra**, Head, Department of Zoology and Environmental Sciences, Gurukula Kangri Vishwavidyalaya, Haridwar has taken his M.Sc. and Ph.D. from Lucknow University, Lucknow. His main areas of specialization include Parasitology, Environmental pollution and Environmental microbiology. He started his career in 1977.

He has published over 100 research papers in reputed national and international journals, has written more than 30 articles and has authored 4 books. He has handled four research projects besides participating in various Indian and International conferences. Prof. Chopra has been chairman/ member of various National expert/academic committees. He is on the panel of referee of various national and international journals and has been actively engaged in various extension and field outreach activities.

He is also a fellow member of Indian Academy of Environmental Sciences, Hardwar and Zoological Society of India. In 2001, Prof. Chopra was awarded Indian Academy of Environmental Sciences Gold Medal for Contribution in the field of Life Sciences and Environmental Sciences. He has also been conferred Nature Conservators, **Kautilaya Gold Medal**-2002 for his outstanding contributions in the field of Microbial Parasitology by Nature conservators, India.

**Dr. G. Prasad**, Reader in the Department of Botany and Microbiology, has earned his M.Sc. degree from Kanpur University and Ph.D. degree from Kurukshetra University. He was awarded Senior Research Fellowship and Post-Doctoral Research Fellowship of the CSIR, New Delhi, during the Ph.D.

He has been the Organizing Secretary of National Seminar on Biotechnology: New Trends and Prospects, 1996. He has published nearly 42 research papers in foreign and Indian journals of International repute. He is Coeditor of the Book "Microbes: Agriculture, Industry and Environment, 2000."

# Aquatic Biodiversity in India

## *The Present Scenario*

*— Editors —*

*Dr. D.R. Khanna*

*Dr. A.K. Prasad*

*Dr. G. Prasad*

*— Associate Editors —*

*Dr. R. Bhutiani*

*Dr. R. Rajput*

**2013**

**Daya Publishing House®**

*A Division of*

**Astral International Pvt. Ltd.**

**New Delhi – 110 002**

*First Impression, 2005*
*Reprinted, 2013*
ISBN 978-81-7035-367-6



*Published by* : **Daya Publishing House®**
*A Division of*
**Astral International Pvt. Ltd.**
81, Darya Ganj, Near Hindi Park,
Delhi Medical Association Road,
New Delhi - 110 002
Phone: 011-4354 9197, 2327 8134
Fax: +91-11-2324 3060
E-mail: info@astralint.com
Website: www.astralint.com

*Laser Typesetting* : **Classic Computer Services**
Delhi - 110 035

*Printed at* : **Salasar Imaging Systems**
Delhi - 110 035

PRINTED IN INDIA

# Preface

It is said that there are more species of life on earth than stars in the visible universe. As we know that most of our planet is covered by water, where million of living species are inhibited. Today, in India we are loosing our biodiversity at a greater rate, the reasons are overpopulation, deforestation and pollution. About 55 per cent of Indian fresh water species are threatened. India is facing and alarming danger to the loss of aquatic biodiversity. There is no option except to develop research strategies and public policies, which can help us in conserving the aquatic biodiversity.

In this book an attempt has been made to determine and analyze their conservation measures. The contributors in this book are all eminent scientists of India) together providing vast experience in this area.

We are grateful to the contributors, who have impaired benefits of their research work. Without their support and help, the outcome of this book simple would not have been possible. We are confident that this book will serve the purpose, for which it is published. We express our thanks to all of our contributors and a very special appreciation to Mr. Kapil Mehta, M/s Deepanjali Computer

Graphics, Jwalapur (Hardwar) for typing this manuscript neatly and timely. We are also thankful to Mr. Anil Mittal of M/s Daya Publishing House, New Delhi for the publication of this book.

**Dr. D.R. Khanna**
**Dr. A.K Chopra**
**Dr. G. Prasad**

# Contents


| | | |
|---|---|---|
| | **Preface** | *v* |
| **1.** | **Sustainable Development and Conservation of Fish Genetic Biodiversity of India**<br>*U.K. Sarkar, D. Kapoor and R. Dayal* | **1** |
| **2.** | **Biodiversity of Cyanophytes and Bacteria Associated with Nitrogen Cycling in the Marine Environment**<br>*P.K. Pandey and C.S. Purushothaman* | **18** |
| **3.** | **Ornamental Fish Biodiversity of India**<br>*Archana Sinha* | **38** |
| **4.** | **Biodiversity of Aquatic Fauna of Mizoram: The Present Scenario**<br>*S.N.Ramanujam* | **61** |

5. **Resource Assessment and Potential of Hill Fisheries in Garhwal Himalayan Region of Uttaranchal: A Perspective** 81

*N.K. Agarwal, D.R. Khanna, B.L. Thapliyal and U.S. Rawat*

6. **Present Status of Biodiversity and Strategies for Development of Wetlands in South Konkan in India** 98

*S.G. Yeragi and S.S. Yeragi*

7. **Algal Communities in Papnash Pond Bidar Karnataka, India** 107

*N. Shiddamallayya, S.B. Angadi and P.C. Patil*

8. **Intertidal Region and Biodiversity** 123

*Amita Saxena*

9. **Fish Fauna of the District Udham Singh Nagar with Some Recommendations for Improvement of Fisheries** 146

*S.P. Badola, Smita Badola and D.R. Khanna*

10. **Role of Water Hyacinth and Other Aquatic Weeds in Wastewater Treatment: An Eco-friendly Approach Having Vast Potential** 154

*V. Singhal and A. Kumar*

11. **Biotic Communities in Aquatic System** 184

*N. Rai, S.A. Ansari, S. Chauhan, Mukesh Ruhela and R. Bhutiani*

12. **Aquatic Diversity in Jharkhand with Reference to Macro Zoobenthos** 202

*Arvind Kumar and Chandan Bohra*

13. **Study of Cultural Eutrophication in Relation to Plant Diversity of Wetland: Ratheshwar in Central Gujarat** 243

*Nirmal Kumar, J.I, Rita N. Kumar and Ira Bhatt*

14. Status of Aquatic Biodiversity of Bhimtal Lake in Kumaon Region (Uttaranchal) 285
*Davendra S. Malik*

15. Effect of Pulp and Paper Mill's Effluent on Microbial Diversity of River Gola 297
*Anil Kumar, Rajeev Rajput, R. Bhutiani, Mukesh Ruhela and V. Singhal*

16. Latitudinal Variation in Phytoplankton Assemblage 304
*A. Wanganeo, S. Gagroo, A.R. Yousuf and R. Wanganeo*

17. Studies on Biotic Diversity of Macro Invertebrates in Bihar River 328
*U. Awasthi and Ashok Awasthi*

18. Coastal Biodiversity of Maharashtra: An Insight 332
*G.N. Kulkarni*

19. Biodiversity: A Present Scenario 339
*Pramod Kumar Joshi*

20. Biodiversity and Conservation of Algae 343
*Narendra Mohan and Jitendra Mohan*

21. Similarity Index of Phytoplankton Species of River Kunda at Khargone, Madhya Pradesh 350
*S.K. Mahajan*

22 Limnological Assessment of Phytoplankton in a Reservoir of Khargone, Madhya Pradesh 354
*S.K. Mahajan*

Index 359

**Aquatic Biodiversity in India: The Present Scenario, 2005** **1–17**
*Edited by*: D.R. Khanna, A.K. Chopra & G. Prasad
*Published by*: Daya Publishing House, New Delhi

# 1

# Sustainable Development and Conservation of Fish Genetic Biodiversity of India

**U.K. Sarkar, D. Kapoor and R. Dayal**
*National Bureau of Fish Genetic Resources, Canal Ring Road, P.O. Dilkusha, Lucknow–226002, U.P.*

India is well known for its mega-biodiversity of biological wealth, harboring over 12 per cent of the shell and fin fishes known. The fisheries sector of India has grown very rapidly in terms of production contribution to GNP as well as foreign exchange in spite of low investments. These resources are still being exploited from wild whereas other resources are largely exploited through cultivated germplasm. As a result the rate of aquatic habitat loss and degradation has been very high. Rapid urbanization, rural and industrial development after the independence resulted a severe deforestation and shrinkage of forest cover, embankments of rivers, extraction of river water for irrigation, siltation of river beds, industrial and domestic water pollution, heavy use of insecticides and pesticides in agriculture, heavy exploitation of fishes using small mesh size nets, chemicals, poisons and explosives has drastically reduced the fish population in natural waters. The condition has become so severe that it has deteriorated the breeding and feeding grounds of fishes. The population of all the fishes in general and in some particular fishes have greatly declined that some of them have completely vanished from some geographical areas so that a noticeable number of fishes have become endangered (EN) in some of the areas where it was found in large number and big sizes in the yester years. The condition has become so severe that it draws the attention of the scientist, researchers and conservationist to work on their conservation and rehabilitation for the survival of fishery resources. Recent report of existing freshwater resources by "Pilot Analysis of Global Ecosystems (PAGE) of World Resource Institute, USA reveals that more than 20 per cent of the world's known 10,000 freshwater fish species have become extinct, been threatened, or endangered in recent decades. In the United States, which has the most comprehensive data on freshwater species, 37 per cent of freshwater fish species, 67 per cent of mussels, 51 per cent of crayfish and 40 per cent of amphibians are threatened or have become extinct (WRI Report, 2000).

Out of about 24,600 fishes of the world (Nelson, 1994), nearly 11 per cent have so far been recorded from India by National Bureau of Fish Genetic Resources (NBFGR), Lucknow, India. Though Indian Fisheries Act of 1879 is a landmark in the conservation of fishes of India no remarkable impact in this regard has yet to be established. In order to manage aquatic genetic resources research input with well defined policies are required (Bartley and Pullin, 1999). In many countries the existing policies for conservation and sustainable utilisation of the aquatic genetic resources are poorly developed (Harvey *et al.*, 1998). Research and policy requirements for conservation of fish germplasm resources are basically based on other resources; however, conservation of fish germplasm resources must also take into consideration the distinct and diverse nature of water resources (Ponniah, 2001). The conservation of fish germplasm resources should be considered as an integrated activity of culture and capture based fisheries and there is need to compete with many other human activities for water, which is increasingly becoming scarce. Inland fisheries and conservation of aquatic species are not adequately taken into consideration, when planning any development project related to water resources. Taking these in to considerations research and policy requirements for sustainable utilisation of waterbodies and fish germplasm resources are discussed in the present communication.

## Aquatic Habitat Potential

India is endowed with vast natural habitats, including snow covered Himalayas, the Indo Gangetic plains, the Deccan plateau, deserts of Rajasthan, coastal regions and the vast seas. Such areas support a broad extent of water resources including cold, warm, brackish and marine waters inhabiting varied types of fishes. The potential inland and marine resource potential of India is presented in Table 1.1.

**Table 1.1: Diversity of Fisheries Resources of India**

| *Resource* | *Area* |
|---|---|
| Freshwater Ponds and tanks | **2.254 ha** |
| Beels, oxbow lakes and swamps | **1.3 m ha** |
| Reservoirs | **2.90 m ha** |
| Rivers and canals | **1,73,287 Km** |
| Brackishwater | **1.235 m ha** |
| Marine | **8,129 Km** |

(*Source*: Report of the working group for ninth five year plan for fisheries, Ministry of Agriculture, Govt. of India, April 1996).

## Cataloguing of Fish Genetic Resources

Knowledge of species and communities can reveal crucial facts necessary to the management of ecosystems and habitats as well as to the identification of important genomes and genes. Identification, listing and prioritization of species are one of the important tasks in conservation. The Convention on Biological Diversity (CBD) has made it obligatory on part of India to document our resources. For equitable sharing of benefits derived from biodiversity, a database is essential. Also NBFGR's mandate reflects its scientific commitment to the cause of fish diversity which includes collection, classification and evaluation of information on fish genetic resources of the country.

The review of literature indicate that a comprehensive and authoritative account of Indian fishes was worked out by Francis

Day in 1889 for the first time. Later on over the next 100 years, a number of researchers and taxonomists *viz.* (Annandale,1922; Mukherji,1931; Hora,1921a, 1936, 1936a, 1937a; Mishra,1976a, 1976b; Silas,1958, 1959; Menon,1974, 1987; Jayaram,1976,1991; Talwar and Jhingran,1991) and many other have reported the fishes from various ecosystem of India. Some of the compilations are of Francis Day who reported 1340 freshwater fishes, Jayaram (1991) reported 742 fresh and brackish water fishes, Talwar and Jhingran (1991) compiled 930 fresh and brackish water fishes. Menon (1974) have published a list of 380 fishes from river Ganga. CMFRI, Cochin (1983) have published a list of 698 marine fin fishes while Talwar and Kacker (1984) reported 548 marine fishes.

In order to fulfill the mandate of the Bureau,NFBGR has made a comprehensive and detailed database of 2118 finfish species found in various ecosystems (Kapoor *et al.,* 2002). This contains information on systematic, habitat and distribution of 2118 finfishes. On the basis of recent publications, reports and proceedings of the workshops the database entitled "Fish Biodiversity of India" was further checked by deleting synonyms and adding more species. As many as 2118 finfish species belonging to 36 orders, 55 sub-orders and 209 families, inhabiting different ecosystems, have been compiled. Some of the species are common to different ecosystems. This book will serve as reference inventory of the fishes inhabiting different ecosystems of the Indian subcontinent. The database has been customized to make them more user friendly and the format provides comprehensive and illustrative output with a provision to append and insert additional parameters (Fig. 1.1). The compilation has been made after consulting many published records and eliminating synonyms. However, this number does not match with many others; for example, the total number of Indian fish species recorded in the IUCN database is 1,424 only (Fish Base, 1998). NBFGR's database on native finfish resources of the country indicate that 2,163 finfish species are distributed in different ecosystems Recent upgradation (2004) of existing fish database of NBFGR contains total 2163 fish out of which 157 belongs to coldwater, 454 freshwater, 182 brakishwater and 1370 marine water (Table 1.2). Recently 21 new species have been reported by the NBFGR–NATP programme from western ghats and north east and these will be added in the NBFGR data base (Anon, 2003-2004). Since most of the surveys are based on fisherman's catch or have been carried out

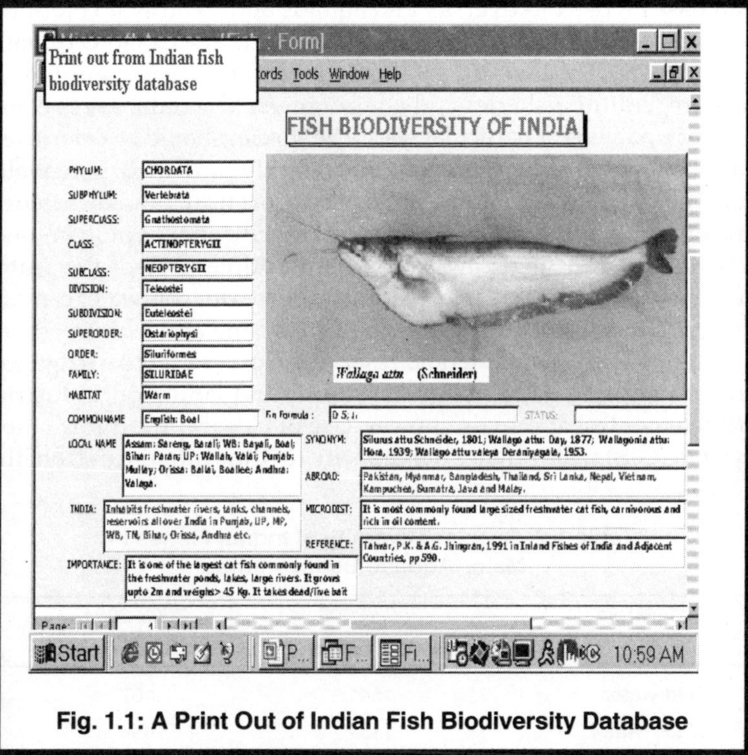


**Fig. 1.1: A Print Out of Indian Fish Biodiversity Database**

with nets normally used for fishing, thus providing a distorted picture of available biodiversity. Another lacuna observed is that most workers, who have carried out biodiversity survey, have only listed the species but not their abundance. Therefore, the research requirements under cataloguing of aquatic resources are a renewed focus on biodiversity survey and taxonomy. As a part of the water bodies database, the digital base map of India showing about 330 rivers and 20 lakes have been extracted from the survey of India map hill ranges and rivers of India at 1: 5 million scale.

Development of a detailed micro level plan for conserving the country's biological diversity is an obligation on part of India under Convention of Biological Diversity (CBD). Also for equitable sharing of benefits derived from biodiversity, fish biodiversity database is essential. NBFGR's efforts have been made to develop an integrated database on fish biodiversity and aquatic bodies on a GIS platform with capacity to add data on environment and socio-economic parameters. This would facilitate integrating programmes in capture, culture fisheries and environment. The database is being further updated for particular uses. The species should be prioritized based on those supporting a fishery, presently cultivated, potentially cultivable, potential ornamental fishes, sport fishes and threatened and endemic species. The next step of prioritization would involve prioritizing and finalizing the single multiple uses of the water bodies as per (a) utilization of natural resources for coldwater capture fishery, sport fishery, culture based fishery and aquaculture, (b) as hill stream fish sanctuaries and reserves (c) other uses like irrigation and drainage, flood control water supply and sanitation, industrial effluent control and hydro power. Use of Geographic Information System (GIS) will enhance the capacity to draw inferences from the database.

**Table 1.2: Species Diversity of Indian Fishes in Major Aquatic Ecosystems**

| *Ecosystem* | *2002-2003* *Fish species (No.)* | *2003-2004* *Fish Species (No.)* |
|---|---|---|
| Cold water | 154* | 157 |
| Warm water | 433# | 454 |
| Brackish water | 171** | 182 |
| Marine | 1360 | 1370 |
| Total | 2118 | 2163 |

* Of these, 34 fishes are common to cold and warm water.

\# Of these, 67 fishes are common to warm and brackish water.

** Of these, 16 fishes are found only in brackish water, 73 are common to warm, brackish and marine water and another 82 are common to brackish and marine water.

## Conservation Status

The preliminary assessment of NBFGR (1992-93) identified 4 endangered, 21 vulnerable, 2 rare and 52 indeterminate fishes covering different ecosystem of the country. Out of 154 cold water fishes 17 fishes were identified as threatened in which 1 categorised as endangered, 4 as vulnerable and 12 as indeterminate (Table 1.2). The NBFGR's NATP programme "Germplasm inventory, evaluation and gene banking of fresh water fishes" with local collaborators is currently generating data from North East and Western Ghats would help in both conservation as well as sustainable commercial utilization of the prioritised food and ornamental species.

**Table 1.3: Threatened fishes of India as per NBFGR**

| | *Endangered* | *Vulnerable* | *Rare* | *Indeterminate* | *Total* |
|---|---|---|---|---|---|
| Freshwater | 3 | 13 | 2 | 28 | 46 |
| Cold water | 1 | 4 | — | 12 | 17 |
| Brackish water | — | 2 | — | 4 | 6 |
| Marine | — | 2 | — | 8 | 10 |
| Total | **4** | **21** | **2** | **52** | **79** |

## Habitat Assessment and Life History Studies

The review of literature indicates that a very little information is presently available on the utilization of stream microhabitat by different life stage of cold-water fishes. The fish use a variety of microhabitats during different developmental stages of their life cycle. Many of the stream fish require specific substrate for their spawning while a good number of cold-water fish shows hiding behaviour under shade. Recently, NBFGR have arranged habitat arrangement data of endangered golden mahseer *Tor putitora* on a geographical information system (GIS) in which the microhabitat of Ladhiya and Kosi river in the Kumaun Himalaya have been critically assessed. This will greatly help in planning macro and microlevel location specific conservation programme. In Bureau's programme on *in situ* conservation, greater focus has been given to understanding habitat requirements of endangered species. Research should be focused on fish life history strategies in relation to microhabitat features.

## Restoration of Endangered Fish Species

Determination of species endangered status of identified fish is very much for taking conservation. In the absence of historical data on its abundance, the only alternative is to draw inferences from the decrease in distribution to fix its endangered status. There is a need to distinguish between local endangerment and species endangerment. There is a possibility that a species might support a good fishery in one location and may be endangered in another location. This is more applicable for large countries like India and for species with wide distribution. In case of local endangerment, a prioritization exercise has to be made to determine whether the life history traits (*i.e.* ecological) and genetical (*i.e.* evolutionary) of the endangered stocks sufficiently differ form those of other population to warrant conservation strategies. There is increasing evidence that intra species variation in life history traits has a strong genetic basis (Parsons, 1997).

Maintenance of a captive population of the endangered species should be attempted, only if there is a reason to believe that the endangered species will become extinct in the wild if not held in captivity. Successful live gene bank programmes are those being operated by the Dexter Endangered Fish Farm for many North American endangered fishes Johnson and Jenson (1991), the Norwegian programme for Atlantic salmon (Walso, 1998). All the above-mentioned programmes have the advantage of being linked either to an active programme of repopulating the natural environment or with ongoing aquaculture activities. In the Indian context live gene banks should be developed only as part of an immediate end objective like rehabiliting the natural ecosystem.Cryo gene banks can serve as safeguards against species extinction and can be utilized in genetic upgradation and ranching programmes. (Ponniah, 1999). NBFGR has presently perfected cryopreservation protocol for 12 endangered/cultivable species and cryopreserved milt of some of the species are being maintained in mini gene bank. Under the recently initiated National Agricultural Technology Project (NATP) the gene bank activities are further intensified to include all prioritized species from western ghats and north east India.

## Sustainable Fisheries Resources Management

According to CGIAR (1998), sustainable agriculture is the successful management of resources for agriculture to satisfy the changing human needs, while maintaining or enhancing the quality of the environment and consigning the natural resources. Fishing adversely affects the natural stocks in many ways. Developing management measures to mitigate these effects would require an awareness, quantification of impact planning management measures. In India, awareness about many of the adverse effects of fishing on biodiversity is low. Greater attention will need to be given in aquaculture to management for sustainability and a system approach has been suggested by (Muir,1996). In India, ornamental fishes are being collected from the freshwaters of western ghats and north east India. Some of these species are endangered Ponniah and Gopalakrisnan (2000) and such collection can lead to their further endangerment. Research and policy support for commercial sustainable utilization of the resources is required. There is also need to safeguard against bio-piracy since freshwater fishes are traded by ornamental industry of many developing countries. In Iceland, the ranching programme for salmon was initiated in the mid eighties and the proportion of ranched salmon in the Icelandic salmon catch increased from less than 20 per cent in eighties to more than 80 per cent in the early nineties (Isaksson, 1998). Codes of practices for transfer of stocks have been drawn for all enhancement programmes (ICES, 1996). In the coldwater habitat, economic sustainability of such stocking practice and their long term impact need to be studied.

## Management of Water Resources on Community Basis

Understanding the existing socio-economic conditions and ensuring the participation of the local community are important prerequisites for better sustainable utilization of natural resources. The participatory management of reservoir fishery in Brazil self-imposed non-fishing periods of protected areas and seasonal ban on certain nets result increased income for all (Christensen *et al.*, 1995). In traditional, community-managed systems there are often unwritten and informal management and conservation guidelines enforced by the community. The 'Sustainable Livelihood Approach' Maine and Coutts (2000), which is linked to the Code of Conduct for responsible Fisheries (CCRF) and which puts people at the centre of development should be incorporated into community based management of aquatic resources. There should be a strong research support for community based management so that management action can be guided and their impact demonstrated. Public perceptions about conservation issue influence to a large extent directly or indirectly policy formulations and implementation. Many conservation actions like area closure or limiting fishing or banning exotic fish are seen by the fisherman or fish farmers as working against their economic interests. Peoples awareness should be one of the main focus to ensure public acceptance and cooperation for such conservation programmes.

## Creating Fish Sanctuaries and Protected Area

The conservation of threatened fish through creation of endangered fish sanctuaries is well developed in countries like USA (Miller and Campbell, 1994; Pearse, 1998).Unlike sanctuaries for terrestrial wildlife, the available sanctuaries for aquatic resources are limited and creation of sanctuaries of protected areas is need of the hour (Menon and Pillai,1996, Musick *et al.*,2000, Ponniah *et al.*, 2002). Even within wildlife sanctuaries, there is less protection for fishes and no scientific management practices are undertaken. The management of marine protected areas (MPAs) has evolved mainly based on terrestrial protected areas. The appropriateness of the legal and institutional mechanisms in support of MPAs based on terrestrial models have been questioned (Leria, 1996). However, aquatic sanctuaries not only help in conserving biodiversity but also rejuvenate fisheries. Marine coral reserves provide protection to spawners ensuring recruitment. Protected aquatic areas have helped to restore numbers of exploited species (Bohnsack, 1993). The sanctuary concept for freshwater fishes is yet to get policy and research support in India. Recently, the status of fish germplasm resources in four water bodies under wildlife protected area in the North East and northern region has been studied by NBFGR, Lucknow through its collaborative network. The studies indicated that sanctuaries can play significant role in maintaining or enhancing recruitment in unprotected areas. The other important observation was that many fishes which have become endangered in other places were available in these waterbodies indicating that this can serve as potential aquatic sanctuary.

## Regulation of Exotic Species Introduction

Exotic species are known to enhance aquaculture production. However, it is essential that before any new exotic is introduced, an evaluation is carried out to assess its capacity to establish itself in nature. Silver carp, which was introduced for polyculture along with Indian major carps, did not establish itself in many parts of India, but it was able to establish in Govindsagar. Therefore, in a large country like India with a wide range of climatic conditions, there is a need for clearance of exotics, as per location. The white spot disease syndrome, which had very adverse effect on Indian shrimp farming, has been attributed to use of imported diseased hatchery seed. Now even wild brood stock are testing positive for the white spot disease indicating its spread from local hatcheries to the wild. Definite policy guidelines including legislation have to be framed to safeguards against this danger. In India the following exotics have been introduced in different strata of uplands waters: *Salmo truto fario* (Brown trout), *S. gairdnery gairdneri* (Rainbow trout), *Cyprinus carpio* var. *specularis* (Miror carp), *C. carpio nudus* (Leather carp), *Ctenopharyngodon idella* (Grass carp), and *Carassius auratus* (Gold fish).

## Potential Species for Breeding, Aquaculture and Biotechnology

The hill stream aquaculture practices may be intensified by utilizing the water bodies at lower order streams and in cold water lakes. Though species diversity of fish is very high, the number of species utilized in culture is limited even worldwide. In India efforts made for the breeding and aquaculture of potential cold water fish have been very encouraging for fishes like *Tor putitora, Tor khudree, Schizothorax* sp., *Labeo dyocheilus* and *Labeo dero*. Successful breeding and hatching of *L. dyocheilus* has been done in the natural conditions of Kosi river (Sarkar *et al.*, 2001). However many more fishes may be brought under cold water aquaculture and ornamental fish farming. In North East India and Western Ghats, there are as many potential cultivable and ornamental species with high consumer preference (Ponniah and Sarkar, 2000; Ponniah and Gopalakrishnan, 2000). Recently, 19 prioritized threatened food fishes and 26 ornamental fishes were bred in the captivity under NBFGR.NATP project. The development of new candidate species for aquaculture and ornamental trade should follow the emphasis, as given in the new Agricultural Policy of the government 'to evolving new location-specific and economically viable improved varieties of agricultural and horticultural crops, livestock species and aquaculture as also conservation and judicious use of germplasm and other biodiversity resources'.

From the conservation point of view, two major issues in aquatic biotechnology need to be tackled. One relates to the utilization of unique genes found in aquatic organisms and the other relates to the effect genetically modified organisms (GMO) on natural stocks (Ponniah, 2001). At present the number of useful genes isolated from aquatic organisms are limited mostly to growth hormone and anti-freeze genes, However, in future, more genes may be isolated from aquatic organisms. Due to their genetic and physiological diversity, aquatic organisms especially from the marine environment would be the source of unique genetic information for new classes of chemicals, which can be utilized in a variety of fields (Ponniah, 2001). Gene banks and tissue repositories are on mechanism to ensure that these materials are available for future biotechnology research programmes.

## Conservation of Aquatic Biodiversity

The conservation of fish genetic resources of the country can be possible with concerted efforts. Taking into consideration the various thrust areas of conservation, the required integration with capture, culture fisheries and environment programmes, and the need for increased research focus, the following action points related to research and policy matters for conservation of coldwater fish genetic resources have been summarized and presented.

1. In order to conserve all aquatic flora and fauna, prioritization of species and particular habitat of the water bodies (streams/rivers, lakes and reservoirs) is important
2. Intensification of soil and water conservation measures, habitat restoration programmes and afforestation
3. There is a need to invest more on research and capacity building in the field of taxonomy, habitat morphology, ecology and life history strategies of fishes
4. The research should cover the temporal and spatial scales adequately with focus on critical early life stages as well as identify genetic units
5. Seed ranching of the endangered fishes and assessment of their performance.
6. Policy should be framed to support development of sustainable farming practices to minimize the impact of farm on natural biodiversity.

7. Research and policy support should be given for domestication of potential cultivable and ornamental fishes as well as for developing biotechnological and pharmaceutical products from aquatic organisms.
8. Peoples awareness on the long term impact of unsustainable culture and capture fishery practices has to be increased. Government supported fisheries cooperative societies can be established.
9. The national water policy being framed by National Water Resource Council (NWRC) under Water Resources Ministry should take into consideration the requirement of fisheries sector and conservation of aquatic resources
10. There is urgent need to establish a network of research institute/local college, NGOs and local communities so that species and location specific research on sustainable utilization aquatic resources can be taken up
11. Legislation covering utilization of water bodies and other aquatic resources should be formulated
12. The fields of aquatic taxonomy, ecology and biology are being neglected by both scientists and funding organizations. Unless input for these fields is coupled with other fields like molecular biology, the science of conservation biology required for sustainable management of our resources would not develop.

## References

Annandale, N. (1922). Fish and fishing in the Inle Lake. *J Bombay Nat. Hist. Soc.* 28 (4): 1038–1044.

Anon (2003-2004). National Agricultural Technology Project annual report, National Bureau of Fish Genetic Resources, Lucknow.

Bartley, D. M. and R. S. V. Pullin (1999). Towards policies for aquatic genetic resources. In Pullin, R. S. V., D. M. Bartley and 1. Kooiman (eds)., Towards policies for conservation and sustainable use of aquatic genetic resources. ICLARM Conf. Proc. 59: 1-16.

Bohnsack, J. A. (1993). Marine reserves–they enhance fisheries, reduce conflicts and protect resources. Oceanus, 36: 63-72.

Christensen, V. and D. Pauly (1995). Fish production, catches and the carrying capacity of the world oceans. Naga, ICLARM 18: 34-40.

CMFRI (Central Marine Fisheries Research Institute. (1983). A code list of common marine living resources of Indian sea. CMFRI supplementary publ.

Day, F. (1889). The fauna of British India, including Ceylon and Burma. Fishes. Vols. 1 and 2, 548 and 509pp. Fish Base (1998). Fish Base 98, CDROM, ICLARM, Manila.

Harvey, B., Ross, C. Greer D. and J. Carolsfeld (1998). Action before extinction: an international conference on conservation offish genetic diversity. Publisher World Fisheries Trust. 259 pp.

Hora, S. L. (1921a). Fish and fisheries of Manipur with some observations on those of the Naga Hills. Rec. Indian Mus., 22 (3): 165-214.

Hora, S. L. (1936). Siluroid fishes of India, Burma and Cylon. 2. Fishes of the genus *Akysis* Bleeker. 3. Fishes of the genus *Olyn* Mellanland. 4. On the use of the generic name *Wallago* Bleeker. 5. Fishes of genus *Heteropneustus* Muller. Rec. Indian Mus., 38 (2): 199-209.

Hora, S. L. (1936a). Notes on fishes in the Indian museum. XXVI. On a small collection of fish from the Chitaldurg district. Mysure, Rec. Indian Mus., 38 (1): 1-7.

Hora, S. L. (1937a). Notes on fishes in the Indian museum. XXXI. On a collection of fish from Nepal. Rec. Indian Mus., 39 (1): 43-46.

ICES. (1996). Report of the working group on introductions and transfer of marine organisms. ICES C. M. 1996/Env: 8.

Isaksson, A. (1998). The status of Icelandic salmonid resources, with special reference to genetic conservation policy. pp 115–127. In Harvey, B., Ross, C. Greer D. and J. Carolsfeld (eds)., Action before extinction: an international conference on conservation of fish genetic diversity. Publisher World Fisheries Trust. 259 pp.

Jayaram, K. C. (1976). Contributions to the study of bagrid fishes. 13. Interrelationships of Indo African catfishes of the family bagridae. Matsya. 2: 47-53.

Jayaram, K. C. (1991). Revision of the genus *Puntius* (Hamilton) from the Indian region (Pisces: Cypriniformes, Cyprinidae, Cyprininnae). Rec. Zool. Survey India. Oce. Pap. 135-178pp.

Johnson, E. J. and B. L. Jenson (1991). Hatcheries for endangered freshwater fishes. In Battle against extinction W. L. Minckley and J. Deacon (eds). . Univ. Arizona Press. 199-217.

Kapoor, D, R. Dayal and A. G. Ponniah (2002), Fish Biodiversity of India, NBFGR Publication 1–710 pp.

Leria, C. (1996). Marine parks and reserves: a brief legal overview. Report of the Regional Workshop on Fisheries Monitoring, Control and Surveillance, 16-20 December 1996. Albion, Mauritius, F AO, Rome.

Maine, R. A. and R. Coutts (2000). Linking of Code of Conduct and Sustainable Livelihoods. In Bay of Bengal News, March, 2000: 24-27.

Menon, A. G. K. (1974). A check list of fishes of the Himalayan and the Indo-Gangetic plains. Indian Fisheries Society of India. Sp. Publ. 1. 136pp.

Menon, A. G. K. (1987). The fauna of India and adjacent countries. Pisces Vol. IV. Teleostei-Cobitoidea. Part I. Homalopteridnae. Zoological Survey of India, Calcutta. 259p.

Menon, N. G. and C. S. G. Pillai (1996). Marine Biodiversity Conservation and Management. Central Marine Fisheries Research Institute, Cochin.

Miller, J. A and C. E. Campbell. (1994). A marine geographic information system for the channel Islands National Marine Sanctuary. In: The fourth California islands symposium Update on the status resources (eds. W. L. Halvorson and G. J. Maender) Santa Barbara museum of natural history, pp135-139.

Misra, K. S. (1976a). The fauna of India and adjacent countries. Pisces (Second edition). 2. Telestomi: Clupeiformes, Scopeiformes and Ateleipiformes. Delhi. xxvii+438pp.

Misra, K. S. (1976b). The fauna of India and adjacent countries. Pisces (Second edition). 3. Teleostomi: Cypriniformes, Siluri. Xxi+387pp.

Mukherjee, D. D. (1931). On a small collection of fish from the Bhavani river (S. India). *J. Bombay Nat. Hist. Soc.* 35 (1): 162-171.

Musick, J. A., S. A. Berkelet, G.M. Cailliet, M. Camhi, G. Huntsman, M. Nammack and M. L. Warren Jr. (2000). Protection of marine fish stocks at risk of extinction. Fisheries (Bethesda), 25 (3): 6-8.

Muir, J. F. (1996). A system approach to aquaculture and environment management, P. 19-49. In D. Baird et al. (eds) Aquaculture and water resource management. Oxford, Black well Science. 219 P.

Nelson, J. S. (1994). Fishes of the world. 3rd Ed. New york. John Wiley and Sons. Inc. 600 pp.

Parson, K. E. (1997). Contrasting patterns of heritable geographic variation in shell morphology and growth potential in the marine gastropod *Bombicium vitatum:* evidence from field experiments. Evolution. 51: 784-796.

Pearse, J. S. (1999). Biodiversity of the rocky intertidal zone in the Monterey bay National Marine Sanctuary: A 24 year comparison. California sea grant coll. programme, La Jolla, USA, pp57-62

Ponniah, A. G. (1999). An integrated approach to gene banking of India's fish germplasm resources. In Harvey, B., Ross, C. Greer D. and J. Carolsfeld (eds)., Action before extinction: an international conference on conservation offish genetic diversity. Publisher World Fisheries Trust, 129-140 pp.

Ponniah, A. G. and A. Gopalakrishnan (2000). Fish biodiversity of Western Ghats. NBFGR's NATP special publication no. 1. National Bureau of Fish Genetic Resources, Lucknow. Pp 1-228

Ponniah, A. G. and U. K. Sarkar. (2000). Fish biodiversity of north east of India. NBFGR's NATP special publication no. 2. National Bureau of Fish Genetic Resources, Lucknow. Pp 1-347

Ponniah, A. G. (2001). Research and policy requirements for sustainable utilization of India's fish germplasm resources. In sustainable Indian fisheries, Pandian, T. J. (eds) National academy of agricultural sciences, New Delhi. pp. 213-224.

Ponniah, A. G, U. K. Sarkar and A. K. Pathak (2002). Need and Approach for Creation of Freshwater Aquatic Sanctuary to Save Endangered Fish Genetic Diversity in India, NAGA (in press).

Pullin, R. S. V., D. M. Bartley and J. Kooiman (eds). . (1999). Towards policies for conservation and sustainable use of aquatic genetic resources. ICLARM Conf. Proc. 59: 277 pp.

Sarkar, U. K., R. S. Patiyal and S. M. Srivastava (2001). Successful induced spawning and hatching of hill stream carp *Labeo dyocheilus* in Kosi river, Indian J Fish 48 (4): 413-416.

Sehgal, K. L., J. P. Shukla and K. L. Shah (1971). Observation on fishery of Kangra Valley and adjacent areas with special reference to mahseer and other indigenous fishes. *J. Inland Fish. Soc. India*, 3: 63-71.

Silas, E. G. (1958). Studies on the cyprinid fishes of the Oriental genus Chela Hamilton. J. Bombay Nat. Hist. Soc., 55 (1): 54-99.

Silas, E. G. (1959). On the natural distribution of the Indian cyprinodont fish Horaichthys setnai Kulkarni. J. Mar. Biol. Assn. Indian 1 (2): 256.

Talwar, P. K. and A. G. Jhingran. (1991). Inland Fishes of India and Adjacent Countries, Vol. 1 and 2. Oxford and IBH Publishing Co. Ltd. 1062pp.

Talwar, P. K. and R. K. Kacker (1994). Commercial Sea Fishes of India. Hand book Zoological Survey of India, Calcutta 997 pp.

Walso, Q. (1998). The Norwegian gene bank program for Atlantic salmon (*Salmo safar*). 97–103 pp. *In:* Harvey, B., Ross, C. Greer D. and J. Carolsfeld (eds). Action before extinction: an international conference on conservation of fish genetic diversity. Publisher World Fisheries Trust. 259 pp.

WRI (World Resource Institute) (2000). Report on "*Pilot Analysis of Global Ecosystems (PAGE): Freshwater Systems*, Washington, DC 20002 USA.

**Aquatic Biodiversity in India: The Present Scenario, 2005** **18–37**
*Edited by*: D.R. Khanna, A.K. Chopra & G. Prasad
*Published by*: Daya Publishing House, New Delhi

# 2

# Biodiversity of Cyanophytes and Bacteria Associated with Nitrogen Cycling in the Marine Environment

**P.K. Pandey and C.S. Purushothaman**
*Central Institute of Fisheries Education, ICAR, Versova, Mumbai–400061*

India is the home of a rich microbial diversity with many of the forms not found anywhere else in the world. Marine bacteria and other microbes form the base of the ocean's food chain and play an important role in the planet's nutrient cycles, but scientists know relatively little about their life cycles. The relatively new field, "ecological genomics" may be able to provide more information on how the organisms' genes work to keep them functioning. The oceans are inhabited by a vast number of bacteria and other microbes, many of which have been difficult to grow under laboratory conditions.

The recent researches have brought about the discovery of several new links that have ultimately changed our understanding of the components of the marine nitrogen cycle. Much of our basic information on the nitrogen cycle is derived from measurements of transformation rates or from experiments with cultivated isolates. Generalization from the behaviour and physiology of cultivated isolates can be misleading, since it appears that many marine microorganisms *in situ* are yet to be obtained in culture (Giovannoni and Rappe, 2000). The rapid increase in the knowledge of genes and molecular biology has had an enormous impact on our understanding of the nitrogen cycle by making it possible to study the ecological underpinnings and diversity of microorganisms involved in specific nitrogen cycle components. Genetic and biochemical investigations have also changed our understanding of processes such as nitrification and denitrification, which were thought to be restricted to very specific habitats and microbes in fact, are more widely distributed.

Oxidation-reduction reactions are involved in nitrogen cycle, many of which are used in the energy metabolism of microbes. Specific enzymes catalyze many of these reactions, and the enzymes and genes are useful targets for studying microbial processes such as assimilatory nitrate reduction, dissimilatory nitrate reduction and nitrogen fixation. Knowledge of the genes encoding enzymes involved in biogeochemical transformations provides useful tools not only for assaying gene expression, but also for determining the diversity of microorganisms involved in specific nitrogen cycle transformations. We still know very little about the link between the ecology of the nitrogen cycle, and the redundancy of microbes and genes in the marine environment. Understanding these links is important for determining the role of microbial diversity in ecosystem processes and the sensitivity of the environment to perturbations.

## Utilization of Nitrogen

General assumption is that most microorganisms use inorganic nitrogen in the form of nitrate, nitrite and ammonium ions. Particularly in the oligotrophic open-ocean gyres, low concentrations of these compounds can limit the rate of productivity in the surface layer (0 to 200 m depth), but nitrogen can regulate productivity even in coastal upwelling regions (Kudela and Dugdale, 2000). In some regions–including coastal regions and the high-nutrient, low-chlorophyll regions–upwelling or runoff can supply nitrogen in concentrations that exceed phytoplankton demand. Thus, there are large geographical variations in the sources and fluxes of nitrate and ammonium ions.

The first comprehensive view of the nitrogen cycle in the surface ocean proposed that inorganic nitrogen was taken up by phytoplankton and that the nitrogen was subsequently recycled from phytoplankton cells by heterotrophs, both large grazers (*e.g.*, planktonic invertebrates) and microbial decomposers. The death and decay of the phytoplankton, either after ingestion by "herbivores" or because of physiological stressors liberated nitrogen in the form of dissolved organic nitrogen or ammonium ions, which were collectively termed "regenerated" nitrogen (Dugdale and Goering, 1967). Biological nitrogen fixation, the reduction of atmospheric dinitrogen gas ($N_2$) to ammonia was thought to be insignificant in the open ocean and essentially all pelagic nitrogen fixation was ascribed to two genera of nitrogen-fixing microbes. Nitrate was believed to be supplied to the upper ocean primarily by mixing, advection and diffusion from deep ocean water, or terrestrial runoff. Because nitrifying bacteria are inhibited by light, it was assumed that nitrification proceeded only in deep water; therefore, the only source of nitrate in surface waters was water mixing from the deep ocean reservoir.

While it was previously assumed that eukaryotic phytoplankton dominated both photosynthesis and dissolved inorganic nitrogen (DIN) assimilation, it is now known that two major groups of small unicellular cyanobacteria (*Synechococcus* and *Prochlorococcus* groups) are extremely abundant in surface waters and contribute a large fraction of photosynthesis and DIN demand (Cavender-Bares *et al.*, 2001). The high–and low-light-adapted *Prochlorococcus* spp. (divinyl chlorophyll *a*-containing cyanobacteria common in oceanic waters) differ in their abilities to use nitrogen sources (Rocap *et al.*, 2001). Earlier, it was assumed that microorganisms generally could use either nitrate or ammonium and that organisms differed in kinetics of nitrate or ammonium uptake and utilization (Eppley *et al.*, 1969) but now, it is clear that some microorganisms may not be able to use nitrate.

The major source of nitrate in the ocean surface is diffusion and upwelling of nitrate-rich deep ocean water. In recent years, it has been shown that even this process, which would appear to be driven largely by physical forcing, has an important microbiological component. Many large oceanic diatoms that sometimes form large mats migrate to great depth to obtain nutrients from the nutrient-rich deep water (Moore and Villareal, 1996), only to return to the surface carrying with them nitrate (Villareal *et al.*, 1999). Migrating *Rhizosolenia* mats may transport an average of 20 per cent, ranging up to 78 per cent, of the upward diffusive flux of nitrate (Villareal *et al.*, 1999). Thus, biological controls are involved even in the upward movement of nitrate from deep water.

## Heterotrophic Bacteria

During the decomposition of organic matter, the primary role of bacteria in the nitrogen cycle was presumed to be the release of inorganic nitrogen ($NH_4^+$), thereby recycling nitrogen (and other nutrients) to phytoplankton. The role of bacteria in trophic transfers through a micrograzer (protozoan) food chain is well appreciated. These food chains result in the transfer of carbon, nitrogen and other nutrients from dissolved organic matter into the food web. However, this picture is complicated by the recent findings regarding the metabolism of heterotrophic marine bacteria, and the composition and sources of organic matter (McCarthy *et al.*, 1997, 1998). Bacteria can use dissolved organic nitrogen as well as organic matter and thus, might even compete with phytoplankton for inorganic nitrogen (Kirchman and Wheeler, 1998). Whether or not bacteria take up inorganic nitrogen probably depends on the C: N ratio of the substrates being used for growth (Goldman and Dennett, 2000). Bacterial regeneration of nitrogen during the mineralization of organic matter also depends on the C: N ratio of cell material relative to substrate availability (Kirchman, 2000) and thus, whether bacteria provide nitrogen or compete with primary producers for nitrogen depends substantially on space and time (Kirchman, 2000). Interestingly, growth rates of bacteria are consistent with the uptake of nitrogen from dissolved free amino acids and ammonium compounds, indicating that the larger pool of dissolved organic nitrogen is not a major source of nitrogen for growth (Kirchman, 2000). Bacteria can take up dissolved inorganic nitrogen while simultaneously liberating ammonium ions through decomposition (Tupas *et al.*, 1994). Thus, bacteria in the ocean can be competing for or regenerating ammonium ions, or both. It is unclear how both processes are occurring simultaneously.

Dissolved organic nitrogen (DON) compounds include a wide range of chemical compounds varying in size, complexity and resilience to degradation. DON can be a large pool in the oceans–3 to 7 μ*M* (Capone, 2000) –and an even larger one in coastal waters (Sharp, 1983). Important but usually minor constituents of DON are amino acids and urea, which are readily used by bacteria and some phytoplankton. Amino acids, either dissolved or free, or combined in oligopeptides, are important sources of organic carbon and nitrogen for bacteria (Kirchman, 2000). Although rates of urea production and catabolism have been measured in marine environments, relatively little is known about the microbiology of urea metabolism in marine systems. Bacteria can be a source or a sink for urea. Urease is a nickel-containing multi-subunit metalloprotein encoded by the *ure* genes (Hausinger *et al.*, 2001), which has been characterized in eukaryotes, cyanobacteria and heterotrophic bacteria, and recently reported in autotrophic nitrifiers. Studies on the diversity of urease genes in the environment are likely to provide interesting information on the distribution of urea utilization capabilities in natural assemblages.

## Nitrogen Fixation

The reduction of atmospheric nitrogen to ammonia (biological nitrogen fixation) is catalyzed by a diverse set of microorganisms. In the marine environment, nitrogen fixation rates are the highest in a few specific habitats such as benthic cyanobacterial mats (Herbert, 1999). In the oceanic biome, there are relatively few known nitrogen fixers despite the fact that dissolved organic nitrogen concentrations are extremely low in many parts of the ocean. Certain species of *Azotobacter* are responsible for nitrogen fixation. Blooms of filamentous nitrogen-fixing cyanobacteria often exploit nitrogen-limiting conditions in lakes and sometimes estuaries, and so, it is a curious paradox that there are so few obvious nitrogen-fixing microorganisms in the ocean (Howarth and Marino, 1988). In the open ocean, the filamentous nonheterocystous cyanobacterium *Trichodesmium* is common in tropical and subtropical waters (Capone *et al.*, 1997; Orcutt *et al.*, 2001). This organism is particularly interesting as it forms macroscopic aggregates of filaments, is buoyant due to gas vacuoles and fixes nitrogen only in the light. Most cyanobacteria segregate oxygen-sensitive nitrogen fixation from oxygen evolved through photosynthesis by fixing nitrogen during the night or in specialized cells called heterocysts, where photosystem-II activity is reduced or absent. *Trichodesmium* is one of the few known genera that evolve oxygen simultaneously with nitrogen fixation without an obvious mechanism to avoid oxygen inactivation. *Trichodesmium* fixes nitrogen only during the day and this cycle is regulated by the synthesis of nitrogenase under the control of a circadian clock (Chen *et al.*, 1998). There are various theories and hypotheses regarding the mechanisms involved in simultaneous nitrogen fixation and photosynthesis, including the possible division of labour among cells that are morphologically similar (Janson *et al.*, 1994; Fredriksson and Bergman, 1997), but the mechanisms whereby *Trichodesmium* fixes nitrogen aerobically are still not completely understood (Capone *et al.*, 1997). The *Trichodesmium* nitrogenase protein is similar phylogenetically to that of other cyanobacterial diazotrophs and is not likely to be more oxygen resistant than other nitrogenases (Zehr *et al.*, 1999).

Other most abundant diazotrophs in oceanic waters are the heterocyst-forming cyanobacterial symbionts of diatoms (Villareal and Carpenter, 1989). These symbionts have not been successfully maintained in culture for extended periods of time and so, relatively little is known about the biology of the symbiotic interactions between the diatoms and cyanobacteria. However, the symbiont-containing diatoms can form large aggregates that can be abundant in oligotrophic waters. These diatoms can form extensive blooms (Carpenter *et al.*, 1999) that can be significant sources of nitrogen in the mixed layer of the ocean.

It was believed that *Trichodesmium* and the symbionts of diatoms were the major nitrogen fixers in the open ocean. However, a number of recent studies have highlighted imbalances in nitrogen budgets that indicate that higher rates of nitrogen fixation are occurring in the open-ocean than was previously estimated (Lipschultz and Owens, 1996; Michaels *et al.*, 1996; Gruber and Sarmiento, 1997). This conclusion is based on biogeochemical calculations rather than direct measurements of nitrogen fixation rates or observed distributions of microorganisms, but has led to a re-evaluation of nitrogen fixation in the sea. Evidence of diverse bacterial and cyanobacterial nitrogen-fixing microorganisms in the Atlantic and Pacific oceans based on amplification of nitrogenase genes from bulk water samples has recently been reported (Zehr *et al.*, 1998, 2000).

It is not well understood what controls the distribution and activity of diazotrophs in the sea. The distributions of *Trichodesmium* and some other cyanobacteria appear to be correlated with water temperature (Carpenter, 1983). It could be that diazotrophs are limited by the availability of iron, a metal that is a component of many proteins, in addition to nitrogenase. Iron distributions in the world's oceans are controlled to a large extent by aeolian transport of dust. Temporal and spatial variation in iron supply may result in oscillation between nitrogen and phosphorus limitations of the oceans through its effect on nitrogen fixation (Wu *et al.*, 2000).

## Nitrification

Isolations of nitrifying and denitrifying bacteria using conventional enrichment techniques have provided extensive culture collections on which our understanding of the biochemistry and ecology of these processes is based. Chemolithoautotrophic nitrification is rather restricted in its occurrence and is represented in culture by 25 species of ammonia oxidizers in the beta and gamma subdivisions of Proteobacteria and by eight species of nitrite oxidizers in the alpha, beta and gamma subdivisions of Proteobacteria (Koops and Pommerening-Roser, 2001). Molecular phylogeny has supported this generalization, showing that both ammonia-oxidizing bacteria and nitrite-oxidizing bacteria belong to a small number of coherent groups (Teske *et al.*, 1994).

The first ammonia-oxidizing bacterium to be cultivated from the marine environment was *Nitrosocystis oceanus* (Watson, 1965), now called *Nitrosococcus oceani*. Strains of *N. oceani* have been obtained from several locations (Carlucci and Strickland, 1968), and it has been detected in sea water by immunofluorescence (Ward and Carlucci, 1985). Members of the *Nitrosomonas/Nitrosospira* group appear to dominate most terrestrial and aquatic environments including marine sediments (Kowalchuk *et al.*, 1998; Whitby *et al.*, 1999; Nold *et al.*, 2000; Ward *et al.*,2000). Purkhold *et al.* (2000) surveyed all published 16S rRNA and *amoA* sequences from both cultures and environmental clones, and concluded that although much diversity among ammonia-oxidizing bacteria remains to be cultured, it is unlikely that entirely novel species will be discovered, at least in the case of the beta-subdivision ammonia-oxidizing bacteria. However, this does not preclude the existence of entirely novel ammonia-oxidizing bacteria that are not detected by probes based on known groups.

Nitrite oxidizers have received somewhat less attention and culture collections are dominated by *Nitrobacter* strains. Recent work in various waste waters, however, has shown that *Nitrospira* is the dominant group in these environments (Gieseke *et al.*, 2001). For both ammonia-oxidizing bacteria and nitrite-oxidizing bacteria, phylogeny and functionality appear to be well correlated, making these groups attractive for molecular phylogeny studies despite their slow and fastidious growth habits in culture. Research on nitrifers, particularly ammonia-oxidizing bacteria, based on 16S rRNA genes and functional genes has proliferated in recent years and was recently reviewed (Kowalchuk and Stephen, 2001).

## Denitrification

Denitrifying bacteria are the opposite of nitrifers in many ways; denitrification ability found in heterotrophic opportunists and chemoautotrophs is widespread among bacteria and Archaea, and has even been reported in Eukarya. Environmental research on the diversity of denitrifers has focused, therefore, on the functional genes involved in the denitrification pathway, mainly nitrite reductase (Braker *et al.*, 2000, 2001) and nitrous oxide reductase (Scala and Kerkhof, 1998, 2001).

Dissimilatory reduction of nitrate to ammonium is often ignored in the marine realm, but could be important in sediments in which fermentative bacteria, with whose metabolism it is often associated, are likely to be found. Christensen *et al.* (2000) found it to be a significant nitrate sink only in sediments with very high organic carbon loading. Bonin *et al.* (1998) suggested that it could be important in coastal sediments and that it is unlikely to occur in the water column. Thus, we refer here mainly to respiratory denitrification in which nitrate is reduced sequentially to nitrite, nitric oxide, nitrous oxide and nitrogen gas.

Chemical distributions in marine sediments and the water column indicate that nitrification is an obligatory aerobic process and that denitrification is an obligatory anaerobic process. Although the oxygen requirements and tolerances vary among isolates, these requirements are reflected in the physiology of cultivated nitrifers and denitrifers, which are predominantly obligate aerobes and facultative anaerobes, respectively. Nitrification and denitrification are often coupled across oxic/anoxic interfaces in both sediments and suboxic waters leading to the loss of fixed nitrogen via mineralization, oxidation and denitrification. As the deep ocean contains high nitrate concentrations, it was long assumed that nitrification occurs in that environment. However, direct rate measurements using $^{15}N$ tracer techniques have consistently shown that most water column nitrification occurs in the lower portion of the euphotic zone and that the nitrate produced there can supply a large fraction of phytoplankton nitrate demand (Ward *et al.*, 1989; Dore and Karl, 1996).

The recent description of aerobic denitrification by denitrifying bacteria introduces a new link in the nitrogen cycle. Aerobic denitrification was first described in *Paracoccus pantotrophus* (Rainey *et al.*, 1999) and like others possessing this ability, *P. pantotrophus* is also a heterotrophic nitrifer. The nitrite generated by the oxidation of ammonia can be released into the medium or denitrified to nitrogen gas. Denitrification of nitrite or nitrate to nitrogen can occur with atmospheric levels of oxygen (Robertson *et al.*, 1995). *P. pantotrophus* was originally isolated from waste water; its ability to denitrify aerobically in batch culture as well as that of several other conventional heterotrophic denitrifers has been confirmed (Robertson *et al.*, 1999). Su *et al.* (2001) reported aerobic denitrification by a strain of *Pseudomonas stutzeri* at rates and with oxygen tolerance greatly exceeding those reported for *P. pantotrophus*. The process may be common in isolates, but its significance in the environment remains uncertain and even in culture, questions still remain. For example, the enzymology of the aerobic pathway is unknown. While *P. pantotrophus* expresses the nitrite reductase gene (*nirS*) under anaerobic conditions, it is not expressed under aerobic conditions (Moir *et al.*, 1995) and thus, the mechanism for aerobic reduction of nitrite remains unknown.

Ammonia-oxidizing bacteria also perform a subset of the conventional set of denitrifying reactions, reducing nitrite to NO and $N_2O$. The process occurs aerobically, but is apparently enhanced at low oxygen concentrations (Goreau *et al.*, 1980; Lipschultz *et al.*, 1981). These gases are also intermediates in denitrification, but could derive from nitrification under a low or nearly zero oxygen concentration. Both $N_2O$ and NO are important in atmospheric processes; they contribute to greenhouse warming and to catalytic destruction of stratospheric ozone. Thus, understanding which processes are responsible for their production could prove to be important in understanding or potentially regulating their fluxes.

It appears that chemoautotrophic ammonia-oxidizing bacteria produce $N_2O$ and NO by using a pathway that is essentially identical to the classical denitrification pathway. First, $NH_3$ is oxidized to nitrite and some of the nitrite is reduced to $N_2O$. The reductions are catalyzed by enzymes that are encoded by nitrite reductase and NO reductase genes that are homologous to the *nirK* and *norB* genes of conventional denitrifying bacteria (Casciotti and Ward, 2001). Even methylotrophs to which some marine nitrifers are related possess genes with homology to *nirK* and *norB* (Wu *et al.*, 2000). The significance of this denitrifying metabolism to the physiological ecology of nitrifying bacteria is unknown. However, its discovery casts uncertainty on the roles of "nitrifers" and "denitrifers" in trace gas metabolism in the oceans and may complicate the use of molecular approaches for studying and detecting their presence and activity in the environment.

Suboxic and anoxic waters and sediments tend to have large fluxes in and sometimes, large accumulations of the gaseous intermediates of nitrification and denitrification. Trace levels of $N_2O$ in oxic ocean waters are thought to arise from nitrification (Yoshinari, 1976) and show a stoichiometric relationship to oxygen utilization. Oxygen minimum zones show depth zones of $N_2O$ accumulation and depletion. $N_2O$ is depleted in the core of the oxygen minimum zone, where denitrification rates are thought to be the greatest. $N_2O$ maxima typically occur both above and below the minimum. Stable-isotope measurement of $N_2O$ from oxygen-depleted waters in the Arabian Sea indicates that both nitrification and denitrification may contribute to the signal (Naqvi *et al.*, 1998). The surface waters of the ocean are generally slightly supersaturated with $N_2O$ and the ocean constitutes a significant source of atmospheric $N_2O$, especially in regions containing oxygen minimum zones (Naqvi and Naronha, 1991; Rahn and Wahlen, 2000).

Autotrophic nitrifying bacteria exhibit some abilities in anaerobic metabolism as well. Enrichment cultures under chemolithotrophic conditions and with very low oxygen concentrations catalyzed the net removal of ammonium ions as nitrogen gas (Muller *et al.*, 1995). Schmidt and Bock (1997) have shown that *Nitrosomonas eutropha* produces gaseous products, mainly NO and nitrogen during growth on $NO_2$ and $NH_3$. The process proceeds at a lower rate than ammonia oxidation in the presence of normal air atmosphere and supports cell growth. Additions of $NO_2$ and NO enhanced the complete removal of nitrogen in the form of $NH_3$ and organic nitrogen without the addition of organic carbon substrates (Zart and Bock, 1998). The relevance of these observations to marine nitrogen cycling is unknown, but the phylogenetic homogenity of the beta-subdivision ammonium oxidizers suggests that analogous metabolic capabilities may exist in marine strains as well.

A completely novel process in which ammonia and nitrite are converted anaerobically to dinitrogen gas has recently been reported from anaerobic wastewater systems (Van Loosdrecht and Jetten, 1998; Strous *et al.*; 1999), and the organisms responsible for this novel metabolism have been identified as relatives of *Planctomyces* (Jetten *et al.*, 2001). Referred to as "anammox", the process probably involves a consortium of the planctomycete organisms and an autotrophic ammonia oxidizer such as *Nitrosomonas europaea* or *N. eutropha*. The planctomycete oxidizes ammonium ions to nitrogen using nitrite, which is produced by the conventional ammonia oxidizer as an oxidant. Both oxygen and nitrite concentrations are maintained at nearly undetectable levels by the metabolism of the members of the consortium, and while both organisms grow quite slowly (generation times for the planctomycete of two weeks or more are reported), the net removal of ammonium ions occurs at a rate 25 times faster than that reported (Stepanauskas *et al.*, 1999) for nitrogen removal by *N. eutropha* growing anaerobically in pure culture (Van Loosdrecht and Jetten, 1998). Still, it can require months to establish an anammox enrichment and the consortium is stable only in bioreactors with long retention times; thus, it seems unlikely that this anammox consortium is active in natural environments (Van Loosdrecht and Jetten, 1998).

Anammox would constitute a shortcut in the conventional nitrogen cycle in which nitrification and denitrification are linked at the level of nitrite, without going through nitrate. Therefore, natural environments where ammonium and nitrite both occur in the presence of low oxygen concentrations might be suitable habitats for anammox-like reactions. Such environments include oxic/anoxic interfaces such as those found at sediment/water interfaces in hemipelagic and shallow sediments, and in stratified lakes and water columns of stratified basins such as the Black Sea and the Cariaco Basin. Upon initial consideration, anaerobic ammonium oxidation seems unlikely to dominate processes in these environments based on observation of chemical distributions. Microaerophilic autotrophic nitrification linked to anaerobic denitrification across the oxic/anoxic interface has been used to interpret the chemical distributions, which typically show depletion of oxygen and nitrate above the interface and accumulation of ammonium ions below it.

In addition to the unconventional activities of "conventional" nitrifers and denitrifers, and the discovery of novel nitrogen metabolic pathways in new organisms, it has also been proposed recently that a short circuit of the nitrification/denitrification couple can also be accomplished abiotically. In marine sediments, which typically contain relatively high manganese levels, nitrogen gas can be produced by the oxidation of $NH_3$ and organic nitrogen by $MnO_2$ in air (Luther *et al.*, 1997) or linked in series to anoxic organic matter oxidation through several biogeochemical reductants, including iron and $H_2S$ (Hulth *et al.*, 1999). Anoxic $NH_3$ oxidation, whether it results directly in nitrogen formation (as in anammox) or in nitrate production (when linked to manganese reduction), would introduce new links into the aquatic and sediment nitrogen cycles. Failure to account for anoxic $NH_3$ oxidation might lead to an underestimate of ammonium ion removal, because the products do not accumulate; they are either lost to the atmosphere immediately or rapidly reduced by the next step in the anaerobic cycling of organic matter.

## Conclusion

We have learnt much about the role of bacteria in nitrogen cycle of marine systems; it must be remembered that our knowledge is superimposed on a dynamic and spatially variable biome. Knowledge gained by long-term studies demonstrates that our previous models of nitrogen cycling in the ocean are insufficient, and that there are changes that have occurred and are occurring with implications on the nitrogen cycle in the oceans. For example, the conceptual model of Dugdale and Goering (1967) was developed during an era when nitrification and nitrogen fixation were assumed to be minor nitrogen fluxes in surface waters. These ecosystem shifts are linked to climatic physical forcing events because physical processes determine the depth of the mixed layer and upwelling or mixing with nutrient-rich deep water. The consequences of the ecosystem shifts are reflected in the linked ecosystem properties of community structure and nutrient fluxes, and ultimately determine the constraints on ocean-atmosphere carbon flux. One of the features of this change seems to be an increased importance of nitrogen fixation as a source of nitrogen. These dynamic ecosystem changes highlight the importance of microbes in the nitrogen cycle of the marine environment.

## References

Bonin, P., P. Omnes and A. Chalamet (1998). Simultaneous occurrence of denitrification and nitrate ammonification in sediments of the French Mediterranean Coast. *Hydrobiologia,* 389: 169-182.

Braker, G., H. L. Ayala-del-Rio, A. H. Devol, A. Fesefeldt and J. M. Tiedje (2001). Community structure of denitrifers, *Bacteria,* and *Archaea* along redox gradients in Pacific Northwest marine sediments by terminal restriction fragment length polymorphism analysis of amplified nitrite reductase (*nirS*) and 16S rRNA genes. *Appl. Environ. Microbiol.,* 67: 1893-1901.

Braker, G., J. Z. Zhou, L. Y. Wu, A. H. Devol and J. M. Tiedje (2000). Nitrite reductase genes (*nirK* and *nirS*) as functional markers to investigate diversity of denitrifying bacteria in Pacific Northwest marine sediment communities. *Appl. Environ. Microbiol.,* 66: 2096-2104.

Capone, D. G. (2000). The marine microbial nitrogen cycle. *In:* D. L. Kirchman (ed)., Microbial Ecology of the Oceans. Wiley-Liss, New York, pp. 455-494.

Capone, D. G., J. P. Zehr, H. W. Paerl, B. Bergman and E. J. Carpenter (1997). *Trichodesmium*: a globally significant marine cyanobacterium. *Science,* 276: 1221-1229.

Carlucci, A. F. and J. D. H. Strickland (1968). The isolation, purification and some kinetic studies of marine nitrifying bacteria. *J. Exp. Mar. Biol. Ecol.,* 2: 156-166.

Carpenter, E. J. (1983). Nitrogen fixation by marine *Oscillatoria* (*Trichodesmium*) in the world's oceans. *In:* E. J. Carpenter and D. G. Capone (ed)., Nitrogen in the Marine Environment. Academic Press, New York, pp. 65-103.

Carpenter, E. J., J. P. Montoya, J. Burns, M. Mulholland, A. Subramaniam and D. G. Capone (1999). Extensive bloom of a $N_2$ fixing symbiotic association in the tropical Atlantic Ocean. *Mar. Ecol. Prog. Ser.*, 185: 273-283.

Casciotti, K. L. and B. B. Ward (2001). Nitrite reductase genes in ammonia-oxidizing bacteria. *Appl. Environ. Microbiol.*, 67: 2213-2221.

Cavendar-Bares, K. K., D. M. Karl and S. W. Chisholm (2001). Nutrient gradients in the western North Atlantic Ocean: relationship to microbial community structure and comparison to patterns in the Pacific Ocean. *Deep-Sea Res.*, 48: 2373-2395.

Chen, Y. -B., B. Dominic, M. T. Mellon and J. P. Zehr (1998). Circadian rhythm of nitrogenase gene expression in the diazotrophic filamentous nonheterocystous cyanobacterium *Trichodesmium* sp. strain IMS 101. J. *Bacteriol.*, 180: 3598-3605.

Christensen, P. B., S. Rysgaard, N. P. Sloth, T. Dalsgaard and S. Schwaerter (2000). Sediment mineralization, nutrient fluxes, denitrification and dissimilatory nitrate reduction to ammonium in an estuarine fjord with sea cage trout farms. *Aquat. Microb. Ecol.*, 21: 73-84.

Dore, J. E. and D. M. Karl (1996). Nitrification in the euphotic zone as a source for nitrite, nitrate and nitrous oxide at Station ALOHA. *Limnol. Oceanogr.*, 41: 1619-1628.

Dugdale, R. C. and J. J. Goering (1967). Uptake of new and regenerated forms of nitrogen in primary productivity. *Limnol. Oceanogr.*, 12: 196-206.

Eppley, R. W., J. N. Rogers and J. J. McCarthy (1969). Half-saturation constants for uptake of nitrate and ammonium by marine phytoplankton. *Limnol. Oceanogr*. 14: 912-920

Fredriksson, C. and B. Bergman (1997). Ultrastructural characterisation of cells specialised for nitrogen fixation in a non-heterocystous cyanobacterium, *Trichodesmium* spp. *Protoplasma*, 197: 76-85.

Gieseke, A., L. Purkhold, M. Wagner, R. Amann and A. Schramm (2001). Community structure and activity dynamics of nitrifying bacteria in a phosphate-removing biofilm. *Appl. Environ. Microbiol.*, 67: 1351-1362.

Giovannoni, S. and M. Rappe (2000). Evolution, diversity and molecular ecology of marine prokaryotes. *In:* D. L. Kirchman (ed)., Microbial Ecology of the Oceans. Wiley-Liss, New York, pp. 47-84.

Goldman, J. C. and M. R. Dennett (2000). Growth of marine bacteria in batch and continuous culture under carbon and nitrogen limitation. *Limnol. Oceanogr.*, 45: 789-800.

Goreau, T. J., W. A. Kaplan, S. C. Wofsy, M. B. McElroy, F. W. Valois and S. W. Watson (1980). Production of $NO_2$-and $N_2O$ by nitrifying bacteria at reduced concentrations of oxygen. *Appl. Environ. Microbiol.*, 40: 526-532.

Gruber, N. and J. L. Sarmiento (1997). Global patterns of marine nitrogen fixation and denitrification. *Global Biogeochem. Cycles*, 11: 235-266.

Hausinger, R. P., G. J. Colpas and A. Soriano (2001). Urease: a paradigm for protein-assisted metallocenter assembly. *ASM News*, 67: 78-84.

Herbert, R. A. (1999). Nitrogen cycling in coastal marine ecosystems. *FEMS Microbiol. Rev.*, 23: 563-590.

Howarth, R. W. and R. Marino (1988). Nitrogen fixation in freshwater, estuarine and marine ecosystems. 2. Biogeochemical controls. *Limnol. Oceanogr.*, 33: 688-701.

Hulth, S., R. C. Aller and F. Gilbert (1999). Coupled anoxic nitrification/manganese reduction in marine sediments. *Geochim. Cosmochim. Acta*, 63: 49-66.

Janson, S., E. J. Carpenter and B. Bergman (1994). Compartmentalisation of nitrogenase in a non-heterocystous cyanobacterium: *Trichodesmium contortum*. *FEMS Microbiol. Lett.*, 118: 9-14.

Jetten, M. S. M., M. Wagner, J. Fuerst, M. C. M. van Loosdrecht, G. Kuenen and M. Strous (2001). Microbiology and application of the anaerobic ammonium oxidation ("anammox") process. *Curr. Opin. Biotechnol.*, 12: 283-288.

Kirchman, D. L. (2000). Uptake and regeneration of inorganic nutrients by marine heterotrophic bacteria. *In:* D. L. Kirchman (ed)., Microbial Ecology of the Oceans. Wiley-Liss, New York, pp. 261-288.

Kirchman, D. L. and P. A. Wheeler (1998). Uptake of ammonium and nitrate by heterotrophic bacteria and phytoplankton in the sub-arctic Pacific. *Deep-Sea Res.* I 45: 347-365.

Koops, H. -P. and A. Pommerening-Roser (2001). Distribution and ecophysiology of the nitrifying bacteria emphasizing cultured species. *FEMS Microbiol. Ecol.,* 37: 1-9.

Kowalchuk, G. A., P. L. E. Bodelier, G. H. J. Heilig, J. R. Stephen and H. J. Laanbroek (1998). Community analysis of ammonia-oxidising bacteria, in relation to oxygen availability in soils and root-oxygenated sediments, using PCR, DGGE and oligonucleotide probe hybridisation. *FEMS Microbiol. Ecol.,* 27: 339-350.

Kowalchuk, G. A. and J. R. Stephen (2001). Ammonia-oxidizing bacteria: a model for molecular microbial ecology. *Ann. Rev. Microbiol.,* 55: 485-529.

Kudela, R. M. and R. C. Dugdale (2000). Nutrient regulation of phytoplankton productivity in Monterey Bay, California. *Deep-Sea Res.,* II 47: 1023-1053.

Lipschultz, F. and N. J. P. Owens (1996). An assessment of nitrogen fixation as a source of nitrogen to the North Atlantic Ocean. *Biogeochemistry,* 35: 261-274.

Lipschultz, F., O. C. Zafiriou, S. C. Wofsy, M. B. Melroy, F. W. Valois and S. W. Watson (1981). Production of NO and $N_2O$ by soil nitrifying bacteria. *Nature,* 294: 641-643.

Luther, G. W., B. Sundgy, G. L. Lewis, P. G. Brendel and N. Silverberg (1997). Interactions of manganese with the nitrogen cycle: alternative pathways to dinitrogen. *Biochem. Biophys. Acta,* 61: 4043-4053.

McCarthy, M. D., Hedges, J. I., Benner, R. (1998). Major bacterial contribution to marine dissolved organic nitrogen. *Science* 281 231-234.

McCarthy, M. D., Pratum T, Hedges J., Benner R. (1997). Chemical composition of dissolved organic nitrogen in the ocean. Nature 390 150-234.

Michaels, A. F., D. Olson, J. L. Sarmiento, J. W. Ammerman, K. Fanning, R. Jahnke, A. H. Knap, F. Lipschultz and J. M. Prospero (1996). Inputs, losses and transformations of nitrogen and phosphorus in the pelagic North Atlantic Ocean. *Biogeochemistry*, 35: 181-226.

Moir, J. W. B., D. J. Richardson and S. J. Ferguson (1995). The expression of redox proteins of denitrification in *Thiosphaera pantotropha* grown with oxygen, nitrate and nitrous-oxide as electron-acceptors. *Arch. Microbiol.*, 164: 43-49.

Moore, J. K. and T. A. Villareal (1996). Size-ascent rate relationships in positively buoyant marine diatoms. *Limnol. Oceanogr.*, 41: 1514-1520.

Muller, E. B., A. H. Stouthamer and H. W. van Verseveld (1995). Simultaneous $NH_3$ oxidation and $N_2$ production at reduced $O_2$ tensions by sewage sludge subcultured with chemolithotrophic medium. *Biodegradation*, 6: 339-349.

Naqvi, S. W. A. and R. J. Noronha (1991). Nitrous-oxide in the Arabian Sea. *Deep-Sea Res.*, 38: 871-890.

Naqvi, S. W. A., T. Yoshinari, D. A. Jayakumar, M. A. Altabet, P. V. Narvekar, A. H. Devol, J. A. Brandes and L. A. Codispoti (1998). Budgetary and biogeochemical implications of $N_2O$ isotope signatures in the Arabian Sea. *Nature*, 391: 462-464.

Nold, S. C., J. Zhou, A. H. Devol and J. M. Tiedje (2000). Pacific Northwest marine sediments contain ammonia-oxidizing bacteria in the ß-subdivision of the *Proteobacteria*. *Appl. Environ. Microbiol.*, 66: 4532-4535.

Orcutt, K. M., F. Lipschultz, K. Gundersen, R. Arimoto, A. F. Michaels, A. H. Knap and J. R. Gallon (2001). A seasonal study of the significance of $N_2$ fixation by *Trichodesmium* spp. at the Bermuda Atlantic Time-series Study (BATS) site. *Deep-Sea Res.*, II 48: 1583-1608.

Purkhold, L., A. Pommerening-Roser, S. Juretschko, M. C. Schmid, H. P. Koops and M. Wagner (2000). Phylogeny of all recognized species of ammonia oxidizers based on comparative 16S and *amoA* sequence analysis: implications for molecular diversity surveys. *Appl. Environ. Microbiol.*, 66: 5368-5382.

Rahn, T. and M. Wahlen (2000). A reassessment of the global isotopic budget of atmospheric nitrous oxide. *Global Biogeochem. Cycles,* 14: 537-543.

Rainey, F. A., D. P. Kelly, E. Stackebrandt, J. Burghardt, A. Hiraishi, Y. Katayama and P. M. Wood (1999). A re-evaluation of the taxonomy of *Paracoccus denitrificans* and a proposal for the creation of *Paracoccus pantotrophus* comb. nov. *Int. J. Syst. Evol. Bacteriol.*, 49: 645-651.

Robertson, L. A., T. Dalsgaard, N. -P. Revsbeck and J. G. Kuenen (1995). Confirmation of aerobic denitrification in batch cultures, using gas chromatography and $^{15}N$ mass spectrometry. *FEMS Microbiol. Ecol.*, 18: 113-120.

Rocap, G., S. Stilwagen, J, Lamerdin, S. Chisholm and F. Larimer (2001). Comparative genomics of marine cyanobacterium Prochlorococcus: from base pairs to niche differentiation. In ASM and Tigr conference on Microbial Genomes, Montery.

Scala, D. J. and L. J. Kerkhof (1998). Nitrous oxide reductase (*nosZ*) gene-specific PCR primers for detection of denitrifiers and three *nosZ* genes from marine sediments. *FEMS Microbiol. Lett.*, 162: 61-68.

Scala, D. J. and L. J. Kerkhof (2001). Horizonal heterogeneity of denitrifying bacterial communities in marine sediments by terminal restriction fragment length polymorphism analysis. *Appl. Environ. Microbiol.*, 66: 1980-1986.

Schmidt, I. and E. Bock (1997). Anaerobic ammonia oxidation with nitrogen dioxide by *Nitrosomonas eutropha. Arch. Microbiol.*, 167: 106-111.

Sharp, J. H. (1983). The distribution of inorganic nitrogen and dissolved and particulate organic nitrogen in the sea. *In:* E. J. Carpenter and D. G. Capone (ed)., Nitrogen in the Marine Environment. Academic Press, New York, pp. 1-36.

Stepanauskas, R., H. Edling and L. J. Tranvik (1999). Differential dissolved organic nitrogen availability and bacterial aminopeptidase activity in limnic and marine waters. *Microbial Ecol.* 38: 264-272.

Strous, M., G. Kuenen and M. S. M. Jetten (1999). Key physiology of anaerobic ammonium oxidation. *Appl. Environ. Microbiol.*, 65: 3248-3250.

Su, J. -J., B. -Y. Liu and D. -Y. Liu (2001). Comparison of aerobic denitrification under high oxygen atmosphere by *Thiosphaera pantotropha* ATCC 35512 and *Pseudomons stutzeri* SU2 newly isolated from the activated sludge of a piggery wastewater treatment system. *J. Appl. Microbiol.*, 90: 457-462.

Teske, A., E. Alm, J. M. Regan, S. Toze, B. E. Rittmann and D. A. Stahl (1994). Evolutionary relationships among ammonia–and nitrite-oxidizing bacteria. *J. Bacteriol.*, 176: 6623-6630.

Tupas, L. M., I. Koike, D. M. Karl and O. Holmhansen (1994). Nitrogen metabolism by heterotrophic bacterial assemblages in Antarctic coastal waters. *Polar Biol.*, 14: 195-204.

Van Loosdrecht, M. C. M. and M. S. M. Jetten (1998). Microbiological conversions in nitrogen removal. *Water Sci. Technol.*, 38: 1-7.

Villareal, T. A. and E. J. Carpenter (1989). Nitrogen fixation, suspension characteristics and chemical composition of *Rhizosolenia* mats in the central North Pacific Gyre. *Biol. Oceanogr.*, 6: 387-405.

Villareal, T. A., C. Pilskaln, M. Brzezinski, F. Lipschultz, M. Dennett and G. B. Gardner (1999) Upward transport of oceanic nitrate by migrating diatom mats. *Nature*, 397: 423-425.

Ward, B. B. and A. F. Carlucci (1985). Marine ammonia–and nitrite-oxidizing bacteria: serological diversity determined by immunofluorescence in culture and in the environment. *Appl. Environ. Microbiol.*, 50: 194-201.

Ward, B. B., K. A. Kilpatrick, E. Renger and R. W. Eppley (1989). Biological nitrogen cycling in the nitracline. *Limnol. Oceanogr.*, 34: 493-513.

Ward, B. B., D. P. Martino, C. M. Diaz and S. B. Joye (2000). Analysis of ammonia-oxidizing bacteria from hypersaline Mono Lake, California, on the basis of 16S rRNA sequences. *Appl. Environ. Microbiol.*, 66: 2873-2881.

Watson, S. W. (1965). Characteristics of a marine nitrifying bacterium. *Nitrosocystis oceanus* sp. n. *Limnol. Oceanogr.*: R274-R289.

Whitby, C. B., J. R. Saunders, J. Rodriguez, R. W. Pickup and A. McCarthym (1999). Phylogenetic differentiation of two closely related *Nitrosomonas* spp. that inhabit different sediment environments in an oligotrophic freshwater lake. *Appl. Environ. Microbiol.*, 65: 4855-4862.

Wu, J., W. Sunda, E. A. Boyle and D. M. Karl (2000). Phosphate depletion in the western North Atlantic Ocean. *Science,* 289: 759-762.

Yoshinari, T. (1976). Nitrous oxide in the sea. *Mar. Chem.*, 4: 189-202.

Zart, D. and E. Bock (1998). High rate of aerobic nitrification and denitrification by *Nitrosomonas eutropha* grown in a fermentor with complete biomass retention in the presence of gaseous $NO_2$ or NO. *Arch. Microbiol.*, 169: 282-286.

Zehr, J. P., E. J. Carpenter and T. A. Villareal (2000). New perspectives on nitrogen-fixing microorganisms in tropical and subtropical oceans. *Trends Microbiol.*, 8: 68-73.

Zehr, J. P., B. Dominic, Y–B. Chen, M. Mellon and J. C. Meeks (1999). Nitrogen fixation in the marine cyanobacterium *Trichodesmium*: a challenging model for ecology and molecular biology. *In:* G. A Peschek, W. Loffelhaedt and G. Schmetterer (ed)., The Phototrophic Prokaryotes. Plenum Publishing Corp; New York, pp 485-500.

Zehr, J. P., M. T. Mellon and S. Zani (1998). New nitrogen-fixing microorganisms detected in oligotrophic oceans by the amplification of nitrogenase (*nifH*) genes. *Appl. Environ. Microbiol.*, 64: 3444-3450.

**Aquatic Biodiversity in India: The Present Scenario, 2005** **38–60**
*Edited by*: D.R. Khanna, A.K. Chopra & G. Prasad
*Published by*: Daya Publishing House, New Delhi

# 3

# Ornamental Fish Biodiversity of India

**Archana Sinha**
*Central Institute of Fisheries Education* (*ICAR*) *Kolkata Centre 32-G.N. Block, Sec-V, Salt Lake City, Kolkata – 700 091*

Ornamental fishes are emerging as a major resource for diversification of aquaculture.The ornamental fish trade is growing by leaps and bounds with vast resources of freshwater and marine ornamental fishes. About 600 fish species of ornamental nature have been reported worldwide from various aquatic habitats. Ornamental fishes have attractive colors, peaceful nature, tiny sizes, suitable for keeping in captivity and adaptable for living within confined space. Their commercial rearing can generate revenue domestically and earn considerable foreign exchange when exported, besides creating employment to men and women alike in both rural and urban areas. Breeding and culture of ornamental fishes has a bright prospect in India, because India is one of the few countries with many indigenous colorful fishes for the ornamental fish trade.

## History of Ornamental Fish Culture

Ornamental or aquarium fishes form an important commercial component of fisheries, providing for aesthetic requirements and upkeep of the environment. Little information exists that dates the origin of ornamental fish culture, but it can be assumed that it was developed in China, where the goldfish was cultured traditionally as ornamental fish (believed to have been somewhere in the year 2000 B.C.). Modern aquarium keeping of fish began in 1805 with the first public display aquarium opened at Regent's Park in England in 1853. Development of aquaria picked up further and by 1928 there were 45 display aquaria open to public, with over 500 aquaria presently functioning worldwide. However, the market for ornamental fish in the world for public aquaria is less than 1 per cent at present and 99 per cent of the market for ornamental fish is still confined to hobbyists.

## World Trade of Ornamental Fishes

During the last four decades, there has been considerable growth and diversification in the international trade of ornamental fishes, which is valued at about US $ 700 million. USA is the largest market of ornamental fish, followed by European Union and Japan. On a global basis, India ranks one among the top few countries with maximum ornamental fish reserves. Despite rich fantastic reserves, favorable climatic factors, available manpower, India is a marginal player in the global trade, which is involved at about Rs. 2.3 crores. Marine Products Export Development Agency (MPEDA) has estimated that India can earn about US $ 5 billion as foreign exchange by export of ornamental fishes.

The major countries with maximum contribution to world market are:

| | |
|---|---|
| Singapore | 30 per cent |
| Hong Kong | 7.6 per cent |
| Thailand | 7.0 per cent |
| Indonesia | 6.9 per cent |
| Philippines | 5.1 per cent |
| Malaysia | 3.3 per cent |
| Japan | 2.5 per cent |

## Ornamental Fishes of India

Ornamental fishes are numerous, geographically distributed widely and reflect great taxonomic and ecological diversity. This diversity is displayed in an astonishing variety of colors, body shapes, locomotion, behavioral pattern, reproductive tactics, feeding strategies and other unique environmental adaptations. Over 1000 freshwater species in about 100 families are represented in the ornamental fish trade. Despite this diversity, only about 250 species in 30-35 families are in great demand and account for the largest volume in the trade. Indian sub-continent holds rich resources of freshwater and marine ornamental fishes. Kolkata, Mumbai and Chennai have emerged as major ornamental breeding centers in the country due to their ideal climatic conditions. West Bengal is a pioneering State in respect of ornamental fish breeding and fish production. It contributed 90 per cent of earnings from export of ornamental fishes from India in 2001-02. Approximately 7-10 thousand people are engaged as part time breeders of ornamental fishes in the State. At the same time, capture sector continues lead in the ornamental fish export trade. Most of these fishes are sourced from wild habitats, especially from indigenous stock of north -eastern States.

## Fresh Water Ornamental Fishes

Ornamental fishes are mainly grouped in to two categories these are

1. Live bearers and
2. Egg layers

## Common Livebearer Fishes

### Guppy (*Poecilia reticulata*)

Its origin is in south America, north of Amazon, but now it is enjoying worldwide distribution. It devour mosquito larvae thereby helping in control of mosquitoes. These are tiny fishes with bright colours, looking very beautiful in groups. Male fish is more colourful than females and may reach upto 2.5 to 3.5 cm in length, while the female are usually larger in size when fully grown. They grow better in the water having temperature ranges 20-25 °C.

Fig. 3.1: Guppy (*Poecilia reticulate*)

**Swordtail (*Xiphophorus helleri*)**

It is originated from Central and North-Eastern South America. The most striking feature is the magnificent sword like extension

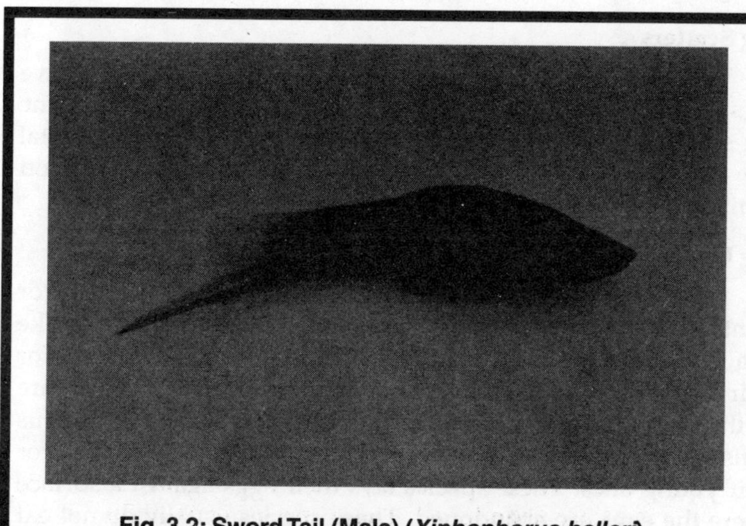

Fig. 3.2: Sword Tail (Male) (*Xiphophorus helleri*)

formed by the lower rays of the caudal fin in the male fish, which serves the purpose as an adornment. The fish prefers slightly saline water and voraciously devours live feed. The usual length of the female fish is 12 cm while that of male is 8 cm. Phenomenon of sex reversal is observed in this species.

### Platy (*Xiphophorus maculatus*)

Platys are of many types-namely red platy, orange platy, green platy and duckcido platy according to the colour. It is also originated from Central and North-Eastern South America. The usual length of male platy ranges from 4–4.5 cm and that of female from 5 to 5.5 cm. They bred once in three weeks and deliver about 75 young ones every time.

### Mollies (*Poecilia latipinna*)

Origin of the fish is same as that of platy and Swordtail. These fishes are easily bred and the usual length reaches 9-10 cms. They prefer saline water and breed every month delivering around 250 youngones at a time. They reach marketable size with in two months.

## Common Egg Layer Fishes

Most of the aquarium species are egg layers with external fertilization. Within this group fishes can be divided in to five sub-groups

### Egg Scatters

These species simply scatter their adhesive or non-adhesive eggs to fall to the substrate into plants or float to the surface. The egg-scatters either spawn in pairs or in groups. There is no parental care given and even they eat their own eggs. They produce good number of eggs. *e.g.* Goldfish (*Carrasius auratus*).

### Egg Depositors

In this case the eggs are either laid on a substrate, like a stone or plant leaf or even individually placed among fine leafed plants like java moss. Egg depositors fall into two groups those that care for their eggs and those that don't care. Among egg depositors that care for their eggs are cichlids and some catfishes. Cyprinids and various catfishes make up the majority of egg-depositors that do not care for their young ones. These species lays their eggs against a surface where the eggs are abandoned. These species usually do not eat their eggs. *e.g.* Angel (*Pterophyllum scalare*), Discuss (*Symphysodon discuss*).

### Egg Buriers

Fishes in this group usually inhabit waters that dry up at some time of the year. The majority of eggs burriers are annual killfish, which lay their eggs in mud. The parents mature very quickly and lay their eggs before dying when the water dries up. The eggs remain in a dormant stage until rains stimulate hatching. *e.g.* Annual killfish

### Mouth Brooders

Mouth-brooders carry their eggs or larvae in their mouth. Mouth brooders can be broken up into ovophiles and larvophiles. Ovophiles or egg loving mouth brooders lay their eggs in a pit, which are sucked up into the mouth of the female. The small number of large eggs hatch in the mother's mouth and the fry remain there for a period of time. Many cichlids and some labyrinth fish are ovophile mouth-brooders. Larvophile or larvae loving mouth-brooders lay their eggs on a substrate and guard them until the eggs hatch. After hatching the female picks up the fry and keeps them in her mouth. When the fry can feed for themselves, they are released. *e.g.* Cichlids

### Nest Builders

Many fish species build some sort of nest for their eggs. The nest ranges form a simple pit dug into the gravel or the elaborate bubble nest formed with saliva coated bubbles. Nest builder practice brood care.*e.g.* Gouramis, Anabantids, some catfishes

## Indian Wild Caught Fish

Since several rivers and streams flowing through dense forests and mountain terrains in India, wild aquarium fishes are in abundant in these rivers and streams and these varieties of fishes constitute the bulk of export made from the country. This source has so far been exploited only to a minimum extent and is a good potential to explore further. West Bengal has identified some trash fishes which are having potential as ornamental fishes. Meghalaya has several species of ornamental fishes, dominated by several *Danio* spp. *Channa* spp. and loaches. A colour changing *Puntius* spp. is also observed in the different streams (Sumer, Umgoi) and paddy fields of Meghalaya. In Manipur, 93 species of fishes have been identified so far as potential ornamental fishes.The freshwater habitats of Kerala abound in more than 50 varieties of ornamental fishes.The natural resources of wild ornamental fish varieties of Kerala especially in the western ghats region are rich and diversified when compared to the north-eastern region of the country and other tropical countries like Singapore, Indonesia, Sri Lanka, North Africa etc. (Table 3.1).

**Table 3.1: Common Freshwater Ornamental Fishes**

| *Scientific name* | *Common Name* |
|---|---|
| *Acanthophthalmus spp.* | Kuhlii Loach |
| *Aequidens maronii* | Keyhole |
| *Aequidens pulcher* | Blue Acara |
| *Anostomus spp.* | Headstander |
| *Aphyocharax spp.* | Bloodfin Tetras |
| *Aphyosemeion spp.* | Killie Fish |
| *Apistogramma spp.* | Dwarf Cichlid |
| *Aplocheilus spp.* | Panchax |
| *Apteronotus albifrons* | Black Ghost Knife Fish |
| *Apteronotus leptorhynchus* | Long Nose Brown Ghost Knifefish |
| *Arnoldichthys spilopterus* | Arnold's Characin, Red-eye Characin |
| *Astronotus ocellatus* | Oscar |
| *Only Albino form of Astyanax fasciatus mexicanis "jordani"* | Blind Cave Fish |
| *Aulonocara nyassae* | African Peacock Cichlid |
| *Aulonocara spp.* | African Cichlids |
| *Bagrichthys hypselopterus* | Black Lancer Catfish |
| *Balantiocheilus melanopterus* | Silver Sharkminnow |
| *Barbodes everetti* | Clown Barb |
| *Barbodes fasciatus* | Striped Barb |
| *Barbodes hexazona* | Tiger Barb |
| *Barbodes lateristriga* | Barb |
| *Barbodes pentazona* | Banded Barb |
| *Bedotia geayi* | Madagascar Rainbow |
| *Benthochromis tricoti* | Benthochromis Tricoti |

*Contd...*

**Table 3.1–Contd...**

| *Scientific name* | *Common Name* |
|---|---|
| *Betta spp.* | Fighting Fish |
| *Boehlkea fredcochui* | Chochui's Blue Tetra |
| *Botia macracantha* | Clown Loach |
| *Brachydanio albolineatus* | Pearl Danio |
| *Brachydanio frankei* | Leopard Danio |
| *Brachydanio kerri* | Kerr's Danio |
| *Brachydanio nigrofasciatus* | Spotted Danio |
| *Brachydanio rerio* | Zebra Danio |
| *Brachygobius spp.* | Bumble Bee Fish |
| *Brochis spp.* | Blue Catfish |
| *Brycinus longipinnis* | African Tetra |
| *Campylomormyrus cassaicus* | Double-nose elephant nose |
| *Campylomormyrus rhynchophorus* | Double-nose elephant nose |
| *Capoeta arulius* | Longfin Barb |
| *Capoeta oligolepis* | Checker |
| *Capoeta partipentazona* | tiger barb |
| *Capoeta semifasciolatus* | Golden Barb |
| *Capoeta tetrazona* | tiger barb |
| *Capoeta titteya* | Cherry Barb |
| *Carassius auratus* | Goldfish |
| *Carnegiella spp.* | Hatchet Fish |
| *Chalinochromis brichardi* | Lake Tanganyika Cichlid |
| *Chalinochromis spp.* | Lake Tanganyika Cichlids |
| *Chanda spp.* | Perchlets |
| *Chilodus punctatus* | Spotted Headstander |
| *Chilotilapia rhoadesii* | Rhoadesii Cichlid |
| *Cichlasoma nicaraguense* | Nicaraguan Cichlid |
| *Coelurichthys microlepis* | Croaking Tetra |
| *Colisa chuna* | Honey Dwarf Gourami |
| *Colisa fasciata* | Giant Dwarf Gourami |
| *Colisa labiosa* | Thick-lipped Gourami |
| *Colisa lalia* | Dwarf Gourami |

*Contd...*

**Table 3.1–Contd...**

| *Scientific name* | *Common Name* |
|---|---|
| *Copeina arnoldi* | Splash Tetra, Characin, Jumping Tetra |
| *Copeina guttata* | Red Spotted Copeina |
| *Corydoras spp.* | Armoured Catfish |
| *Corynopoma riisei* | Swordtail Characin |
| *Crenicara filamentosa* | Checkerboard Lyretail |
| *Crenicara maculata* | Checkerboard Cichlid |
| *Cyathopharnx furcifer* | Thread Fin Furficer |
| *Cyprichromis leptosoma* | Yellowtail Cyprichromis |
| *Cyrtocara moorii* | Lake Malawi Cichlid |
| *Danio devario* | Bengal danio, Sind danio |
| *Danio malabaricus* | Giant danio, Malabar danio |
| *Dermogenys pusillus* | Half Beak |
| *Dianema urostriata* | Stripe Tailed Catfish |
| *Epalzeorhynchus kallopterus* | Flying Fox |
| *Epalzeorhynchus siamensis* | Siamese Flying Fox |
| *Epiplatys spp.* | Killie Fish |
| *Eretmodus cyanostictus* | Dwarf Goby Cichlid |
| *Eretmodus maculatus* | Tangyanikan Clown Cichlid |
| *Esomus malayensis* | Flying Barb |
| *Farlowella acus* | Twig Catfish |
| *Gasteropelecus spp.* | Hatchet Fish |
| *Gnathochromis permaxillaris* | African cichlid |
| *Gnathonemus macrolepidotus* | Elephant nose |
| *Gnathonemus petersi* | Elephant nose |
| *Gymnocorymbus ternetzi* | Black Widow Tetra |
| *Gyrinocheilus aymonieri* | Sucking Asian Catfish |
| *Hasemania nana* | Silver Tip Tetra |
| *Helostoma rudolfi* | Pink Kissing Gourami |
| *Helostoma temminckii* | Green Kissing Gourami |
| *Hemigrammopetersius caudalis* | Yellow-tail Congo Tetra |
| *Hemigrammus spp.* | Tetras |

*Contd...*

**Table 3.1–Contd...**

| *Scientific name* | *Common Name* |
|---|---|
| *Hemiodopsis sterni* | Striped Hemiodopsis |
| *Homaloptera orthogoniata* | Indonesian Lizard Fish |
| *Hyphessobrycon spp.* | Tetras |
| *Inpaichthys kerri* | Blue Emperor Tetra |
| *Iodotropheys sprengerae* | African Cichlid |
| *Julidochromis spp.* | Dwarf Cichlid |
| *Kryptopterus bicirrhis* | Glass Catfish |
| *Kryptopterus macrocephalus* | Poormans Glass Catfish |
| *Labeo bicolor* | Redtail Shark |
| *Labeo erythrurus* | Red Fin Shark |
| *Labeo frenatus* | Rainbow Shark |
| *Labeo variegates* | Variegated Shark |
| *Laetacara curviceps* | Curviceps |
| *Laetacara dorsigerus* | Cichlid |
| *Laubuca laubuca* | Indian Hatchet Fish |
| *Leiocassis siamensis* | Siamese Catfish, Bumble Bee Catfish |
| *Lepidarchus Adonis* | Flagtail Tetra, Adonis Tetra |
| *Leporinus arcus* | Lipstick Leporinus |
| *Leporinus fasciatus* | Banded Leporinus |
| *Leporinus maculatus* | Spotted Leporinus |
| *Leporinus multifasciatus* | Multi-banded Leporinus |
| *Loricaria filamentosa* | Whiptail Catfish |
| *Macrognathus aculeatus* | Spiny Eel |
| *Macropodus opercularis* | Paradise Fish |
| *Megalamphodus spp.* | Tetras |
| *Melanochromis auratus* | Auratus |
| *Melanochromis simulans* | Auratus |
| *Only non-Albino form of Mesonauta festivus* | Festivum |
| *Metynnis spp.* | Silver Dollars |
| *Moenkhausia spp.* | Tetras |
| *Monodactylus argenteus* | Angel Mono, Malayan Mono, Batfish |

*Contd...*

**Table 3.1–Contd...**

| *Scientific name* | *Common Name* |
|---|---|
| *Monodactylus sebae* | African Mono |
| *Morulius chrysophekadion* | Black Shark |
| *Myleus rubripinnis* | Red Hook |
| *Nannacara anomala* | Golden Dwarf Acara |
| *Nannacara aureocephalus* | Golden Head Cichlid |
| *Nannacara taenia* | Dwarf Lattice Cichlid |
| *Nannostomus spp.* | Pencil Fish |
| *Nematobrycon spp.* | Emperor Tetra |
| *Neolamprologus brichardi* | Princess of Burundi |
| *Neolamprologus cylindricus* | Tanganyikan Cichlid |
| *Neolamprologus leleupi* | Lemon Cichlid |
| *Neolamprologus meeli* | African cichlid |
| *Neolamprologus mustax* | Mustax, Mask Lamprolagus |
| *Neolamprologus ocellatus* | African cichlid |
| *Ophthalmotilapia spp.* | Blacknosed Threadfin Cichlid |
| *Oryzias latipes* | Golden Medaka |
| *Osteochilus hasselti* | Bony lipped barb |
| *Osteochilus vittatus* | Bony lipped barb |
| *Otocinclus arnoldi* | Sucker Catfish |
| *Oxygaster oxygastroides* | Glass Barb |
| *Pantodon buchholzi* | Butterfly Fish |
| *Papiliochromis altispinosa* | Bolivian Butterfly Cichlid |
| *Papiliochromis ramirezzii* | Ram |
| *Paracheirodon axelrodi* | Cardinal Tetra |
| *Paracheirodon innesi* | Neon Tetra |
| *Paracyprichromis nigripinnis* | Blue Neon Cyprichromis |
| *Parosphromenus deissneri* | Licorice Gourami |
| *Pelvicachromis pulcher* | Kribensis |
| *Pelvicachromis subocellatus* | Kribensis |
| *Pelvicachromis taeniatus* | Kribensis |
| *Petitella georgiae* | False Rummy Nose |
| *Petrochromis trewavasae* | 'Texas' Cichlid, White Spotted Peerchromis |

*Contd...*

**Table 3.1–Contd...**

| *Scientific name* | *Common Name* |
|---|---|
| *Phenacogrammus interruptus* | Congo Tetra |
| *Pimelodella pictus* | Pictus Catfish |
| *Pimelodus ornatus* | Catfish |
| *Poecilia latipinna* | Sailfin Mollie |
| *Poecilia reticulate* | Guppy |
| *Poecilia sphenops* | Black Mollie |
| *Poecilia velifera* | Yucatan Sailfin Mollie |
| *Poecilocharax weitzmani* | Shining Tetra |
| *Prionobrama filigera* | Glass Bloodfin |
| *Pristella maxillaris* | Pristella |
| *Pseudogastromyzon myersi* | Dwarf Stone Sucker |
| *Pterophyllum spp.* | Angel Fish |
| *Puntius asoka* | Asoka Barb |
| *Puntius bimaculatus* | Two Spot Barb |
| *Puntius conchonius* | Rosy Barb |
| *Puntius cumingi* | Cummings Barb |
| *Puntius filamentosus* | Black Spot Barb |
| *Puntius lineatus* | Striped Barb |
| *Puntius nigrofasciatus* | Ruby Barb |
| *Puntius ticto* | Ticto Barb |
| *Puntius vittatus* | Kooli Barb |
| *Rasbora argyrotaenia* | Silver Rasbora |
| *Rasbora borapetensis* | Red Tail Rasbora |
| *Rasbora caudimaculata* | Red Tail Rasbora |
| *Rasbora dorsiocellata* | Emerald Eye Rasbora |
| *Rasbora dusonensis* | Yellow Tail Rasbora |
| *Rasbora einthoveni* | Blue Line Rasbora |
| *Rasbora elegans* | Two Spot Rasbora |
| *Rasbora hengelii* | Harlequin rasbora |
| *Rasbora heteromorpha* | Harlequin rasbora |
| *Rasbora kalochroma* | Clown Rasbora |
| *Rasbora leptosoma* | Copper Striped Rasbora |
| *Rasbora maculata* | Dwarf Spotted Rasbora |

*Contd...*

**Table 3.1–Contd...**

| *Scientific name* | *Common Name* |
|---|---|
| *Rasbora pauciperforata* | Red Line Rasbora |
| *Rasbora sarawakensis* | Sarawak Rasbora |
| *Rasbora steineri* | Gold Line Rasbora |
| *Rasbora taeniata* | Blue Line Rasbora |
| *Rasbora trilineata* | Black Scissortail |
| *Rasbora vaterifloris* | Flame Rasbora |
| *Rhodeus amarus* | Bitterling |
| *Rhodeus sericeus* | Bitterling |
| *Semaprochilodus insignis* | Prochilodus |
| *Semaprochilodus taeniurus* | Flagtail Prochilodus |
| *Spathodus erythrodon* | Blue Spotted Goby Cichlid |
| *Sphaerichthys osphronemoides* | Chocolate Gourami |
| *Sturisoma panamense* | Armoured Catfish |
| *Symphysodon spp.* | Discus |
| *Synodontis decorus* | Catfish |
| *Synodontis multipunctatus* | African Catfish |
| *Synodontis nigriventris* | Upsidedown Catfish |
| *Tanganicodus irsacae* | Goby Cichlid |
| *Tanichthys albonubes* | White Cloud |
| *Telmatherina ladigesi* | Celebes Rainbow |
| *Thayeria spp.* | Hockeystick Tetra |
| *Thoracocharax spp.* | Hatchet Fish |
| *Toxotes jaculator* | Archer |
| *Trichogaster leeri* | Pearl Gourami |
| *Trichogaster microlepis* | Moonbeam Gourami |
| *Trichogaster trichopterus* | Golden Gourami |
| *Trichopsis pumilus* | Gourami |
| *Trichopsis vittatus* | Gourami |
| *Trinectes maculates* | Flounder |
| *Triportheus spp.* | False Hatchet |
| *Tropheus spp.* | African Cichlids |
| *Xiphophorus helleri* | Swordtail |
| *Xiphophorus maculatus* | Platy |
| *Xiphophorus variatus* | Variegated Platy |

## Marine Ornamental Fishes

Although many of the freshwater ornamental fish sold to the public are farm raised, essentially all of the marine reef products are collected from the wild. The Asian/pacific region is the global centre of marine diversity, it supports more species of coral and fish than does any other region on earth. This region is home to over 4000 species of reef fish and more than one-third of the world's coral reef. More than 300 species from marine habitats are represented in the ornamental fish trade. With a long coastal line and several Islands stretching around with lagoons and coral reefs in our country, there are innumerable varieties of colourful marine ornamental fishes therein. But, the difficulty in care and rearing. Some of the marine ornamental fishes are conserved in coastal lagoons and sanctuaries in lakshadweep islands, Wandoor national park in Andamans, marine park in Gulf of mannar and in a few selected places in the coastal regions.

The most commonly available marine ornamental fishes are shown in Table 3.2.

**Table 3.2: Common Marine Ornamental Fishes**

| *Scientific name* | *Common Name* |
|---|---|
| General (not species specific) | |
| All species of the family Acanthuridae | Surgeon fish |
| All species of the family Anomalopidae | Flashlight fish |
| All species of the family Apogonidae | Cardinal fishes |
| All species of the family Balistidae | Triggerfish |
| All species of the family Brotulidae | Eel-Pouts |
| All species of the family Bythitidae | Cusk Eels |
| All species of the family Callionymidae | Dragonets |
| All species of the family Carapidae | Pearlfish |
| All species of the family Centriscidae | Razor fish |
| All species of the family Chaetodontidae | Butterfly fish |
| All species of the family Cirrhitidae | Hawk fish |
| All species of the family Ephippididae | Batfish |
| All species of the family Grammatidae | Grammas |
| All species of the family Holocentridae | Squirrel fish, Soldier fish |

*Contd...*

**Table 3.2–Contd...**

| *Scientific name* | *Common Name* |
|---|---|
| All species of the family Labridae | Wrasses |
| All species of the family Malacanthidae | Blanquillos |
| All species of the family Monocentrididae | Pineapple fish |
| All species of the family Mugiloididae | Weevers |
| All species of the family Mullidae | Goatfish |
| All species of the family Muraenidae | Moray eels |
| All species of the family Ostraciidae | Box fish |
| All species of the family Pegasidae | Seamoths |
| All species of the family Pempherididae | Sweepers |
| All species of the family Pholidichthyidae | Convict blennies |
| All species of the family Plesiopidae | Longfins |
| All species of the family Pomacanthidae | Angel fish |
| All species of the family Pomacentridae | Damsel fish |
| All species of the family Priacanthidae | Bullseyes |
| All species of the family Pseudochromidae | Dottybacks |
| All species of the family Scaridae | Parrotfish |
| All species of the family Syngnathidae | Pipe fish, Seahorses |
| All species of the family Zanclidae | Tangs |
| Species (in alphabetic sequence) | |
| *Alectis spp.* | Trevally |
| *Anthias spp.* | Rock Cods |
| *Brachirus spp.* | Scorpion fish |
| *Canthigaster spp.* | Puffer fish |
| *Cirripectes stigmaticus* | Blenny |
| *Cromileptes spp.* | Groupers |
| *Dendrochirus spp.* | Scorpion fish |
| *Ecsenius axelrodi* | Blenny |
| *Ecsenius bicolor* | Blenny |
| *Ecsenius graveri* | Blenny |
| *Ecsenius melarchus* | Blenny |
| *Ecsenius midas* | Blenny |
| *Ecsenius pulcher* | Blenny |

*Contd...*

**Table 3.2–Contd...**

| *Scientific name* | *Common Name* |
|---|---|
| *Gobiodon spp.* | Gobies |
| *Lipophrys nigriceps* | Blenny |
| *Lobotes spp.* | Jumping Cod |
| *Lythrypnus spp.* | Gobies |
| *Macolor spp.* | Snapper |
| *Meiacanthus astrodorsalis* | Blenny |
| *Meiacanthus grammistes* | Blenny |
| *Meiacanthus ovalauensis* | Blenny |
| *Nemateleotris spp.* | Gobies |
| *Plotosus lineatus* | Eel-tailed catfish |
| *Ptereleotris spp.* | Gobies |
| *Pterois spp.* | Scorpion fish |
| *Rhinopias spp.* | Scorpion fish |
| *Scolopsis bleekeri* | Spine-cheek |
| *Scolopsis bilineatus* | Spine-cheek |
| *Siganus spp.* | Rabbit fishes |
| *Signigobius spp.* | Gobies |
| *Symphorichthys spp.* | Snapper Hifin |
| *Valenciennea strigata* | Goby |

## Scope for Entrepreneurs in Ornamental Fish Trade

### Culture of Ornamental Fishes

For the breeding of ornamental fish, the needed infrastructure facilities have to be set up supported by the application of relevant technical know how. Rearing of commercial ornamental species can be undertaken in recirculation and flow-through water systems designed and established to maintain good water quality and to stimulate natural running water conditions. Different types of live feeds and artificial feeds are available in the market to rear ornamental fishes. Several workers pursue research work on production of indigenous feed for these fishes on a continuing basis. While, in every major metropolitan city there are aquarists who own few small ponds/cement tanks where they breed many freshwater ornamental fishes exclusively for domestic markets, this industry needs to be adequately popularized. Colourful hand books on aquarium fish keeping and maintenance are available for the hobbyists but by need is to provide literature on breeding and rearing of ornamental fishes to identified farmers in the regional languages so that they can follow the technologies and produce healthy ornamental fishes to promote their trade.

## Marketing of Accessories

In addition to the breeding, rearing and export of ornamental fishes, this trade has generated an ancillary business of abroad. For beautification and maintenance of aquaria, rocks and gravels, artificial toys, natural and artificial plants, dry feed, live feed, aerators, filters are in use. There is a great demand for all these accessories. Different types of decorative toys with beautiful colorations, attractive shapes that are non-toxic to fishes are gaining popularity in the market. Submerged varieties of simulated aquatic ornamental plants the natural habitat of ornamental fishes for placement in aquaria, have a developing market. There are many aquatic plants for aquaria and some of them are costlier than ornamental fishes. Commonly available attractive aquatic plants are ribbon grass (*Vallisneria* spp.), arrow weed (*Sagittaria* spp.), spike rush (*Acorus* spp.), lace plant (*Aponogeton* spp.), faneard (*Cabomba* spp.), Indian water fern (*Ceratopteris* spp.), hornwort (*Certophyllum* spp.), Amazon Sword Plant (*Echinoderus* spp.), Hydrilla (*Hydrilla* spp.), Mint (*Ludwigia* spp.), Water Star (*Hygrophila* spp.), etc. Most of these plants can be grown and multiplied under controlled conditions. Artificial, non-toxic plants are also available in the market and are now increasingly attracting customers due to their blended colours and durability. Apart from plants, a number of decorative toys are available for imparting an attractive look to an aquarium. They include plastic bubblers in the shape of mermaid, underwater diver, oyster shell, angler human skull, tortoise, frog etc. These can be efforts to improve the material used for the manufacture of these, and also quality, texture, colour and material of these toys so that their utility can be enhanced, thereby providing a diversified activities status to the trade.

## Breeding of Ornamental Fishes

The demand for ornamental fishes in domestic as well as International market is increasing rapidly. As such, sustainable

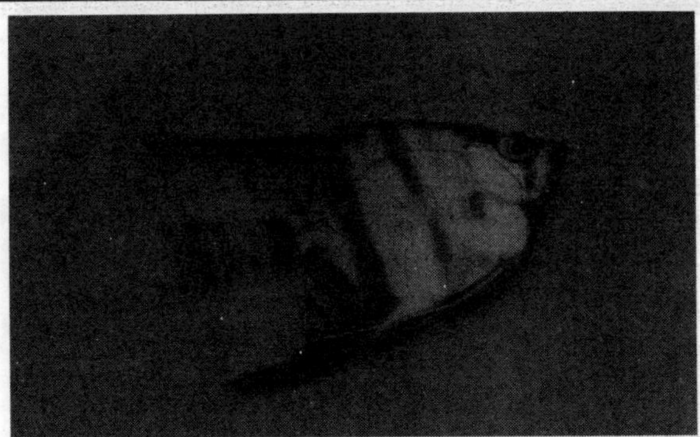

**Fig. 3.3: Angel (*Pterophyllum scalare*)**

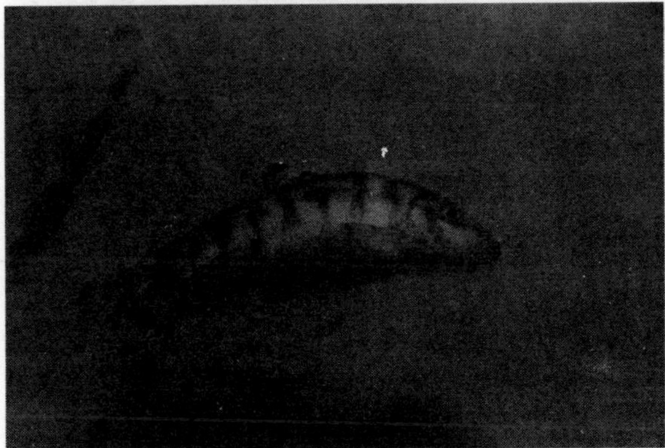

**Fig. 3.4: Banded loach (*Naemacheilus savana*)**

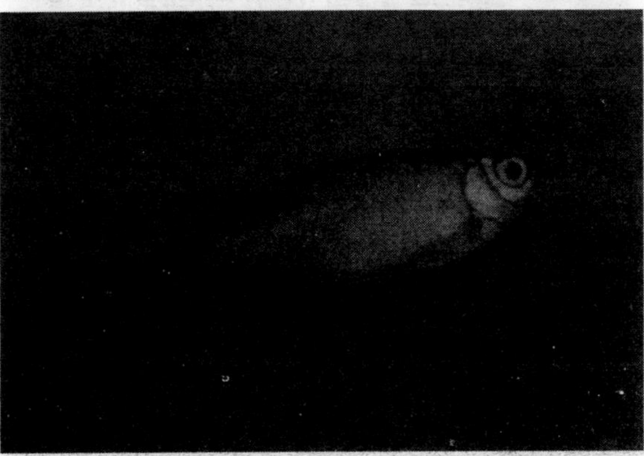

**Fig. 3.5: Blue danio (*Brahydanio albolineatus*)**

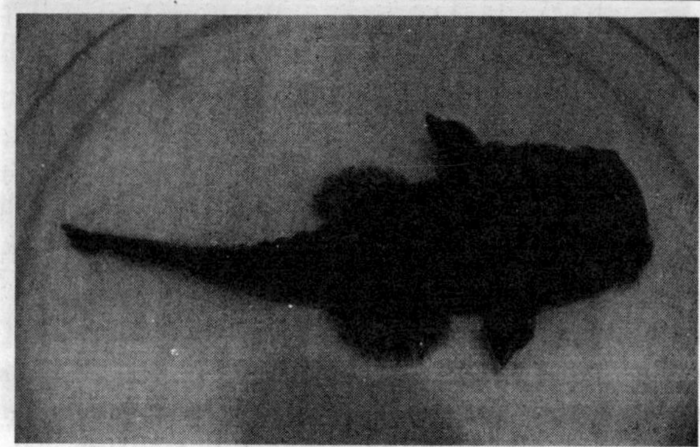

**Fig. 3.6: Devil Cat Fish (*Chaca chaca*)**

**Fig. 3.7: Eel (*Lycodontis tile*)**

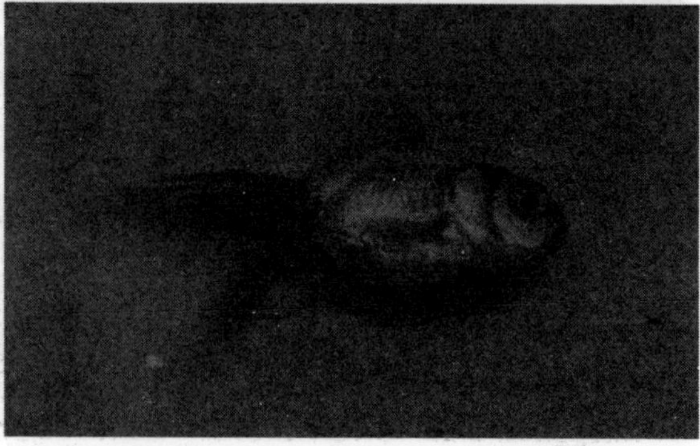

**Fig. 3.8: Gold Fish (*Carasius auratus*)**

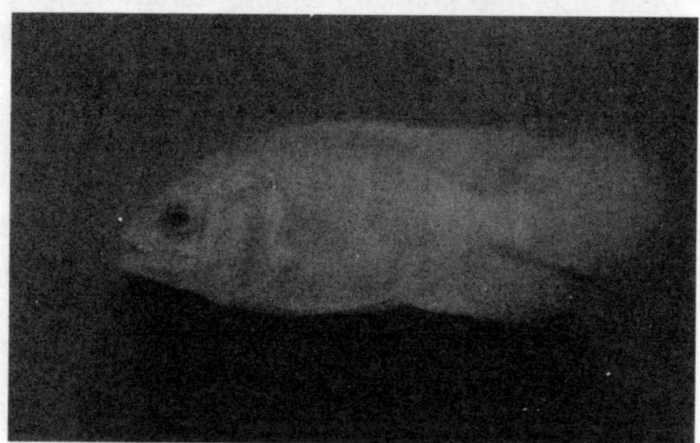

**Fig. 3.9: Oscar (*Astronotus ocellatus*)**

**Fig. 3.10: Siamese fighter (*Betta splendens*)**

**Fig. 3.11: Sucker Fish (*Glyptothorax sp.*)**

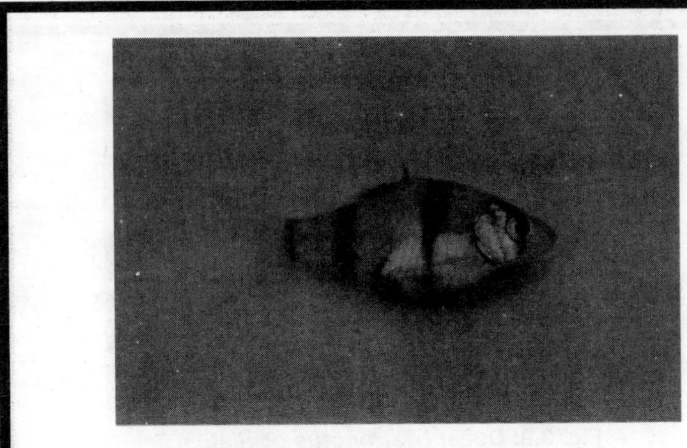

**Fig. 3.12: Tiger barb (*Puntius tetrazona*)**

exploitation of wild stocks of these fishes will not be able to meet the increasing demand. It is therefore essential to evolve appropriate breeding and rearing technology to produce both marine and freshwater ornamental fishes under controlled conditions in land-based infrastructural facilities. The technologies of breeding different varieties of ornamental fishes have now been established to such an extent that most of the aquarium fishes can not be bred as a household activity, both in rural and urban areas. Most of the aquarists breed only the common varieties of aquarium fishes like gold fish, guppys, platys, mollys, swordtails, gouramys, tetras, barbs etc., which are easy to breed. In order to enable to householders to upgrade their capabilities, the State Government should come forward to encourage aquarists and interested entrepreneurs to take up farming of these highly priced fishes. Simultaneously technologies on the production of live fish food and nutritionally balanced dry feed in various forms such as pellets, powder, flakes, microcapsules etc., should be developed up by technologists so that they can be extended to the hobbyists and entrepreneurs.

## Export of Ornamental Fishes

In spite of having immense natural ornamental fish resources, and technology for breeding and rearing them, not much of headway has been in the country in the matter of export of ornamental fishes to foreign countries. So as to move ahead in this endeavours, MPEDA, Kochi has prepared a directory of ornamental fish exporters in which they have identified 25 ornamental fish exporters in India especially, in Kolkata, Mumbai, Chennai and Kochi. The farmers and exporters have to be brought together, for the purpose of integrating the production and export activities in a manner that would be mutually beneficial. The establishment of such a relationship would push up the level of exports of ornamental fishes from the country, particularly to USA, Europe and Japan. It has been reported that 8 per cent of the estimated 86 million houses in USA keep aquaria in their homes, 14 per cent of the estimated 21 million houses in Great Britain, 4 per cent of homes in Belgium and Holland and 5 per cent of German and 20 per cent of Dutch houses keep fish. China, South Africa and several other countries too have the hobby of ornamental fish keeping. In view of the huge demand for export of ornamental, it is possible to undertake mass production of ornamental fish by farmers, to be made available to exporters. In fact producers can become exporters so as to have the advantage of earning foreign exchange themselves.

## Conclusion

It is sure that there are quite a large numbers of ornamental species found in our country, However, most of the potential ornamental fish resources are seen as highly unprotected and there is injudicious exploitation. The present need is to enforce regulations as provided under the Act, and immediate steps should be taken for mass awareness on the need for ecological restoration. A study of possibilities of ornamental fish trade in India shows that the resources which we have should be properly utilized for earning more of foreign exchange and for upgrading livelihood of ornamental fish farmers and entrepreneurs. The trade has a sizeable potential for employment too. This has to be harnessed by way of training more and more manpower in this trade. Strengthening and expanding research and extension in this field is very essential for refining and expanding technology basis of this line of fishery industry

## References

Chapman, F. A. (2000). Ornamental fish culture, freshwater. In: Stickney, R. R. (Ed). Encyclopedia of aquaculture, John Wiley and Sons, New York, pp. 602-610.

Kishori Lal Tekriwal and Andrew Arunava Rao (1999). Ornamental and aquarium fish of India. Kingdom Books, P. O. box 15 Waterloo PO 76 B Q England, 144p.

Srivastava L. S. (1994). Ornamental fish-New export opportunities. *Yojana*, Nov. 15, 22. p.

Talwar, P. K. and Jhingran, A. G. (1991). Inland fishes of India and adjacent countries. Oxford and IBH Publishing Co., New Delhi.

**Aquatic Biodiversity in India: The Present Scenario, 2005** **61–80**

*Edited by*: D.R. Khanna, A.K. Chopra & G. Prasad

*Published by*: Daya Publishing House, New Delhi

# 4

# Biodiversity of Aquatic Fauna of Mizoram: The Present Scenario

**S.N.Ramanujam**

*Dept. of Zoology, Pachhunga University College, Aizawal – 796 001, Mizoram, India*

Biodiversity is the variability of organisms on planet earth which is a mega ecosystem. Variation is one of the features in the process of evolution and is occurring continuously in nature. Biodiversity provides the basic biotic resources on which human race is sustaining. The world Commission on Environment and Development (WCED) constituted by General Assembly of United Nations published its report in 1987 which provided a major boost and endorsement to the need for conserving the World's rich biodiversity, particularly that of the tropical areas. The convention was attended by representatives from 170 countries and the proceedings ratified by 104 countries.

The term biodiversity includes three different aspects *viz.* Genetic diversity, Species diversity and Ecosystem diversity which are closely related to each other. The criteria often used while evaluating priorities of ecosystem or biogeographic area are–species richness, taxonomic distinctiveness, centres of density, representativeness, complimentarity and irreplaceability. The diversity of organisms is the source of all germplasm.

Biodiversity can be studied at three different levels: genetic, species and ecosystem.

## Genetic Diversity

It is concerned with the variation in genes within a particular species. Many billion individuals are produced through sexual reproduction. Each differs from the other in its hereditary constitution or the genetic information contained in its genes. It is this difference which has given us beautiful butterflies, roses, parakeets or coral in a myriad hues, shapes and sizes. Examples: Brown Bear, Sloth Bear, Himalayan Black Bear; Himalayan Tahr, Nilgiri Tahr.

## Species Diversity

It refers to the variety of living organisms on earth. Species differ from one another markedly in their genetic makeup and do not interbreed in nature. Closely related species however have much of their hereditary material in common. For instance, humans and chimpanzees have about 98.4 per cent of the same genes.

## Ecosystem Diversity

This refers to the variety of habitats. A habitat is the sum total of the climate, vegetation and geography of a region. There are several kinds of habitats around the world. Change in climatic conditions is accompanied by a change in vegetation as well. For instance cactus and thorny scrub replace grassland or forests as the climate changes from temperate to hot and dry. Each species is adapted to a particular kind of environment. As the environment changes, species best adapted to that environment become predominant. Thus the variety or diversity of species in the ecosystem is influenced by the ecosystem itself. Geographical changes within an ecosystem lead to genetic diversity or diversity within the same species. In India the Himalayan Tahr is found in the Eastern Himalayas, while the Nilgiri Tahr inhabits the slopes of the southern Western Ghats. Examples: Corals, Grasslands, Wetland, Desert, Mangrove, Tropical rain forests, etc.

There are about 5 -20 million species of living forms estimated to be present on this planet and out of it only about 1.5 million species have been identified. It includes 3,00,000 species of green plants and fungi, 8,00,000 species of insects, 40,000 species of vertebrates and 3,60,000 species of micro organisms. The total number of species of insects which includes both aquatic and terrestrial forms is more than the total number of species in all other groups of animal put together. The tropical environment encourages the species diversity both among plants and animals. It is mainly because of two reasons. 1. In tropics the conditions for evolution were optimum and for extinction fewer. 2. The species biodiversity was conserved over geological time due to low rate of extinction. Biological diversity is the result of interaction between climate, organisms, topography, parent soil material, time and heredity. Unlike the terrestrial organisms the biodiversity of aquatic organisms depend upon the drainage pattern and barriers between them.

As the scenario stands, the biodiversity studies in India are still at its infancy. India is the seventh largest country in the world and Asia's second nation with an area of 32,87,263 sq. km. Indian main land stretches from 8.4' to 37.6' N latitude and from 68.70' to 97.25'E longitude. It has a land frontier of around 15,200 km. and a coast line of 7516 km. The richness in the biodiversity found in the country (Table 4.1) is because of its vast landscape with variation in climate and altitude and hence a variety of ecological niches. They include humid tropical North-Eastern hill région and Western Ghats, hot deserts of north-west, cold deserts and the snow covered mountains of Himalayas, the warm coasts of peninsular India and a variety of aquatic bodies. The Indian gene centre is among the twelve mega diversity regions of the world.

In India too as in rest of the world the plant biodiversity is studied in much more detail as compared to animals. Among the animals the least importance is given to aquatic biodiversity. There are no clear estimates about the marine biota though the coast line is 7000 km. long with a shelf zone of 4,52,460 sq. km. and extended economic zone of 2,03,410 sq. km. There is an abundance of seaweeds, fish crustaceans, molluscs, corals, reptiles and mammals. The variation in the geographical and climatic condition has helped to harbour a rich variety of aquatic fauna.

**Table 4.1: Number of recorded Fauna in India**

| *Taxon* | *No. of Species* |
|---|---|
| Protozoans | 2577 |
| Porifera | 519 |
| Cnidiria | 237 |
| Ctenophora | 10 |
| Platyhelminthes | 1622 |
| Nematoda | 2350 |
| Rotifera | 310 |
| Kinoryncha | 10 |
| Gastrotricha | 88 |
| Acanthocephala | 110 |
| Sipuncula | 38 |
| Mollusca | 5042 |
| Echiura | 33 |
| Annelida | 1093 |
| Onychophora | 1 |
| Arthropoda | 57525 |
| Phoronida | 3 |
| Bryozoa | 170 |

The North-east India is one of the mega biodiversity centres in the Indian sub-continent. It is also one of the 10 distinct biogeographic zones of the country (SAARC, 1992). The region has a variety of ecosystems like tropical wet evergreen moist deciduous, sub-alpine forest and grasslands to numerous freshwater lakes, rivers, swamps and marshy wetlands. It is also having the maximum number of endemic plants and animals found in the country

Mizoram having an area of 21,081 sq. km. lies between 21°56′ N and 24°31′N latitude and 92°16′ E and 93°26′E longitude. The state has international borders of Bangladesh in the west and Myanmar in the east and south. On the northern side it is bordering with three states of India *viz.* Tripura in North West, Assam in north and Manipur in north east. The hilly state is covered with tropical and subtropical forest cover. The gorges are steep with north-south trending mountains. The average temperature varies between 11° C and 21° C in winter and climbs upto 20° C and 29° C in summer months. The monsoon season stretches from mid May to mid October. The annual rainfall is about 250 cm. The relative humidity is quite high during monsoon season. The Aizawl, the capital of the state lies just north of Tropic of Cancer.

The drainage pattern of Mizoram (Fig. 4.1) appears to be parallel. Most of the rivers either flow northwards or southwards and a watershed is formed in the middle of the state. The important rivers include Tuirial, Tuirini, Tlawng, Tuival which run from south to north entering the Barak river; Mat, Kolodyne which run southward and Karnafuli (Tlabung) towards the southwest. The total length of the rivers in the state is 1100 km. occupying an area of 6000 hectares. Tlawng is the longest river (102 km.) by length whereas Kolodyne is

Fig. 4.1: Drainage Map of Mizoram

largest by volume. The important lakes of the state include the Palak lake, Rengdil lake and Tamdil lake. They are known to possess a rich faunal biodiversity.

Ecologically the state is located within one of the identified 'hotspots' of the world harbouring endemic and endangered species of fauna and flora.Though the state is rich in flora and fauna, the biological diversity which is characterized by Indo-Malayan and Burmese elements is still largely unmapped. Very little has been explored so far aquatic fauna is concerned. There are few reports on biodiversity of fish fauna (Ramanujam and Harit, 2002; Ramanujam and Razi,2002) and the chelonians (Choudhury, 1998; 2001, Pawar and Choudhury, 2000).

## Biodiversity of Lower Invertebrates

Lower organisms like invertebrates are very important. But, biodiversity of many of them, especially insects, are still not fully understood. Apart from being bio-indicators, they play a vital role in controlling insect pests. Lower vertebrates serve as an important source of food to most carnivorous animals including humans.

The number of lakes, pools and ponds which are perennial in the state is very much negligible as the soil quality is very porous and terrain is hilly. Many of the small streams gets almost dried up during months from November to March. Several groups of lower invertebrates are known to be present in the water bodies of Mizoram. They include microorganisms, parasitic organisms, insects and their larval stages, crustaceans and their larval stages which are economically important as food animals and also disease causing organisms. There is no literature available regarding the aquatic invertebrate fauna of the state. There are no records of the systematic studies done on the micro invertebrates in the water bodies of the state inventory studies have revealed the presence of certain zooplankton listed in the Table 4.2. The macro invertebrates reported to be present in the rivers and some lakes is given in the Table 4.3. Insects with an aquatic larval forms but a winged adult are often restricted to a particular water body as compared to entirely aquatic species. They form a richest group in terms of diversity and adaptability. A detailed alpha taxonomic studies are required to know the biodiversity of both micro and macro invertebrates. A few of their larval forms may be used as food organisms in pisciculture which is coming up in the state.

**Table 4.2: Zooplankton found in the Water Bodies of Mizoram**

| *Phylum* | *Taxa* |
|---|---|
| Cladocera | *Daphnia sp.* |
| | *Moina sp.* |
| Copepoda | *Cyclops sp.* |
| | *Neodiaptomus sp.* |
| Ostracoda | *Cypris sp.* |

**Table 4.3: Macroinvertebrate Taxa found in the Water Bodies of Mizoram**

| *Phylum* | *Taxa* |
|---|---|
| Arthropoda | *Macrobrachium sp.* (freshwater shrimp) |
| | *Brotia costula Proturus pelagicus* (freshwater crab) |
| | *Culex* sp. (larva) |
| | *Chironomus sp.* (larva) |
| | *Baetis sp.* (larva) |
| | *Crocothemis sp.* (larva) |
| | *Nepa sp.* (Adult) |
| | *Ranatra sp* (Adult) |
| Mollusca | *Brotia costula* |
| | *Brotia costulata* |
| | *Belanocochlis glandiformis* |

One species of crab, *Proturus pelagicus* locally known *as* Chakai, a species of shrimp, *Macrobranchium* sp. locally called as Kaikuang and three species of freshwater snails, *viz. Brotia costula, Brotia costulata* and *Balanocochlis glandiformis,* having the same local name Chengkawl are consumed by local people on a large scale. These Gastropods are collected from rivers and streams during the low tide season. The presence of the same species have also been reported from freshwater systems of Bangladesh and Malaysia.

## Biodiversity of Chelonian Fauna

There are approx 294 species of chelonians alive today. Nearly 40 -45 species are found in India. They all share a basic appearance, with a hard shell made up of scutes (hard, bony skin) that can be described as either a hard or soft shell. Chelonians are a well-known group and can be found in ponds, streams, lakes, and oceans. Several species are endangered. There are three main groups of chelonians: turtles (medium-sized freshwater chelonians and large saltwater chelonians), tortoises (large land dwelling chelonians sometimes found in the desert), and terrapins (small turtles, in North-America considered to be one species with hollow plates on the carapace, and in Great Britain considered to be a number of species of pond turtles). There are 11 families among Chelonians. Very little is known about the chelonian fauna of the region (Choudhury,1990, 1996 a,b). Recently seven sp. of chelonians from Barak valley region of Assam has been recorded (Gupta, 2003).

The chelonian fauna of Mizoram is virtually unknown except for the recent work of (Choudhury, 1998,2001 and Harit and Ramanujam, 2002). The survey of these chelonians is mostly carried out in the western, southern and south-eastern parts of the state covering Dampa sanctuary, Ngenpui river, a tributary of the Kolodyne river towards the northern fringe of Ngengpui wildlife sanctuary and Phawngpui National park (50 sq. km.). The park ranges from 1,100 to more than 2,000 meters above msl. and has the highest peak in Mizoram (Blue Mountain–2,157 m above msl).

Table 4.4 gives the different species of Turtles and Tortoises available in the State. The tortoise is represented by a single family Testudinidae and the turtles by two families *viz.* Emydidae and Trionchidae. Six species of the turtles out of seven reported belong to family Emydidae.The Assam roofed turtle *Kachuga sylhetensis* is kept under critically endangered category, the Asian brown tortoise, *Monouria amys* as vulnerable. Three species *viz. Kachuga tentoria, Pyxidea mouhatii* and *Melanochelys trijuga* found in the state are endemic as per IUCN record, 1997.

All species of chelonians are eaten by the local tribals making conservation efforts difficult. They are being smuggled out to Far East countries through Myanmar and China as they are used in traditional Chinese medicine. The turtles are pouched by the neighbouring Myanmarese and are shared with the local people. However, some are protected in the notified wildlife sanctuaries and national parks which account for a meager 4.8 per cent of the total area of the state.

**Table 4.4: Chelonian Fauna of Mizoram**

| *Order* | *Common Name* | *Zoological Name* |
|---|---|---|
| Chelonia | Elongated tortoise | *Indotestudo elongate* (Blyth) |
| | Brown Hill/Asian brown tortoise | *Monouria emys* (Schlegel and Muller) |
| | Asian leaf turtle | *Cyclemys dentata* (Gray) |
| | Indian roofed turtle | *Kachuga tecta* (Gray) |
| | Indian tent turtle | *Kachuga tentoria* (Gray) |
| | Keeled box turtle | *Pyxidea mouhotii* (Gray) |
| | Asiatic Giant softshell turtle | *Pelochelys cantorii* (Gray) |
| | Indian black turtle | *Melanochelys trijuga* (Schweigger) |
| | Assam roofed turtle | *Kachuga sylhetensis* (Jerdon) |

Short cycles of shifting cultivation, locally known as 'jhuming', due to rapid growth in population is the main cause of their habitat destruction. Usually it is done during the months of Feb.-April burning the hill slopes for cultivation by local farmers. Following it a large number of turtles and tortoises are found either burnt or partly burnt.

## Biodiversity of Fishes

Aquatic resources, mainly fishes form a major part of food for a large population of many developing countries like India, Bangladesh, Myanmar, Nepal and Pakistan. Fishes as a very good source of protein have been exploited by man since time immemorial and most of the economically important fishes are cultured in small and large scale all over the world to meet the ever increasing demand.

Fishes constitute nearly half the number of vertebrate fauna found in the world. Of total 39,900 vertebrate species recognized all over the world over 21,723 are living species of fishes of which 8411 are of freshwater (excluding commonly diodromous fishes that may have landlocked population) and 11,650 marine species. In India total recorded fishes are 2,500 out of which 930 are freshwater and 1,570 are marine. Fishes show a great diversity in habitats, ranging from extreme deep oceans 7,000m below the surface to total dark caves. They tolerate a wide range of salinity and temperature (Jayaraman, 1999).

The state Mizoram has approximately 10,000 acres of water bodies where variety of fishes are found. Out of it nearly 6,000 hectares are riverine and rest consists of lakes, ponds and pools. It is estimated that about 1280 metric tons of fishes are procured from ponds and the average catch from river is 3.5 kg/hectare. The biodiversity studies of fishes are still in alpha taxonomic level. It is estimated that there are about 650 species of freshwater fishes available in India (CAMP workshop, 1997, Lucknow), out of which 80 species have been recorded in the state (Tables 4.5 to 4.12) (Ramanujam and Razi, 2002).

**Table 4.5: Diversity of Fish Fauna in Rivers of Mizoram**

| *Sl.No.* | *Name of the River* | *Number of* | | | |
|---|---|---|---|---|---|
| | | *Orders* | *Families* | *Genera* | *Species* |
| 1. | Tuirial | 6 | 11 | 20 | 34 |
| 2. | Tuirini | 3 | 4 | 9 | 9 |
| 3. | Tlawng | 3 | 3 | 6 | 9 |
| 4. | Tuivai | 1 | 1 | 4 | 4 |
| 5. | Mat | 5 | 6 | 13 | 17 |
| 6. | Kolodyne | 6 | 10 | 19 | 24 |
| 7. | Karnafuli | 6 | 15 | 32 | 38 |

The fishes found in the rivers of Mizoram belong to three main categories *viz.* true hill stream group, torrential fishes and plain water type. Cypriniformes fishes are found widely distributed throughout the state. More than 50 per cent fish species belong to only one family cyprinidae. The fishes belonging to genera *Barilius* and *Puntius* exhibit high degree of diversity with five species each distributed in six of the seven rivers studied. Though the density of Siluriformes and Perciformes fishes are less, they also showed a great degree of diversity represented in six rivers and by six families each. Majority of the Silurian fish species (ten out of thirteen) come under threatened category either endangered or vulnerable. Comparatively very few species of cyprinids (twelve out of twentysix) are threatened.

**Table 4.6: Fish Fauna of Tuirial River**

| *Sl.No.* | *Order* | *Family* | *Genus* | *Species* |
|---|---|---|---|---|
| 1. | Cypriniformes | Cyprinidae | *Barilius* | *barnoides* |
| 2. | | | | *bendelisis* |
| 3. | | | | *shacra* |
| 4. | | | | *tileo* |
| 5. | | | | *vagra vagra* |
| 6. | | | *Crossocheilus* | *burmanicus* |
| 7. | | | | *latius latius* |
| 8. | | | *Danio* | *aequipinnatus* |
| 9. | | | | *naganensis* |
| 10. | | | *Esomus* | *danricus* |
| 11. | | | *Garra* | *annandalei* |
| 12. | | | | *gotyla gotyla* |
| 13. | | | | *lissorhynchus* |
| 14. | | | *Neolissocheilus* | *hexagonolepis* |
| 15. | | | *Puntius* | *conchonius* |
| 16. | | | | *sarana spilurus* |
| 17. | | | | *ticto* |
| 18. | | Cobitidae | *Botia* | *dario* |
| 19. | | | | *rostrata* |
| 20. | | | *Acantoph-thalmus* | *pangia* |
| 21. | | Psilorhynchidae | *Psilorhynchus* | *gracilis* |
| 22. | | | | *balitora* |
| 23. | | Balitoridae | *Acanthocobitis* | *botia* |
| 24. | | | *Balitora* | *brucei* |
| 25. | | | *Schistura* | *rupecula* |
| 26. | | | | *scaturigina* |
| 27. | | | | *vinciguerrae* |
| 28. | Siluriformes | Sisoridae | *Erethistes* | *pussilus* |
| 29. | Beloniformes | Belonidae | *Xenentodon* | *cancila* |
| 30. | Synbranchi-formes | Mastacem-belidae | *Mastacembelus* | *armatus* |
| 31. | Perciformes | Chandidae | *Chanda* | *nama* |
| 32. | | Channidae | *Channa* | *orientalis* |
| 33. | | Nandidae | *Badis* | *badis* |
| 34. | Osteoglossi-formes | Notopteridae | *Notopterus* | *notopterus* |

**Table 4.7: Fish Fauna of Tuirini (Seruli) River**

| Sl.No. | Order | Family | Genus | Species |
|---|---|---|---|---|
| 1. | Cypriniformes | Cyprinidae | *Aspidoparia* | *morar* |
| 2. | | | *Barilius* | *shacra* |
| 3. | | | *Chela* | *laubuca* |
| 4. | | | *Neolissocheilus* | *hexagonolepis* |
| 5. | | | *Puntius* | *chola* |
| 6. | | | *Securicula* | *gora* |
| 7. | Siluriformes | Bagridae | *Mystus* | *vittatus* |
| 8. | Perciformes | Gobiidae | *Glossogobius* | *giuris* |
| 9. | | Chandidae | *Chanda* | *nama* |

**Table 4.8: Fish Fauna of Tlawng**

| Sl.No. | Order | Family | Genus | Species |
|---|---|---|---|---|
| 1. | Cypriniformes | Cyprinidae | *Barilius* | *tileo* |
| 2. | | | | *vagra vagra* |
| 3. | | | *Cirrhinus* | *mrigala* |
| 4. | | | | *reba* |
| 5. | | | *Labeo* | *calbasu* |
| 6. | | | | *rohita* |
| 7. | | | *Securicula* | *gora* |
| 8. | Siluriformes | Schilbeidae | *Clupisoma* | *garua* |
| 9. | Perciformes | Gobiidae | *Glossogobius* | *giuris* |

**Table 4.9: Fish Fauna of Tuivai River**

| Sl.No. | Order | Family | Genus | Species |
|---|---|---|---|---|
| 1. | Cypriniformes | Cyprinidae | *Barilius* | *shacra* |
| 2. | | | *Chagunius* | *chagunio* |
| 3. | | | *Puntius* | *clavatus* |
| 4. | | | *Tor* | *mosal* |

**Table 4.10: Fish Fauna of Mat River**

| *Sl.No.* | *Order* | *Family* | *Genus* | *Species* |
|---|---|---|---|---|
| 1. | Cypriniformes | Cyprinidae | *Barilius* | *bendelisis* |
| 2. | | | *Danio* | *aequipinnatus* |
| 3. | | | | *naganensis* |
| 4. | | | *Esomus* | *danricus* |
| 5. | | | *Garra* | *gotyla* |
| 6. | | | *Neolissocheilus* | *hexagonolepis* |
| 7. | | | *Puntius* | *conchonius* |
| 8. | | | | *Sarana spilurus* |
| 9. | | | *Semiplotus* | *modestus* |
| 10. | | | *Tor* | *tor* |
| 11. | | Balitoridae | *Balitora* | *brucei* |
| 12. | | | | *cavia* |
| 13. | Siluriformes | Sisoridae | *Glyptothorax* | *Sinense* |
| 14. | | | | *telchitta* |
| 15. | Beloniformes | Belonidae | *Xenentodon* | *cancila* |
| 16. | Synbranchi-formes | Mastacem-belidae | *Mastacembelus* | *armatus* |
| 17. | Perciformes | Channidae | *Channa* | *Sp.* |

The river Karnafuli bordering Bangladesh was dominant both in diversity and density. Though Tlawng is the longest river in the state the number of species recorded was much lower compared to other major rivers studied *viz.* Karnafuli, Tuirial and Kolòdyne. The average riverine fish production in the state is 3-5 kgs./hectare, whereas it is as high as 21-30 kgs./hectare in Karnafuli. Temperature and physiography may be the factors limiting the distribution and density of fishes in rivers of Mizoram. The average temperature in the western and northern regions of the state is considerably higher as compared to the average temperature in the eastern region. The average elevation is 700 meters in the west as compared to 1500 meters in the east (Singh, 1994)

The fishes of the state show Malayan affinity. Hora (1944) critically studied Indian fish genera and concluded that not only the Indian fish fauna had a marked Malayan affinity, but was even related to the fauna of Thailand and China. An examination of Indian endemic genera of fishes also shows a close kinship with fauna in countries towards east of India *viz.* Burma, Malaya, Sumatra and Java (Jhingran, 1991). The state of Mizoram, though rich in biodiversity of icthyofauna, does not have many endemics because of the political limits of the country cutting through the eastern Himalayan biogeographic zone.

**Table 4.11: Fish Fauna of Kolodyne River**

| *Sl.No.* | *Order* | *Family* | *Genus* | *Species* |
|---|---|---|---|---|
| 1. | | | *Aspidoparia* | *morar* |
| 2. | | | *Barilius* | *vagra* |
| 3. | | | | *shacra* |
| 4. | | | *Crossochelius* | *latius latius* |
| 5. | | | *Danio* | *naganensis* |
| 6. | | | *Garra* | *annandalei* |
| 7. | | | | *gotyla* |
| 8. | | | | *manipurensis* |
| 9. | | | *Neolissocheilus* | *hexagonolepis* |
| 10. | | | *Osteobrama* | *cotio* |
| 11. | | | *Puntius* | *conchonius* |
| 12. | | | | *Sp.* |
| 13. | | | *Tor* | *masal* |
| 14. | | Balitoridae | *Acanthocobitis* | *botia* |
| 15. | Siluriformes | Bagridae | *Aorichthys* | *aor* |
| 16. | | | | *seenghala* |
| 17. | | | *Batasio* | *batasio* |
| 18. | | Schilbeidae | *Eutropiichthys* | *vacha* |
| 19. | | Sisoridae | *Glyptothorax* | *telchitta* |
| 20. | Beloniformes | Belonidae | *Xenentodon* | *cancila* |
| 21. | Synbranchi-formes | Mastacem-belidae | *Mastacembelus* | *armatus* |
| 22. | Perciformes | Chandidae | *Chanda* | *nama* |
| 23. | | Gobiidae | *Glossogobius* | *giuris* |
| 24. | Anguilliformes | Anguillidae | *Anguilla* | *bengalensis* |

**Table 4.12: Fish Fauna of Karnafuli (Tlabung) River**

| *Sl.No.* | *Order* | *Family* | *Genus* | *Species* |
|---|---|---|---|---|
| 1. | Osteoglossi-formes | Notopteridae | *Notopterus* | *notopterus* |
| 2. | Clupeiformes | Clupeidae | *Gudusia* | *chapra* |
| 3. | Cypriniformes | Cyprinidae | *Amblyphary-ngodon* | *mola* |
| 4. | | | *Barilius* | *Sp.* |
| 5. | | | *Catla* | *catla* |
| 6. | | | *Cirrhinus* | *mrigala* |
| 7. | | | | *reba* |
| 8. | | | *Danio* | *naganensis* |
| 9. | | | *Esomus* | *danricus* |
| 10. | | | *Garra* | *gotyla* |
| 11. | | | *Labeo* | *calbasu* |
| 12. | | | | *gonius* |
| 13. | | | | *rohita* |
| 14. | | | *Osteobrama* | *cotio* |
| 15. | | | *Puntius* | *conchonius* |
| 16. | | | *Salmastoma* | *bacaila* |
| 17. | | | *Securicula* | *gora* |
| 18. | | | *Semiplotus* | *modestus* |
| 19. | Siluriformes | Bagridae | *Aorichthys* | *aor* |
| 20. | | | *Mystus* | *cavasius* |
| 21. | | | | *vittatus* |
| 22. | | Schilbeidae | *Ailia* | *coila* |
| 23. | | | | *punctata* |
| 24. | | | *Eutropiichthys* | *vacha* |
| 25. | | Siluridae | *Ompak* | *bimaculatus* |
| 26. | | | | *padba* |
| 27. | | | *Wallago* | *attu* |
| 28. | | Sisoridae | *Erethistes* | *pussilus* |
| 29. | | | *Nangra* | *nangra* |
| 30. | | Clariidae | *Clarias* | *batrachus* |
| 31. | | Heteropneustidae | *Heteropneustes* | *fossilis* |

*Contd...*

**Table 4.12–Contd...**

| Sl.No. | Order | Family | Genus | Species |
|---|---|---|---|---|
| 32. | Beloniformes | Belonidae | *Xenentodon* | *cancila* |
| 33. | Perciformes | Belontidae | *Colisa* | *fasciatus* |
| 34. | | Chandidae | *Chanda* | *nama* |
| 35. | | | *Parambassis* | *ranga* |
| 36. | | Channidae | *Channa* | *orientalis* |
| 37. | | Gobiidae | *Glossogobius* | *Giuris* |
| 38. | | Sciaenidae | *Johnius* | *coitor* |

At present the state fish supply is dependent to a large extent on the fishes brought from Andra Pradesh and other neighbouring states. Scarce scientific information, lack of conservation measures and old fashioned techniques of fishing has lead to a very meager fish production in the state (Pandey *et al.*, 1993). The availability of species like *Catla, Labeo, Cirrhinus, Puntius* and *Barilius* suggests that culture of these species may be taken up as an alternative to the fishes brought from other state. Further studies on various factor effecting species diversity and dominance indices in different riverine system will help in selecting the appropriate indigenous species for local culture requirement. The schemes for increasing the productivity of crops, livestock and fish single mindedly at the cost of local species and by using inputs have affected the biodiversity. The policy of promoting high yielding varieties and assessment of progress and success on the basis of consumption of fertilizer and plant protection chemicals has led to ignoring the indigenous varieties. The government subsidy and credit policy is instrumental in adopting these schemes.

Fishes like *Hetropneustus fossilis, Anabas testiduents, Channa punctatus, Rasbora daniconius, Apolochilus punchay,* and *Daneo rario* which may be available in the aquatic systems of the state may be a good biological agent for controlling mosquito larvae responsible for malaria which is very prevalent in southern parts of the state.

## Biodiversity of Amphibians

Amphibians are the highly endemic (nearly 50 per cent) as compared to fishes and other aquatic invertebrates. Out of 206 species found in the country, 110 are supposed to be endemic and 3 are threatened. There are many species of Amphibians which are known to occur in the varied ecosystems of the state. The numerous rivers and rivulets originating in the hills, along with other temporary and permanent water bodies, provide a diverse range of habitat conditions for amphibians. Interestingly, all the three living orders of Amphibia, *viz.*, Gymnophiona (limbless amphibians), Caudata (tailed amphibians) and Anura (tailless amphibians) are distributed in North-East India. There are as many as 29 species of Amphibians which are reported to be endemic to North-East India. Where ever the adults live, all Amphibians need aquatic environment for their reproduction and metamorphosis. Amphibians have been identified in six main guilds depending on their activity. They include the diurnal and nocturnal species which may be terrestrial arboreal or arboreo–terrestrial. Pawar (1999) has studied the assemblage of Amphibian fauna in relation to the alteration in their habitat. Among the eight Amphibian genera which are not found outside India two of them *viz. Bufoides* and *Microhylid* are found in Mizoram. Recently three species of frogs have been recorded from Aizawl district which falls in the northern part of the state. They include *Euphylctis cyanophylctis, Limnonectes limnocharis and Pterorana khare,* all belonging to the family Ranidae (Ramanujam and Dey, 2003). The different species of amphibians found in Mizoram are listed in Table 4.13.

## Conservation Measures

The state has certain drawbacks which have to be taken care before implementing the conservation measures effectively. They include the absence of trained taxonomists which is necessary for inventorization studies on biodiversity. It is more so with lower aquatic invertebrates. The school and college teachers maybe trained to undertake such responsibilities. Proper harvesting, storage, distribution and marketing of various products have to be streamlined.

Not all persons concerned with management of biological resources understand the concept of biodiversity in proper perspective. Many of them suffer from biased attitudes. So it is imperative that those who plan, decide and implement the developmental programmes are adequately trained and educated in favour of biodiversity conservation.

**Table 4.13: Amphibian Fauna of Mizoram**

| Family | Zoological Name |
|---|---|
| Icthyophidae | *Icthyophis sp.* |
| Megophryidae | *Megophrys parva* |
| Bufonidae | *Bufo melanostictus* |
| Microchylidae | *Microchyla berdmorei* |
| | *Microchyla ornate* |
| | *Koloula pulchra* |
| | *Uperodon cf. systoma* |
| Rhacophoridae | *Philatus sp.* |
| | *Chirixalus vittatus* |
| | *Polypedates leucomystax* |
| | *Rhacophorus maximus* |
| Ranidae | *Limnonectes cf. limnocharis* |
| | *Haplobatrachus tigerinus* |
| | *Euphlyctis cyanophlyctis* |
| | *Occidozyga sp.* |
| | *Pterorana khare* |
| | *Rana laticeps* |
| | *Rana sp.* |

## Threat to Biodiversity

Habitat destruction by man due to shifting (jhum) cultivation, expansion of agriculture in the valleys, clearance for settlement, encroachment of various kinds, felling of trees, poisoning and dynamiting the rivers for fish, are major threats to the habitat.

Introduction of the exotic species, and monoculture without proper ecological evaluation may have adverse effect on biodiversity of the indigenous species. Indiscriminate popularization of hybrid variety may influence biogenetic resources of the state. Dynamiting and poisoning of the water bodies by natural or synthetic piscicides for fish catching, poaching of turtles for smuggling to China and Far East where they are used as medicines are very prevalent among the tribals of the state. Trade in wildlife and their products, over exploitation of biodiversity beyond sustainable limit and change in food habits of the population of the state due to subsidized food distribution are also factors which threaten the diversified aquatic faunal wealth of the state. Information like studies on ecological factors of aquatic ecosystem, species inventory studies in the inaccessible areas, microbial biodiversity and information on biosphere reserves are to be taken up for optimal utilization and preserving the immense biodiversity of Mizoram.

Creation of new protected areas including the water bodies through out the state, extension of the existing protected area, check on poaching and trade, check on jhum cultivation, survey of all existing water bodies and the most important of all, the awareness campaign are the suggested measures for conservation of aquatic biodiversity of the state. Considering the high literacy in the state, any awareness campaign should be more fruitful in comparison to other states of the north-eastern region.

## References


Anonymous (1997). BCPP CAMP Workshop on Freshwater fishes of India. National Bureau of Fish Genetic Resources, Lucknow, India.

Choudhury, A. U. (1990). Two freshwater turtles of the genus *Kachuga* from Assam. *J. Bombay nat. Hist. Soc.*, 87 (1), 151-152.

Choudhury, A. U. (1996a). New localities for Brown hill tortoise *Manouria emy* (Schlegel and Muller) from Karbi Anglong, Assam. *J. Bombay nat. Hist. Soc.*, 93 (3), 590.

Choudhury, A. U. (1996b). The keeled box turtle *Pyxidea mouhotii* Gray–a new record for Manipur. *J. Bombay nat. Hist. Soc.*, 93 (3), 590-591.

Choudhury, A. U. (1998). *Pyxidea mouhotti* (Gray) in southern Assam and Mizoram. *J. Bombay Nat. Hist. Soc.* 95 (3), 511.

Choudhury, A. U. (2001). Some chelonian records from Mizoram. *J. Bombay Nat. His. Soc*, 98 (2), 184-190.

Gupta, A. K. (2003). Personal communication.

Harit, D. N. and Ramanujam, S. N. (2002). Reptilian fauna of Mizoram, India. *Cobra*, 47, 5-7.

Hora, S. L. (1944). On Malayan affinities of freshwater fish fauna of peninsular India and its bearing on probable age of the Garo-Rajmahal gap. *Proc. nat. Inst. Sci. India*, 10 (4), 423-439.

Jayaraman, K. C. (1999). The Freshwater Fishes of the Indian Region. Narendra Publishing House, Delhi, 551 pp.

Jhingran, V. G. (1991). Fish and Fisheries of India (3rd ed)., Hindustan Publishing Corporation (India). Delhi.

Pandey, A. C., Singh, S. P. and Tiwari, R. P. (1993). Fisheries resources and problems of fisheries in the state of Mizoram. J. *North Eastern Council*, 13 (1), 21-26.

Pawar, S. (1999). Effects of habitat alteration on Herpetofaunal assemblages of evergreen forest in Mizoram, North-Eastern India. *Dissertation of Master's degree, Saurashtra Univ.*, Rajkot, India.

Pawar, S. S. and Choudury, B. C. (2000). An Inventory of chelonians from Mizoram, North East India: New records and some observations on threats. *Hamadryad*, 25, 144-158.

Ramanujam, S. N. and Dey, M. S. (2003). Amphibians of Mizoram (communicated).

Ramanujam, S. N. and Harit, D. N. (2002). Report on the fish fauna of Tiau and Tuipui rivers of Mizoram, India. *Natcon.*, 07, 227-230.

Ramanujam, S. N. and Razi, M. A. (2002). A report on the biodiversity of icthyo fauna and reptilian fauna of Mizoram. Proceedings of Regional Seminar on Biodiversity of the North East, Aizawl, Mizoram.

Singh, S. N. (1994). Mizoram Historical, Geographical, Social, Economic, Political and Administrative. Mittal Publications, New Delhi

**Aquatic Biodiversity in India: The Present Scenario, 2005** **81–97**
*Edited by*: D.R. Khanna, A.K. Chopra & G. Prasad
*Published by*: Daya Publishing House, New Delhi

# 5

# Resource Assessment and Potential of Hill Fisheries in Garhwal Himalayan Region of Uttaranchal: A Perspective

**N.K. Agarwal, D.R. Khanna*, B.L. Thapliyal and U.S. Rawat**
*Deptt. of Zoology, H.N.B. Garhwal University, Srinagar*
** Deptt. of Zoology and Environmental Sciences, Gurukul Kangri University, Hardwar*

The Garhwal Himalayas comprising districts of Uttaranchal, *viz.* Dehradun, Tehri, Pauri, Chamoli, Rudraprayag, Hardwar and Uttarkashi, lie between latitude 29° 40′–31° 45′north and longitude 77° 30′–81° 06′ east, covering an area of 30,090 square km. (Vass, 2002). The district boundaries of Nainital, Almora and Pithoragarh separate it from the East, River Tons from Himachal Pradesh in the West and starting from the South, the region extends through the snow clad peaks, to the Indo-Tibbetan borders in the North (Fig. 5.1).

**Fig. 5.1: Indo-Tibetan Borders**

The Garhwal region is endowed with several cold-water lentic and lotic water bodies, offering the coldwater fisheries activities for self-reliance in animal protein diet. The lentic water bodies are snow fed rivers, their tributaries and several small spring-fed as well as snow-fed streams and rivulets. The lotic water bodies are in the form of natural cold water lakes, vernacularly called "Tals" (Table 5.1) as well as few man made reservoirs (Table 5.2). Keeping in view of large number of available water resources and indigenous fish species of commercial importance, the cold-water fishery of Garhwal has a vast potential for sustainable development and exploitation.

The aquatic resources for fisheries in Garhwal Himalayan region of Uttaranchal are broadly classified as:

**Table 5.1: Garhwal Himalayan Lakes and their Altitudinal Location**

| *Middle Garhwal (1200-1800 m als)* | | | *Upper Garhwal (1800-2400 m als)* | | | *Cold Zone (2400 m als and above)* | | |
|---|---|---|---|---|---|---|---|---|
| *Name* | *Location* | *Altitude* | *Name* | *Location* | *Altitude* | *Name* | *Location* | *Altitude* |
| Gandhiyal | Pauri | 1660 | Deoria (1.6 Ha) | Chamoli Tal | 2395.2 | Sharashat | Uttarkashi | 5745.5 |
| | | | | | | Sahashtra Tal | Tehri | 5326.3 |
| Dhar | | | Tarakund | Pauri | 2395.2 | | | |
| Bhanti Tal | Rudra Prayag | 1800 | Tarkeshwar Kund | Pauri | 2095.8 | Dev Tal | Chamoli | 5299.4 |
| | | | | | | Sidh Tal | Chamoli | 4946.1 |
| Airoli Tal | Pauri | 1800 | Jal Tal (1.6 Ha) | Chamoli | | Narsingh Tal | Uttarkashi | 4940.1 |
| | | | Sukh Tal | Chamoli | | Roop Kund | Chamoli | 4790.4 |
| | | | (1.1 Ha) | | | Matrica Tal | Uttarkashi | 4790.4 |
| | | | Rishi Tal | Uttarkashi | 1880.0 | Ling Tal | Uttarkashi | 4640.4 |
| | | | Tara Tal | Pauri | 2000.0 | Brahm Tal | Chamoli | 3280 |
| | | | Nag Tal | Chamoli | 2200.0 | Hem Kund | Chamoli | 4251.5 |
| | | | | | | Vasuki Tal | Chamoli | 4194.6 |
| | | | | | | Gandhi Sarovar | Chamoli | 4191.6 |
| | | | | | | Apsara Tal | Tehri | 4191.6 |
| | | | | | | Satopanth | Chamoli | 4191.6 |

*Contd...*

**Table 5.1–Contd...**

| *Middle Garhwal (1200-1800 m als)* | | | *Upper Garhwal (1800-2400 m als)* | | | *Cold Zone (2400 m als and above)* | | |
|---|---|---|---|---|---|---|---|---|
| *Name* | *Location* | *Altitude* | *Name* | *Location* | *Altitude* | *Name* | *Location* | *Altitude* |
| | | | | | | Hem Kund | Chamoli | 3952.1 |
| | | | | | | Bheki Tal | Chamoli | 2994.0 |
| | | | | | | Dodi Tal (3.0 Ha) | Uttarkashi | 2991.3 |
| | | | | | | Saptrishil Tal | Uttarkashi | N.A. |
| | | | | | | Masal Tal | Uttarkashi | 2890.0 |
| | | | | | | Kedar Tal | Uttarkashi | 3970.0 |
| | | | | | | Tamba Kund | Tehri | 3440.0 |
| | | | | | | Gaurikund | Chamoli | 3000.0 |
| | | | | | | Nandi Kund | Chamoli | 4450.0 |
| | | | | | | Kashni Tal | | 4680.0 |
| | | | | | | Panya Tal | | 4880.0 |

**Table 5.2: Multipurpose River Valley Hydro-electric Projects and their Status in Garhwal Himalayas**

| *Name* | *River* | *District* | *Status* |
|---|---|---|---|
| Bhairon Ghati Hydel Project | Bhagirathi | Uttarkashi | Proposed |
| Harshil Hydel Scheme | Bhagirathi | Uttarkashi | Proposed |
| Luhari Nagpala Hydel Scheme | Bhagirathi | Uttarkashi | Surveyed |
| Maneri Bhali Hydel Scheme stage I | Bhagirathi | Uttarkashi | Completed |
| Maneri Bhali Hydel Scheme stage II | Bhagirathi | Uttarkashi | Under Construction |
| Pala Maneri Hydel Scheme | Bhagirathi | Uttarkashi | Surveyed |
| Tehri Dam | Bhagirathi | Tehri | Under Construction |
| Koteshwar Dam | Bhagirathi | Tehri | Under Construction |
| Kotli Bahl Dam | Ganga | Tehri | Proposed |
| Veer Bhadra Barrage | Ganga | Dehradun | Completed |
| Vishnu Prayag Project | Alaknanda | Chamoli | Under Construction |
| Karna Prayag Project | Alaknanda | Chamoli | Proposed |
| Srinagar hydel Scheme | Alaknanda | Pauri Garhwal | Initial work started |
| Uttyasu Hydel Scheme | Alaknanda | Pauri Garhwal | Proposed |
| Vishnugad Pipal Koti Hydel Scheme | Alaknanda | Chamoli | Proposed |
| Tapowan Vishnu Gad Hydel Scheme | Dhauli Ganga | Chamoli | Proposed |
| Malan Hydel Project | Malan | Pauri Gahrwal | |
| Bilkhet Vyas Ghat Project | Nayar | Pauri Gahrwal | |
| Dak Patthar Barrage | Yamuna | Dehradun | Completed |
| Yamuna Hydel Stage-II | Asan | Dehradun | Completed |
| Kalagarh Hydel Project | Ramganga | Pauri Garhwal | Completed |
| Lakhwal Hydel Project | Yamuna | Dehradun | Under Construction |

## Riverine Resources

Two principle drainage systems *viz.* (*i*) the Ganga drainage system and (*ii*) the Yamuna drainage system contributes for the rich riverine resources to the region. The Ganga system include two major rivers the Alaknanda and the Bhagirathi and their tributaries. The Alaknanda is the biggest river of Garhwal Himalaya covering a distance of about 240 km. Its main tributaries are Vishnu Ganga, Birahi, Nandakini, Pinder, and Mandakini. The river Bhagirathi is the second largest river of Garhwal, originates from Gomukh glacier, meet to river Alaknanda at Devprayag. Its major tributaries are the Kedar Ganga, Jar Ganga, Asi Ganga and river Bhilangana. The other important rivers and rivulets of Garhwal are, the river Yamuna, Song, Jarganga, Nayar and Khoh. The Yamuna system includes the river Yamuna, which originates from Yamunotri glacier, borders the Uttaranchal state with Himachal Pradesh. Its main tributaries are river Tons, Peber, Rupin and Supin. The river Yamuna flows through district Tehri and Dehradun in Uttaranchal. Beside the natural rivers and their tributaries, the small irrigation channels and diversion channels of hydroelectric projects in the plains of district Dehradun and Pauri also provide good scope for fishery. The main riverine resources and their total length in the region are dipicted in Table 5.3.

**Table 5.3: Main Riverine Resources in the Garhwal Region**

| *District* | *Major Riverine Resources* | *Total Length* |
|---|---|---|
| Chamoli | Alaknanda, Birhi, Nandakini, Vishnuganga, Amrit Ganga, Madmahsehwari Heylang | 102 Km |
| Rudraprayag | Alaknanda, Mandakini | N.A. |
| Uttarkashi | Bhagirathi, Asiganga, Yamuna, Tons, Gaduged, Peber Rupin, Supin | 220 km |
| Tehri | Bhagirathi, Bhilangana, Alaknanda | 215 Km |
| Pauri | Ramganga, East Nayar, West Nayar, Alaknanda, Ganga | 115 Km |
| Dehradun | Ganga, Yamuna, Song, Tons, Sushwa, Asan | 179 km |
| Hardwar | Ganga | 15 Km |

## Lacustrine Resources

Several natural coldwater lakes are situated at different altitudes in Chamoli, Uttarkashi, Pauri and Ruderaprayag districts. Total area of these water bodies is about 100 ha. (Vass, 2002). Most of them are difficult to access due to tough terrain and very low temperature. These high altitude lakes can be stocked with the trout for recreational purpose to attract the trackers and anglers, as is done in Dodi Tal, where fingerlings of brown trout was stocked at experimental level. However, few lakes are situated on comparatively lower altitudes, small in size ranging from 0.12-4.00 ha and can be utilized for culture/sport fishing (Table 5.1). At present, no culture fisheries are being exercised in these lakes.

## Reservoir Resources

A number of river-valley projects have been planned for the purpose of hydroelectric generation by damming the rivers in the Garhwal Himalayas. Due to these major developmental activities, several reservoirs are coming up and some have already come up. The reservoirs in existence are Maneri Bhali Reservoir, Dak Patthar Barrage, Kulhal Hydel Barrage, Veer Bhadra Barrage, etc. But unfortunately, a culture fishery in these water bodies has yet not started. Besides, a large reservoir on the Bhagirathi River, *i.e.*, Tehri Dam Reservoir, is being constructed and will come up in existence very soon. A total water capacity at Full Supply Level (FSL) will be 3,500 million cumecs and 925 million cumecs at Dead Storage. According to a rough estimate 3,277 ha of reservoir area may be available for the fisheries development.

## Ponds

Due to the typical mountanious topography, the pond resources in hill region are very limited. But numerous small watershed/ irrigation tanks for irrigation purpose may be seen in the agriculture fields in hill region. Their diversified use for fish culture at very small level is possible. However in District Dehradun, about 50 ha and in Hardwar, about 284 ha area is covered by village ponds, which are used for irrigation and fish culture. There are also few government farm ponds at Bentwali Mandi in Dehradun. Mahseer stock is been raised here. Seed of common carp and mirror carp is also produced here.

## Fish Biodiversity

The Garhwal region of Uttaranchal state is conferred with valuable indigenous fish germplasm and pristine water resources with tremendous range in their thermal regime. In total, 64 species have been reported from different water bodies of the region (Singh *et al.*, 1987). The principle food fishes of the region belong to schizothoracids (snow trout). After schizothoracids, *Tor* (Mahaseer), *Labeo* etc. also have their own importance in the region. Some minor carps and loaches of less economic importance are also present in the small spring fed tributaries and consumed by the denizens. These belong to the Genera *Barilius, Puntius, Garra, Crossocheilus, Nimachelus, Glyptothorax, Psedechenis* and are very common in the region. The species of *Barilius, Puntius, Nimachelus,* and *Crossocheilus* are found in the small spring-fed streams of low gradient, while *Garra, Glyptothorax, Psedechenis,* occur in the turbulent waters of snow fed as well as spring-fed streams of high gradient in upper reaches. The exotic fish species-brown and rainbow trouts are also present in the Garhwal region, but restricted to the upper reaches of the river Bhilangana and river Alaknanda above 1500m asl, and few show fed rivulets Asiganga, Pinder, Virahi, etc and few lotic water bodies at high altitude *viz.* Dodi Tal etc.

## Fishery Status

### Capture Fisheries

The capture fishery contributes the major chunk in the fish food basket of the Garhwal region of Uttaranchal, particularly in the hill areas. Though, capture fishery is not done in organised way in Garhwal. The poor residents are used to going to nearby hill-streams for fishing for their own consumption and surplus catch are sold by them for their livelihood in local markets. There is hardly any fish-landing centre in the uplands from where the exact contribution of capture fishery could be ascertained. Yet it is estimated that the total fish catch from the river and streams of Tehri district alone is about 85,000 kg per year. This is based on the survey done by the author during 1999 by interviewing the local people engaged in fishing activities and also by the personal records of the shopkeepers/hotel owners who purchased fishes from local inhabitants. Capture fishery resources play vital role in supplementing protein requirement of poor people located in remote Himalayan region and providing source of income to a section of people.

The major component of capture fishery in Garhwal region is snowtrout. The minor carps and loaches also contribute to the fish catch from small streams in upper reaches but are not of much economic importance. Mahseer also registered in good amount in local catches from Nayar and Song river, being one of the principal fish species. Mahseer and *Labeo* are in considerable amount in the river Ganga at foothills also.

**Snow Trout Fishery**

The Snow trout (Schizothoracids), endemic to Himalayas, contributes to 85 per cent in the catches and are commonly occurring in almost all the snow fed streams and rivers of the Garhwal Himalayan region (Singh *et al.*, 1987). Snowtrout play vital role in supplementing nutrition requirement to poor people living in remote and hilly area of the region. In Garhwal hills snow trout fishery also provides income to a weaker section of the society who don't have cultivable land and are unemployed. The *Schizothorax richardsonii, Schizothorax plagiostomus, Schizothoraichthys curvifrons* and *Schizothoraichthys progastus* are the species of snow trout group occurring in the region. *S. richardsonii* is the only species in snow trout groups which is widely distributed from 500m–2000m asl in the region. The middle reaches of all the watershed of the Garhwal Himalayan region have vast potential of this fishery. The *Schizothoraichthys progastus* occur in good number in the catches from the mighty river Ganga at foothills and migrate upstream during summer months (Singh *et al.*,1987). These also showed their presence in the fish haul from river Mandakini and tributaries of river Alaknanda and Bhagirathi of upper reaches (Bhatt, 1999). The snow trout fishing is generally done by conventional fishing methods *viz.* gill nets commonly known as 'Phans' in the Upstream and by cast nets in the foothills.

**Mahseer Fishery**

The Mahseer (*Tor putitora*) is next in capture fishery. It has recreational as well as food value in the region. *Tor putitora* and *Tor tor* are the only species of Mahseer group appearing in the catches. Mahseer is in appreciable number in the river Ganga at foothills–Haridwar and Rishikesh. Being the Hindu pilgrim centre, fishing is banned at these places therefore a 25km stretch of river Ganga from Laxman Jhulla (Rishikesh) to Haridwar including Chilla Barrage at Rishikesh works as natural fish sanctuary for mahseer germ plasm conservation. The mahseer migrate upward in spring and summer months from the foothills -their natural abode, and ascend to the smaller tributaries of river Ganga, Alaknanda, Bhagirathi and Bhilangana for breeding (Nautiyal, 1994). The Alaknanda, Bhagirathi, Bhilangana, Song and Nayar are the important mahseer water and are ideal for the promotion of Mahseer sport fishing centres in the Garhwal Himalaya.

### Trout Fishery

The fish catch from the snow fed rivers, rivulets, and other lotic waters from above 1500 m asl also include few exotic species–Brown trout and Rainbow trout. Trouts were introduced in Garhwal region in the year 1910 by the then state emperor. The efforts of river ranching of rainbow and brown trouts are now showing results in the fish catch from high altitude streams.

### Miscellaneous Fishery

Miscellaneous fishery of the region constitutes species of *Labeo, Barilius, Puntius, Noemacheilus, Pseudoecheneis, Glyptothorax, Botia, Crossochelius* etc. The fish catch from small spring fed streams and rivulets (vernacularly called gads) mainly consist of *Barilius, Puntius* and *Noemacheilus sp.* but of very poor economic and food value due to their small size. However, *Labeo* sp. predominate in the river Ganga and other streams in foothills.

## Culture Fisheries

Culture fishery in the hill districts of Garhwal Himalayan region is very limited and needs to be encouraged. Fish ponds in upland are very few and small in sizes due to terrain. Mainly common carp is culture here owing to the availability of their seed from the govt. owned hatcheries. Trout (Rainbow trout and Brown trout) are also being culture in runoff ponds of a NGO in Chamoli District for demonstration among local people to promote trout culture in the region. Now people in the hills are gaiting interest in fish culture but still at very small level. They are using their small irrigation tanks of agriculture fields for stocking carp seeds. A small number of farmers in hill districts have started to pickup fish culture practices by converting their agriculture land locating nearby perennial water sources into run off fishponds.

However, in the foothills of Garhwal Himalaya, comparatively culture fishery exists in good way. In govt owned Mahseer fish farm at Baintwali Mandi in district Deharadun, Mahseer and common carp are produced. Few fish farms in private sector can be noticed in Hardwar, and Kotdwar districts, where profitable culture of major carps (*viz.* common carp, grass carp, *Labeo, Rohu* and *mirgala*) and cat fishes (*viz., Channa* sp.) is taking place.

## Sport Fishery

The coldwater streams in the high altitude region of Garhwal are very ideal for the sport fishery. The trout and mahseer is very popular among tourists as sport fish. As per ICAR survey of 1974, the length of streams, suitable for trout, is about 400 km (Kumar, 1992).

## Fish Seed Resources

### Hatcheries

For the development of culture fisheries in any region, hatcheries are one of the prerequisite infrastructure facilities. A very little has been done in this direction. Five hatcheries exist in the region and one is likely to be established in Pauri District

#### Kaldyani Hatchery

This oldest trout hatchery was established in 1910 at Kaldyani at an altitude of 1540 m asl, at a distance of 14 km from Uttarkashi along the River Asiganga. The brown trout, *Salmo truta fario* was introduced in the hatchery in 1910 and few thousand brown trout fingerlings were produced every year and stocked in the River Asiganga. This hatchery was badly damaged due to heavy rainfall and land-slides in 1982 and excessive floods in 1984. This needs renovation.

#### Gangori Hatchery

This hatchery was established in 1989-90 at Gangori along the River Bhilangana at a distance of about 5 km from Uttarkashi. The hatchery is fed by ground water, naturally coming out from the mountains. The seed of trout and mirror carp is produced here and used for the stocking in Bhagirathi river. The seed is also supplied to private individuals on a very small scale.

#### Talwari Hatchery

The Talwari Hatchery is located in Pinder Valley at Talwari at an altitude of 1,700 m asl. It is a spring-fed hatchery and it was originally established for trout propagation. The rainbow trout seeds produced at this hatchery are being stocked in the Pinder and the Nandakini Rivers.

### Varangana Hatchery

This trout hatchery was recently established in Varangana, 10 km from Gopeswar. Here, the mirror carp seed and rainbow trout seeds are being produced in good amount.

### Mahseer Hatchery

This hatchery is established at Baintwali Mandi Dehradun. Mahseer seed are produced here. Mahseer farm activity at experimental level is going here.

## Potential of Hill Fisheries

As stated above, there is a vast network of water bodies in form of small tributaries, natural lakes and man made reservoirs. Some more reservoirs (by the completion of hydro electric project) will take shape in coming years. All these water bodies have vast potential for capture as well as culture fisheries development. At present, no attempt has been made to manage these water resources for fishery viewpoint and not even to explore the possibilities. A very preliminary work has been carried out on the fish and fisheries of this region (Badola and Singh, 1981; Singh *et al.*,1987; Singh and Agarwal, 1993). However, several workers have studied the Physico-chemical and Biological analysis of various water bodies (Badola and Singh, 1982; Singh, 1985; Sharma, 1987; Nautiyal *et al.*,1993; Singh *et al.*, 1994; etc.).

These studies are helpful in knowing the pre-impoundment characteristic of fluvial bodies. The Physico-chemical and biological studies in upper reaches of the Ganga (Singh, 1988) provide us the data for the pre-impoundment characteristics of some reservoirs, taking shape on the Bhagirathi and Ganga rivers. A comparative study of some Physico-chemical and biological characteristics of the river Bhagirathi at Gangnani and the reservoir at Maneri is available (Nautiyal *et al.*, 1988). These basic data may help to start the coldwater reservoir fishery. Further emphasis is required on similar lines, because due to the upcoming reservoirs, a vast area of water will be available in future for reservoir fisheries.

## Strategies for Development of Hill Fisheries

The potential of using aquatic resources for the expanding demands of food and employment opportunities is apparent in Asia (Fernando, 1984). This is also true for Garhwal Himalaya. Basic strategy for the development of hill fisheries should be to adopt separate measures and plans for different types of water resources. In this context, water resources are classified as:

1. Large reservoirs
2. Small reservoirs
3. Natural lakes
4. Small ponds and tanks
5. Small perennial streams.

Due to the construction of large dams, the flowing system is getting converted into large semi-lacustrine system with high water column, results in the complete change in Physico-chemical and biological environment. These large water bodies are basically meant for power generation and will face high water level fluctuation. The environmental factors will also vary. Hence it is very necessary to select such fish species for stocking which can easily survive, breed and grow fast by utilising the maximum food available in all the water zones. The species recommended in such large reservoirs for culture fishery management are Mahseer (*Tor putitora*), snow trout (*S. richardsonii, S. curvifrons, Schizothoraichthys progastus*), minor carp (*Labeo dero, L. diochelius*) etc. Apart from these, some exotic carps, such as mirror carp, Grass carp, and rainbow trout can also be considered for culture.

The stocking-restocking pattern for fishery management is a common type of practice in the large reservoirs. But the natural breeding grounds of the indigenous fishes should also be developed and protected near the headwaters of the reservoir. Prior to developing and protecting the breeding grounds, spawning behaviour and ecology of the indigenous fishes should be taken into consideration. The schizothoracids are the main commercially important indigenous food fish of the region. The information on the breeding biology of *Schizothorax* (Agarwal, 1996,2001) and *Tor putitora* (Nautiyal, 1994) are available. However, more emphasis is needed on the breeding aspects of indigenous cultivable fishes.

In big reservoirs and high altitude lakes, there is distinct possibility of cage culture practice by using flexible frames and floating cages. The designing of cheaper cages by using local available material will certainly usher in a new and profitable fish farming system in the region.

In small reservoir, natural lakes and ponds, the stocking–restocking patterns for fish culture is most appropriate. The strict conservative measures, like close season, close area, size limitation of mesh and fish, and regulation of fishing catch limits should be followed. This will ensure high yield of fish from reservoirs and lakes.

The small ponds and tanks (meant for irrigation) may also be used for fish culture. This may be done by providing technical assistant/training to the villagers and creating awareness among them.

Suitable stretches of small perennial streams with low gradient, should be identified to make small running water pools by the side of the stream, with provision to check excess water, inflowing in the pools. In such small pools, stocking–restocking patterns should be adopted. The fingerlings of *Schizothorax* sp. Common carp and grass carp are suitable for stocking these small pools. The streams like Pinder, Birahi, Madmaheshwari, Balganga, Asiganga, Mandakini Nandakini, Bhilangana are suitable for stocking of trout and Schizothoracids seed. These streams are very rich in periphyton and benthic invertebrates, *viz.* larvae and nymphs of several insects (Ephemeroptera, Plecoptera, Trichoptera and Diptera), which serve as a food organisms. The water temperature (range between 8-19 °C) and DO content (9-13.5 ppm) are quite suitable to trout ranching programmes in these streams.

To meet the fish seed requirements and large scale production of fingerlings of snowtrout, mahseer and mirror carps, the hatcheries and fish farms should be set up primarily on the periphery of the reservoirs and others suitable sites. Now technique for artificial breeding and hatching of snowtrout is also available (Agarwal *et al.*,2001). Earlier this was a constraint for snow trout seed production.

The strategies for enriching the food resources at different tropic level of the reservoirs and lakes may be achieved by acclimatising and transplanting choice feed organisms to the fish. The fish food manufacturing units should also be set up in the region for providing cheap artificial feed to the hatcheries and fish farms.

Organisations like fishery co-operative societies should be set up and promoted for proper marketing and transportation of fish catch. This will be helpful in socio-economic development of fishing community.

The efficient conservation measure can be undertaken with the effective implementation of fisheries act. The assistance from forest department and local bodies in this direction may also be a part of the strategies.

Finally, the people involved in fishing should be trained in latest fisheries technically know how. The regular field training and refresher courses for generating trained manpower will certainly work towards the development of fisheries in the region.

## Acknowledgement

The authors are grateful to Prof. H. R. Singh, Head, Department of Zoology, Allahabad University and Prof. Asha Chondola Saklani, Head, Department of Zoology, HNB Garhwal University, Srinagar Garhwal, for encouragement. We are also thankful to ICAR New Delhi [Project No. 4 (22) 95-ASR I] for the financial assistance.

## References

Agarwal, N. K. (1996). Fish Reproduction. A P H Publishing Corporation, New Delhi. Pp. 1-157.

Agarwal, N. K. (1996). Reproductive strategies of Snowtrout (*Schizothorax* sp). with reference to the environment of the river Bhagirathi and Bhilanagana of Tehri Garhwal. Final project report of ICAR Scheme, New Delhi pp 1-91.

Agarwal, N. K, Thapliyal, B. L. and U. S. Rawat (2001). Artificial Breeding of snowtrout, *Schizothorax richardsonii*, inhabiting the Bhilangana river of Garhwal Himalaya. J. Inland Fish Soc. India. 33 (1), 77-80.

Badola, S. P., and H. R. Singh (1991b). Fish fauna of Garhwal hills. Part IV (Tehri district). Ind. J. Zool., 18 (2): 115-118.

Badola, S. P., and H. R. Singh (1991b). Hydrobiology of River Alaknanda of the Garhwal Himalayas. Indian J. Ecol., 8; 269-276.

Fernando, C. H. (1984). Reservoirs and lakes of south east Asia (Oriental region). In: F. B. Taub (ed), Lakes and Reservoirs. Amsterdam Netherlands, Elsevier Science Publishers, pp 411-446.

Kumar, V. (1992). Present status of cold water fisheries of Garhwal Himalayas. In: K. L. Sehgal (ed)., Recent Researches in cold water fisheries. Today and Tomorrow Printers and Publishers, pp 11-18.

Nautiyal, P. (1994). Mahseer, the game fish–Natural history, status and conservation practices in India and Nepal, Rachana publishers, pp A-1-D-77.

Nautiyal, P., N. K. Agarwal and H. R. Singh (1993). Certain abiotic components of the fluvial system and their inter-relationship as evidenced by statistical evaluation. In H. R. Singh (ed), Advances in limnology. Narendra Publishing House Delhi, pp 181-186.

Nautiyal, P., Pokhriyal, R. C., Gautam, A., Rawat, D. S. and Singh, H. R. (1988). Maneri dam on the river Bhagirathi–A lacustrine environment in the making. In: S. K. Kulshreshtha, U. N. Adholia, O. P. Jain and Anita Bhatbagar (eds), Proc. Nat. Symp. on past, present and future of Bhopal Lakes. pp. 175-182.

Sharma, R. C. (1987). Aims and strategies of the fisheries management for the Tehri dam reservoir in Garhwal Himalayas, India. In: Proc. Asian reservoir fisheries management (IDRC Canada), pp 145-153.

Singh, H. R. (1985). The Bhagirathi ecosystem In: Environment and natural resources, eds. V. P. Agarwal and S V. S. Rana. Soc. Biosci., Muzaffarnagar, India. pp 151-155.

Singh, H. R. (1988). Pollution study of the upper Ganga and its tributaries. Final technical report of research project submitted to Ganga Project Directorate, New Delhi, pp 222.

Singh, H. R., Badola, S. P. and Dobriyal, A. K. (1987). Geographical distributional list of ichthyofauna of the Garhwal Himalayas with some new records. J. Bombay Nat. Hist. Society, vol 84 (1): 126-132.

Singh, H. R., Nautiyal, P., Dobriyal, A. K., Pokhriyal, R. C., Negi, M., Baduni, V., Nautiyal, R., Agarwal, N. K., and Gautam, A. (1994).

Water quality of river Ganga (Garhwal Himalayas). Acta. Hydrobiol. 36 (1): 3-15.

Singh, N and Agarwal, N. K. (1993). Organs of adhesion in four hill stream fishes: A comparative morphological study. In: Advances in Limnology eds. H. R. Singh. Narendra Publishing House, New Delhi, 311-316.

**Aquatic Biodiversity in India: The Present Scenario, 2005** **98–106**
*Edited by*: D.R. Khanna, A.K. Chopra & G. Prasad
*Published by*: Daya Publishing House, New Delhi

# 6

# Present Status of Biodiversity and Strategies for Development of Wetlands in South Konkan in India

**S.G. Yeragi and S.S. Yeragi**
*K.J. Somaiya College of Science and Commerce, Mumbai – 400 077*

The undisturbed mangrove forests may act as the sea word barrier and check considerably the coastal erosion and minimize the tidal thrust or strong storm hit from the sea. Davies (1910) defined the mangrove as the plant which live in muddy loose wet soils in tropical tide waters.

Mangroves are the only trees amongst relatively small group of higher plants that have been remarkable successful in colonizing the intertidal zone of the interphose between land and sea".

Aquaculture plan of Government of India was for increasing the production of prawn to get the foreign exchange but this exploitation greatly damages the mangroves hence there was a ban. The conflict between the exploitation of mangrove for human needs, for development and construction of aquaculture farm in intensive or semi intensive manner in the mangrove reclainged zone on one hand and the conservation of the mangrove forest or ecosystem for sustainable yield of the forests and sustenance of mangrove and its adjacent coastal capture fisheries. A correlation between certain penaeid species and mangrove forest appear to hold and one might say no mangroves no prawns.

The wetland literature contains a vast number of terms to designate different kinds of wetlands. This terminology is often used differently by different workers and a number of taxonomical schemes have been put forth. The estimated area of global wet lands is 7 to 8 million $Km^{2.}$ Bogs and Fens occupy 60 of the total estimated area (3.46 million $Km^{2)}$. Wetlands are extremely productive part of the landscape. Wetlands are continuously disappearing throughout the world. Loss rate in USA is 80000 to 1,60,000 hectares per year (2 to 4 lakh hectares/yr).

To meet their demand for fuel, bread and butter, lakhs of these rural people exploit these forest wood from this slow growing mangrove woodlands and harvest or catch fish, prawn, and crabs with the nets. This large-scale exploitation of these natural resources gives little chance for the replenishment of these natural wealth or resources.

Sustainable development is not possible without wetland because of their critical role in water quality and Biodiversity. Therefore sound wetland science is needed to formulate government regulations and policies.

**Table 6.1: Wetland of Indian Continent**

| | | |
|---|---|---|
| River and canals | = | 1.71 lakh/Km |
| Area under Reservoirs | = | – 20.50 lakh/hectare |
| Tanks and ponds | = | 28.55 lakh/hectare |
| Brackish water areas | = | 14.22 lakh/hectare |

## Wetlands Ecosystem

Wet lands are a group of highly evolved halophytes occupying the intertidal zone in estuaries, creeks, lagoons and coastal mud flats. They grow well in poorly oxygenated water logged saline habitats, a seasonal fluctuation with fresh water is also equally important for their life long survival. The undisturbed wetlands act as the seaward barrier and check considerably the coastal erosion and minimize the tidal thrust. Traditionally wetlands are considered to be transitional ecosystem often described as successional links in series from open water to land.

1. The International Union for the Conservation of Nature and Natural Resources (IUCN) treating wetlands as just "wet areas". Wet lands are areas submerged land twice a day in tidal cycle water dominated areas to be considered would include marshes, bogs, swamps, fens, peatlands, estuaries, bays, sounds, lagoons, lakes, rivers and reservoirs.
2. Opposed to this, a biological and relatively narrow conceived definition was adopted during the International and Biological program (IBP). A wetland is an area dominated by specific herbaceous macrophytes the production of which takes place predominantly in the aerial environment above the water level while the plants are supplied with amounts of water that would be excessive for most other higher plants bearing aerial shoots.
3. U.S.Fish and Wildlife Service (USFWS) by states that wetlands are lands transitional between terrestrial and aquatic systems where the water table is usually at or near the surface.

## Scope's Wetland Types

1. Bogs are typically oligotrophic and usually acidic and dominated by *Sphagnum* species.
2. Fens are minerotrophic, meso to eutrophic in which nutrients are rich and dominated by herbaceous plants.
3. Marshes are mainly meso–to eutrophic and moderately saline with dominated by herbaceous plants.
4. Swamps are minerotrophic with or without peat and thick woody shrubs or trees.
5. Shallow lakes are permanent bodies of water may be saline, alkaline or fresh, lot of disturbances, water turbidity high due to high velocity of wind.

6. Floodplains are systems of wetlands associated with rivers and streams, comprising relatively large areas of riparian forests.

## Functions of Wetlands

1. Wet land ecosystem is highly productive zone compared to others.
2. Prawn species breed and complete their life cycle in the shallow mangrove water.
3. The *Scylla serrata* is a dominating species inhabiting with high density muddy ground, dense forest, rich detritus, protection of still roots favour this species.
4. The fish like *Mugil* spp. *Latus calcarifer* are dominated their population density is always high compare to open estuaries and creek. Annual production is high in wetland of Deogad, Mumbai, Mithbav, Achara, and Uran etc.
5. Average annual production for wetland is around 1125 $cm^2 yr^{-1}$.
6. The standing stock of animal biomass the annual animal production is also relatively high in wetland ecosystem.
7. The primary production is high due to high rate of photosynthesis hence secondary production is also high.
8. The wet lands imply an important landscape role in nutrient recycling and storage plant and animal harvest and species conservation.
9. Mycoflora of wetlands help in the formation of insoluble organic polymers in to soluble monomers which continuously add nutrient load in associated area of the creeks and esturies. The mycoflora help in making the wetlands highly nutritious having thick population of resources organisms are present throughout the seasons.
10. Wet lands appeared as safe home for avifauna. Many water, birds present in large number in these areas. Population of birds is decreasing all over the world but wetlands are only survival sources because human interference is very lesser and so safe for their nesting.

The flooded wet lands provide a refuge from predaters and security from the physical and biological threats of channels.

### Wetlands Contribute

1. World's food production.
2. Fisheries development.
3. Preservation of Avifauna.
4. Increasing nutritional load.
5. Pollution control.
6. Protection from land erosion.
7. Recreation.
8. Waterflow protection.
9. Overflow biological Diversity.

## Future Development and suggestions for wetlands in 21st century

In the Indian sub-continent, the mangrove ecosystems along with about 35 true mangrove species and more than 35 mangrove associated flora or back mangals are distributed within the intertidal zone of east and west coast of India.

Untawale and Jagtap (1992) have mentioned 47 mangroves and 37 mangrove associated flora from Indian deltaic mangals. India has very long coastal line and thus long coastal line is very much variable in relation to its physiognomy and ecological features. Indian mainland coast line length is extended to 5,700 Km, within these zones, mangrove areas were estimated only 3,979 $Km^2$ and mud flat 22,961 $km^2$. Since the last five decades,the Indian mangrove ecosystem have been highlighted by several Indian researches. Recently, the mangrove swamps is merged under forest department.

Several of these coastal mangrove ecosystems bear the important 'Gene Pool' for several endangered or threatened flora and fauna. These mangrove forest resources have provided good number of articles of commerce to the rural poor people. Mangrove timbers are hard, durable, free from pests and possess high calorific values and very much used for housing and fuel wood. Local native collect large-scale honey and wax. Ecological engineering of wet lands the development of better techniques for building and constructing wetlands that initiate natural wetlands.

Complete inventories of wetlands. Redefination of wetland taxonomy and evaluation systems.

Development of sustainable economic uses of wetlands while protecting their natural features education of Govt.officials, land owners, engineers and lawyers and biologists on wetland functions, values and technique for management.

Improved international co-operation and information exchange on wetlands. Advanced planning for wetlands and mass awareness to the natives. Improved tools such as ecosystem models and geographic information system for better management of wetlands.

## Problems of South-Konkan Wetlands

The problems of the South-Konkan mangals are identified as:

1. Ecological problems.
2. Management problems
3. Regeneration problems
4. Afforestation Problem
5. Mortality of trees

The mangrove forest ecosystem is dynamic one and changes very frequently. These changes may be caused due to continuous siltation, erosion of the mangal lands, which caused by routine tidal and upstream flow of neighbouring rivers, channels and creeks.In the changing mangrove environment, the succession of the diverse mangrove species also show unique adaptation.The wetlands like *Avicennia* spp, *Sonneratia* spp, *Excoecaria* spp.are the pioneer trees grow as and when the new lands or island come up,the rate of biodiversity amongst these species is high and hence they are present all over the estuaries and creeks of South-konkan.

No proper or adequate research and development have been undertaken in Konkan. The steps and research programmes are as follows:

1. Phonological studies of mangroves.
2. Reproduction techniques should be developed for the mangroves.

3. Nursery development techniques for important mangroves require to develop.
4. Studies on the regeneration of dominant species of mangroves.
5. Studies on natural regeneration status.
6. Studies on growth and yield of major mangroves in relation to salinity.
7. Studies on the effect of hypersaline condition.
8. Studies on the pest.
9. Studies on the deforestation of *Sonneratia apetala* (Buch–Ham)
10. Pathological investigation
11. Mycological studies in relation to decomposition of waste dumped.

Most of these work programmes are based on the forestry management in relation to harvest the timbers and other forest production.

## Conclusion

Though mangroves are saet tolerant species, their better growth observed in less saline zone with much height and more density.A little attention has been made for the conservation of the endangered mangroves and study the mangal ecosystem of the south Konkan.Therefore the wet lands are not Waste lands.The mother nature has given this beautiful landscape to generate resources and protection of life.We don't want lip sympathy, but let us not destroy it if we want to preserve it.

## Qualitative Status of Mangroves in South Konkan

1. *Rhizophora apiculata* (Blume)
2. *R. muconata* (Lanite)
3. *Kandelia candel* (L) Druce
4. *Criceps decandra* (Griff)
5. *Ceriops tugal* (Perr) *Rab*
6. *Bruguiera gymnorrhiza* (L) Lank
7. *B. parviflora* (Raxb)

8. *Avicennia alba* (Blume)
9. *A. marina* (Forsk)
10. *A. officinalis* (L)
11. *Sonneratia alba* (Smith)
12. *S. casedaris* (L)
13. *S. apetala*
14. *Exoecaria agallocha* (L)
15. *Acanthus ilicifolius* (L)
16. *Derris trifoliate* (Lour)
17. *D. scandens* (Benth)
18. *D. indica* (Bennett)
19. *Salvadora persica* (L)
20. *Lumnitzera racmosa* (Wild)
21. *Aegiceras comiculatum* (BI)
22. *Acrostichum aureim* (L)

## References

Babu K. N. (1999). Environmental Studies in relation to mangroves of Uran creek. Ph. D. Thesis Jagtap T. G. (1987). Seasoned Distribution of organic matter in mangrove environment in Goa. India. J. Mar-Sci 16. PP. 103-106.

Kumudrajan Nasker. Ecology and Biodiversity of Indian mangroves. Daya Pub. House Delhi–110035.

Navalkar B. S. (1952). Succession of Mangrove vegetation in Bombay and selected islands. J. Bom. Nat. Hist. Soc. 50 157-166.

Patwardhan, D. D. (1986). Ecology of Mithbav creek. M. Sc. Thesis, University of Mumbai.

Patwardhan, D. D. (1990). A study of estuarine ecology of South Konkan, Ph. D. Thesis, University of Mumbai.

Shet P. B. (1994). Ecology of mangrove in relation to organisms. M. Sc. Thesis, University of Mumbai.

Sidhu S. S. (1963). Studies on the mangroves of India East Godavari region, India. Forestar, 89: 337-351.

Untawale A. G. (1984). Mangroves of India. Present status and multiple use status report submitted to the UNDP/UNESCO. Regional mangrove project for Asia and Pacific.

Vyas M. M. (1991). Ecology of Manori creek. M. Sc. Thesis, of University of Mumbai.

Yeragi A. S. (2002). Biodiversity and environmental studies of Mangroves. Ph. D. Thesis, University of Mumbai.

**Aquatic Biodiversity in India: The Present Scenario, 2005** **107–122**
*Edited by*: D.R. Khanna, A.K. Chopra & G. Prasad
*Published by*: Daya Publishing House, New Delhi

# 7

# Algal Communities in Papnash Pond Bidar Karnataka, India

**N. Shiddamallayya, S.B. Angadi and P.C. Patil**
*Phycology Laboratory, Department of P.G. Studies and Research in Botany, Gulbarga University, Gulbarga – 585 106, Karnataka*

The study of fresh water system with respect to physical, chemical, geological and biological parameters is termed as limnology. Fresh water biology emphasizes the ecology of organisms in relation to their fresh water habitat. It occupies a relatively small portion of the earth surface as compared to marine and land habitat, but their importance to man is far greater than other areas because, they are the most convenient and cheapest source of water for domestic and all other activities of man. They also provide the important sites for waste disposal systems.

Fresh water habitats are isolated from each other by land and sea. The sea organisms with little dispersal mechanism fail to get established even in favourable places. The niches in different rivers, streams, lakes and ponds which are separated from each other will be occupied by different species. The combination and concentration of various factors influence the growth and diversity of phytoplankton. Changes in phytoplankton species composition are a central feature of aquatic ecosystem dynamics. Phytoplankton constitutes the vary basis of nutrient cycle of an aquatic ecosystem. They play a key role in maintaining proper equilibrium between biotic and abiotic components of aquatic ecosystem. Many limnologists have carried remarkable investigation on the phytoplankton of different water bodies and its seasonal variation. Significant among those are that Hutchinson (1944) has worked on the phytoplankton periodicity and chemical changes in lake waters. Niemi (1972) observed the variations in phytoplankton population in eutrophic and non-eutrophic water bodies of Finland. Berman and Pollingher (1974) have studied the annual and seasonal variations of phytoplankton distribution in Lake Kinneret. Ilmavirata *et al.* (1984) reported phytoplanktons in Finnish lakes and emphasized on its species composition and its relations to water chemistry. Arvola (1986) studied the abundance and species composition of phytoplankton in 54 small lakes in southern Finland. Pieterse and Roos (1992) investigated the diurnal changes in physical, chemical, phytoplankton biomass and their interrelationship. Kilham *et al.* (1996) studied the abundance and seasonal changes of planktonic diatoms. Harvey *et al.* (1997) described multivariate analytical methods to analyze the abundance and species composition of phytoplankton in relation to water quality. Noges and Laugaste (1998) studied the seasonal and long-term changes in phytoplankton of Lake Varotsjav. Huden-christiane (2000) investigated phytoplankton assemblages in the St Lawrence river.

In India, many hydrobiologists have carried out a considerable amount of research on seasonal succession, production, phytoplankton abundance and the variations in the different fresh water ecosystems. Gandhi (1956) made an attempt to know the knowledge of fresh water Diatomaceae of sub western India. Govind (1963) worked on the phytoplankton functions with reference to environmental condition and factors regulating the production and succession of algae. Ray and Rao (1964) studied the density of diatoms in relation with the water conditions. Seenayya (1971) studied the phytoplanktonic populations in pond of Hyderabad.

David *et al.* (1974) worked on phytoplankton in relation with limnological studies. Bharati and Bongale (1975) studied the systematic account of fresh water algae of Raichur. Sharma *et al.* (1978) studied the seasonal variations of physico-chemical characteristics in several fresh water bodies around Jaipur and compared with the phytoplankton communities, Hosmani and Bharati (1980) carried out the work on limnological studies of ponds and lakes of Dharwad with reference to phytoplankton ecology of four water bodies. Panth *et al.* (1983) worked on the phytoplankton populations diel variations in sub-tropical lake. Kant and Raina (1985) studied the qualitative and quantitative distribution of phytoplankton. Goel *et al.* (1986) studied the fresh water bodies with special reference to their chemistry and phytoplankton in southern Maharashtra. Ayyappan and Chandrasekhara (1987) investigated the spatial and temporal fluctuations of phytoplankton of Ramsamudra tank in Dakshina Kannada district. Hosmani (1988) studied the seasonal changes in phytoplankton communities in a fresh water pond of Dharawad Karnataka. Singh and Ahmed (1990) studied the comparative status of phytoplankton in relation to limnological characters of a pond. Shastree (1992) discussed on dynamics of phytoplanktonic fluctuations in lentic water body. Rajkumar *et al.* (1994) observed a species (phytoplankton) in relation with environmental factor in urban aquatic ecosystems. Padhi (1995) studied the phytoplankton of polluted and unpolluted fresh water ponds. Hegde and Sujata (1997) studied the distribution of planktonic algae in three fresh water habitat of Dharwad. Kalyani and Charya (1999) worked on phytoplankton diversity, dynamics and abundance in Bhadrakali Lake, Warangal, Andhra Pradesh. Prasad *et al.* (2000) surveyed the summer phytoplankton assemblages of 15 ponds at Varanasi.

Limnological studies of water bodies occurring northern parts of Karnataka was limited to Gulbarga region only. Gaddad *et al.* (1983); Vijayakumar (1992); Vijayakumar and Paul (1994); and Angadi *et al.*, 1999) have conducted studies about the physico-chemical and hydrobiology of water in and around the ponds and lakes of Gulbarga district. But no one had ever conducted limonological studies on the water bodies of Bidar. Hence the present account is an attempt to accumulate an information pertaining to algal community with reference to physico-chemical nature of standing fresh water body from the Bidar.

## The Study Area

The present investigation has been carried out on Papnash pond Bidar which falls under 17°-55′ 'N' latitude and 77°-32′ 'E' longitude and located about 551 meters above the sea level. The temperature ranges from 12 °C to 44 °C and rainfall of 966.9 mm has been recorded during the period (Table 7.1). The pond is situated at the western side of the Bidar city, which's northeast and southwest boundaries are flanked by hills. Fields demarcate eastern and western boundaries. The mean depth of water level in pond is 2.1 meter with a maximum depth of 4 meter during the rainy season and the minimum being 0.75 meter during the dry season. *Acacia, Dalbergia* and other vegetation covers the northern and southern sides of the pond. It is covered with a large quantity of *Ipomea fistulosa.* Mart. This pond is exploited by the pilgrim and surrounding people for various domestic activities, such as bathing, washing of clothes as well as cattle dumping activities.

**Table 7.1: Monthly Total Average Values of Rainfall Data of Bidar**

| *Month* | *Rainfall (mm)* |
|---|---|
| October-1999 | 76.7 |
| November | NIL |
| December | NIL |
| January-2000 | NIL |
| February | 4.00 |
| March | NIL |
| April | 12.7 |
| May | 8.2 |
| June | 231.2 |
| July | 187.7 |
| August | 388.5 |
| September | 57.9 |

Total Rainfall during the period of study was 966.9 mm.

Total monthly average Rainfall during the study period was 80.575 mm.

## Materials and Methods

The investigation of physico-chemical and biological parameters carried out during October–1999 to September–2000. The water samples and algal samples were collected on monthly basis for the study of various physico-chemical parameters and algal community.

All the field activities and observation of the pond were made between 8: 30 to 10: 30 AM through out the study period and samples were brought to the laboratory for further analysis. Standard methods prescribed for the analysis were employed for the estimation of various parameters, according to (APHA, 1995 and Trivedi and Goel, 1986). Algal samples were examined under microscope by using micrometer and preserved in 4 per cent formalin. Identifications of algal taxa were made by the following keys given by (Prescott, 1951; Deshikachary, 1959 and Anand, 1998).

## Results and Discussion

The investigation on the Papnash pond Bidar has been for the first time carried out to reveal the limnological parameters. Some work has been done on the limnological study of lakes, ponds, streams and reservoirs in India. The influence of abiotic factors on water body and its relationship with phytoplankton was noticed. The fresh water body's size, shape, depth, climatic conditions, the region to which they belong, flora, fauna and pollution problems play an important role making it difficult to compare the characteristics of individual water body with one another.

The pond has been polluted to a large extent by disposal of sewage and other human activities. Hence in the present investigation effort was made to correlate the physico-chemical factors with the algal population. Therefore a number of physico-chemical and biological parameters have been analyzed during the study. The results of the physico-chemical characteristics have been illustrated in Table 7.2.

Physico-chemical factors of the water body plays an important role in qualities and distribution patterns of different phytoplankton groups. This is due to various groups of algae have different requirement of nutrients. The fluctuations in phytoplankton periodicity depending upon availability of various nutrients are presented in Table 7.3.

### Myxophyceae

Members of Myxophyceae are ubiquitous in natural waters and many times forms temporary or permanent blooms in polluted water body. It was observed that the blue green algal sps. ranged from 1 to 4 during the study period, *viz. Oscillatoria limosa, O. irrigua, Gleocapsa*

**Table 7.2: Monthly Values of Physico-chemical Parameters of Papnash Pond, Bidar (October 1999 to September 2000)**

| *Month/Year* | *Atmospheric Temperature °C* | *Water Temperature* | *ph* | *$CO_2$* | *Total Alkalinity* | *DO* | *$Cl_2$* | *Dissolved Organic Matter* | *$NO_2$–N* | *$PO_4$–P* | *Total Hardness* | *$SO_4$* |
|---|---|---|---|---|---|---|---|---|---|---|---|---|
| October 1999 | 31.0 | 24.0 | 7.40 | 0.22 | 575 | 10.0 | 05 | 0.80 | 0.015 | 0.010 | 138 | 3.2 |
| November | 30.0 | 21.0 | 7.70 | 0.29 | 750 | 11.4 | 11 | 1.13 | 0.031 | 0.014 | 130 | 1.2 |
| December | 31.0 | 19.0 | 7.40 | 0.36 | 725 | 07.6 | 17 | 1.09 | 0.138 | 0.010 | 148 | 4.0 |
| January 2000 | 30.5 | 22.0 | 7.25 | 0.37 | 740 | 06.0 | 11 | 1.28 | 0.182 | 0.090 | 166 | 8.8 |
| February | 31.0 | 21.5 | 7.20 | 0.55 | 565 | 05.6 | 18 | 1.47 | 0.141 | 0.350 | 168 | 9.3 |
| March | 31.0 | 20.0 | 7.40 | 0.75 | 645 | 09.6 | 12 | 1.24 | 0.019 | 0.464 | 200 | 16.3 |
| April | 32.0 | 22.0 | 7.45 | 0.52 | 720 | 06.4 | 15 | 1.60 | 0.040 | 0.069 | 200 | 9.1 |
| May | 37.0 | 26.5 | 7.60 | 0.20 | 520 | 05.6 | 16 | 1.09 | 0.022 | 0.113 | 216 | 8.6 |
| June | 26.0 | 22.5 | 7.20 | 0.44 | 445 | 05.8 | 17 | 0.70 | 0.017 | 0.034 | 168 | 8.2 |
| July | 29.0 | 25.0 | 7.80 | 0.85 | 345 | 08.6 | 19 | 1.53 | 0.084 | 0.318 | 270 | 7.6 |
| August | 25.0 | 24.0 | 7.60 | 1.05 | 380 | 09.6 | 10 | 0.96 | 0.015 | 0.041 | 058 | 6.0 |
| September | 31.5 | 26.0 | 7.50 | 1.10 | 415 | 09.2 | 23 | 0.83 | 0.016 | 0.019 | 088 | 6.0 |

All the values expressed in $mgl^{-1}$ except temperature and pH.

**Table 7.3: Periodicity of Papnash Pond, Bidar (October 1999 to September 2000)**

| Sl.No. | Name of the Species | 1999 | | | 2000 | | | | | | | | |
|---|---|---|---|---|---|---|---|---|---|---|---|---|---|
| | | Oct. | Nov. | Dec. | Jan. | Feb. | March | April | May | June | July | Aug. | Sep. |
| **Myxophyceae** | | | | | | | | | | | | | |
| 1. | *Oscillatoria limosa* | + | + | + | + | + | + | – | – | – | + | + | + |
| 2. | *O. irrigua* | + | + | + | – | + | + | + | – | + | + | + | + |
| 3. | *Gleocapsa rupestris* | + | – | – | – | – | + | + | + | – | – | – | + |
| 4. | *Merismopodia* sp. | + | – | + | + | + | – | – | – | – | + | + | + |
| | | 4 | 2 | 3 | 2 | 3 | 3 | 2 | 1 | 1 | 3 | 3 | 4 |
| **Chlorophyceae** | | | | | | | | | | | | | |
| 5. | *Ulothrix lonata* | + | + | – | + | + | – | + | + | – | + | + | + |
| 6. | *U. cylindricum* | + | + | + | + | + | – | – | + | + | + | – | – |
| 7. | *Chlorella vulgaris* | + | + | + | – | + | – | + | – | – | + | + | + |
| 8. | *C. fusca* | + | + | + | + | – | + | – | + | + | + | + | + |
| 9. | *Closterium acerosum* | + | + | + | – | + | + | – | – | + | + | + | + |
| 10. | *C. tumidium* | + | + | + | + | + | – | – | – | – | – | + | + |
| 11. | *Spirogyra condensata* | + | + | + | + | + | + | + | + | + | + | + | + |
| 12. | *S. aequinoctialis* | + | + | + | + | + | + | + | + | + | + | + | + |
| 13. | *Oedogonium hatei* | + | + | + | + | – | + | – | – | – | + | + | – |

Contd...

**Table 7.3–Contd...**

| Sl.No. | Name of the Species | 1999 | | | 2000 | | | | | | | | |
|---|---|---|---|---|---|---|---|---|---|---|---|---|---|
| | | Oct. | Nov. | Dec. | Jan. | Feb. | March | April | May | June | July | Aug. | Sep. |
| 14. | *O. desikachary* | – | + | + | + | + | + | + | – | – | – | + | + |
| 15. | *Volvox aureus* | + | + | + | + | + | – | – | – | – | + | + | + |
| 16. | *Mougeotia scalaris* | – | – | + | + | – | – | + | + | + | + | + | + |
| 17. | *Scendesmus incrassatulus* | + | + | – | – | – | – | – | – | – | – | + | + |
| 18. | *Kirchnerilla lunaris* | + | + | + | + | – | – | – | – | – | + | + | + |
| 19. | *Cylindrocapsa geminella* | – | + | + | + | + | – | – | – | – | + | + | + |
| 20. | *Dactylococcopsis* sp. | – | + | + | – | + | – | – | – | – | – | + | + |
| | | 12 | 15 | 14 | 12 | 11 | 7 | 6 | 5 | 6 | 12 | 15 | 14 |
| **Bacillariophyceae** | | | | | | | | | | | | | |
| 21. | *Navicula rilomboidles* | + | + | + | + | + | – | + | + | + | + | – | + |
| 22. | *Frustulia rilomboidies* | + | – | + | + | + | + | – | + | – | – | + | + |
| 23. | *Stauroneies anceps* | + | + | – | + | + | + | + | – | + | + | – | + |
| 24. | *Pinnularia viridis* | + | + | + | + | + | – | + | – | – | + | – | + |
| 25 | *Synedra ulna* | – | + | – | ++ | + | + | – | + | + | – | + | – |
| 26. | *Fragilaria* sp. | – | – | + | + | + | – | + | – | – | + | – | + |
| | | 4 | 4 | 4 | 6 | 6 | 3 | 4 | 3 | 3 | 4 | 2 | 5 |

**Table 7.3–Contd...**

| Sl.No. | Name of the Species | 1999 | | | 2000 | | | | | | | | |
|---|---|---|---|---|---|---|---|---|---|---|---|---|---|
| | | Oct. | Nov. | Dec. | Jan. | Feb. | March | April | May | June | July | Aug. | Sep. |
| **Euglenophyceae** | | | | | | | | | | | | | |
| 27. | *Phacus longicauda* | + | + | + | + | + | + | – | – | – | + | + | + |
| 28. | *P. swirenkoi* | + | + | + | + | + | – | + | – | – | + | + | + |
| 29. | *P. triqueter* | – | + | + | – | + | + | + | – | – | – | + | + |
| 30. | *P. anacoelus* | + | – | + | + | – | + | + | + | – | + | + | + |
| 31. | *P. curvicouda* | + | + | + | – | + | + | – | – | – | + | + | + |
| 32. | *Trachelomonas mammillosa* | + | – | + | + | + | – | – | – | – | + | – | – |
| 33. | *T. armata* | + | + | + | + | + | + | – | – | – | – | – | + |
| 34. | *T. similis* | – | + | + | – | + | + | + | – | – | + | + | – |
| 35. | *T. superba* | + | + | + | + | + | + | + | – | – | + | + | – |
| 36. | *Euglena oxyuris* | – | + | + | + | + | + | – | – | – | + | + | – |
| 37. | *E. minuta* | – | + | – | – | + | + | + | – | – | + | – | – |
| 38. | *E. gracilis* | + | + | + | + | + | – | – | – | – | + | – | + |
| | | 8 | 10 | 11 | 8 | 11 | 9 | 6 | 1 | 0 | 10 | 9 | 7 |
| | | 28 | 31 | 32 | 28 | 31 | 22 | 18 | 11 | 19 | 29 | 29 | 30 |

*rupestris*, and *Merismopodia* sp. The lowest number of sp. 1 was found during May and June, and the highest number of blue-green algae was recorded in October and in September during the study period.

The factors, like high temperature, dissolved oxygen, Sulphate, low nitrite, Phosphorus, and dissolved organic matters, favoured the growth of Myxophyceae in the pond.

Similarly Rao *et al.* (1995) and Singh (1960) had observed the same phenomenon. It was observed that the low levels of phosphorus, nitrite and dissolved organic matters may not be responsible for the luxuriant growth of myxophyceae. In addition, high level of dissolved oxygen, sulphate, temperature play an important role in increasing the blue-green algal distribution in the pond.

The observations made in the present study confirms the observations made by the Yoshimura (1932) and Presscott (1938) emphasized nitrite, phosphate and organic matter often play a role in the distribution of Myxophyceae. Chu and Tiffany (1951) and Zafar (1967) have correlated the high temperature and bright sunshine with blue–green algal population.

## Chlorophyceae

It has been observed through the literature that the Chlorophyceae members may adopt any type of environment. During investigation it was noted that the Chlorophyceae sps. periodicity varies from 6 to 15. *viz. Ulothrix zonata, U. cylindricum, Chlorella vulgaris, C. fusca, Closterium acerosum, C. tumidium, Spirogyra condensata, S. aequinoctialis, Oedogonium hatei, O. desikarcharyii, Volvox aureus, Mogeotia scalaris, Scendesmus incrassatulus, Kirchnerilla lunaris, Cylindrocapsa geminella* and *Dactylococcopsis* sps. The limited number of 6 species were noted during April to June and more number of sps. (15) were in the month of November and August. The fluctuation in periodicity and distribution patterns in Papnash pond can be attributed to high alkalinity, dissolved organic matter, dissolved oxygen and high pH. According to Zafar (1964) and Munawar (1970) low nitrate, phosphate, and high dissolved oxygen are the factors responsible for the maximum occurrence of green algae.

Similarly in the present work the same features were noted. Verma and Mohanty (1995) and Gujarathi and Kanhere (1998) have also made the similar observations as that of earlier workers and concluded that temperature, pH, phosphate and dissolved oxygen play an important role in the growth and development of Chlorophyceae members.

## Bacillariophyceae

Ecology of fresh water diatoms has attracted many workers. During the working period it was perceived that the periodicity of diatoms fluctuated from (3) to (6). *Viz. Navicula rhomboidies, Frustulia rhomboidies, Stauroneies anceps, Pinnularia viridis, Synedra ulna* and *Fragilaria* sp. The minimum number of species (3) during March, May and June, and a maximum 6 in the month of January and February was noted. The factors like temperature, total alkalinity, dissolved organic matter, nitrite, phosphate and sulphate have often emphasize and played a significant role in the distribution of diatoms in the present study. Similarly it was observed by Munawar (1970) and Hosmani and Bharati (1980) in their work.

## Euglenophyceae

Several workers have discussed on the distribution of Euglenophyceae in various types of fresh water environment. In the present work distribution of Euglenophyceae was noted, and that varied from none to (12) sps. during the working period The (12) Species are *Phacus longicauda, P. swirenkoi, P. triquetor, P.anacpelus, P.curvicouda, Trachelomonas mammillosa, T.armata, T.similis, T. superba, Euglena oxyuris, E. minuta* and *E. gracilus*. In the month of May, none of the species were present, but 12 sp. were noticed in the month of December and February.

The factors such as alkalinity, dissolved organic matter, nitrite and phosphate may be responsible for the growth and distribution of Euglenophyceae members. Other workers also noted high pH, temperature and dissolved oxygen favoured the growth of Euglenophyceae in their work by Seenayya (1971) and Hegde and Bharati (1984).

## Summary and Conclusion

The present study was made to understand the algal communities in Papnash pond Bidar for a period of one year. The samples were collected once in a month during the period of investigation. These samples were subjected to analyze various physico-chemical parameters and periodicity of phytoplankton of the water body. It was observed that all physico-chemical factors fluctuated during the investigation. This is because of domestic sewage and other human activities. It was noticed during the study that the phytoplankton was significantly associated with most of the physico-chemical parameters. It was found that dissolved oxygen, free carbon dioxide, sulphates, carbonates, bicarbonates and nitrites were sufficient in quantity for the phytoplankton. All the phytoplankton groups such as Myxophyceae, Chlorophyceae, Bacillariophyceae and Euglenophyceae are positively interrelated with each other. From the obtained results one may conclude that sewage disposal and other domesticated activities were the factors causing fluctuation in physico-chemical parameters, which supports the growth and diversity of algal community in Papnash pond Bidar.

## Acknowledgement

The authors are grateful to The chairman, Department of Botany, Gulbarga University, Gulbarga for providing necessary facilities.

## References

Anand N. (1998): "Indian fresh water micro algae". Centre of advanced study in Botany. Madras (India).

Angadi S. B., Lingannaiah. B. and Eshwaralal Sedamkar (1999): Limnological studies of Jagat tank Gulbarga. *Fresh water ecosystem of India*. Edt. K. Vijaykumar. Daya Pub. New Delhi: pp. 133-159.

APHA. AWWA. WPCF. (1995): "Standard methods for the examination of water and waste water". *American Public Health Association. American wastewater Association and water pollution control Federation*. 19th Edition. Washington. D. C.

Arvola L. (1986): Spring phytoplankton of 54 small lakes in Southern Finland. *Hydrobiologia*. 137: pp. 125-134.

Ayyappan S. and Chandrashekhara G. T. R. (1987): Limnology of Ramasamudra tank-phytoplankton. Mysore. *J. agric. Sci*. 21: pp. 59-64.

Berman T. and Pollingher. U. (1974): Annual and seasonal variations of phytoplankton, chlorophyll and photosynthesis in lake Kinneret. *Limnol. Oceanogr*. 19: pp. 31-54.

Bharati S. G. and Bongale. U. D. (1975): Systematic account of fresh water algae at Raichur, Karnataka state. India. *J. Karnataka Univ.*, 20: pp. 130-141.

Chu Hao-Jan and Tiffany L. H. (1951): Distribution of fresh water Chlorococcaceae in Szechwan. China. *Ecology*. 34 (4): pp. 709-718.

David A. Rao., N. G. S. and Ray P. (1974): Tank fishery resources of Karnataka. *Bull. Cent. Inland fish. Res. Inst.* Barrackpore. 20: p. 87.

Desikachary T. V. (1959): "Cyanophyta" pub. by Indian Council of Agricultural Research. New Delhi. (ICAR).

Gaddad S. M. Nimbargi P. M. and Rodgi S. S. (1983): Ecological studies of two polluted water bodies Of Gulbarga. *Poll. Res.* 2 (2): pp. 49-52.

Gandhi H. P. (1956): A contribution to the knowledge of the fresh water diatomaceae of southwestern India. *J. Indian Bot. Soc.* 35: pp. 194-209.

Goel P. K., Khatavkar. S. D., Kulkarni A. Y. and Trivedy R. K. (1986): Studies on fresh water in southwestern Maharashtra with special reference to their chemistry and phytoplankton. *Poll. Res.* 5 (2) pp. 79-84.

Govind B. V. (1963): Preliminary studies on plankton of the Tungabhadra reservoir. Indian. J. Fish. 10 (1): pp. 148-158.

Gujarathi A. S. and Kanhere R. R. (1998): Seasonal dynamics of phytoplankton population in relation to abiotic factors of a fresh water pond at Barwani (M. P). *Poll. Res.* 17 (2): pp. 133-136.

Harvey M., Therriault J. C. and Simard N. (1997): Late-summer-distribution of phytoplankton in relation to water mass characteristics in Hudson Bay and Hudson Strait (Canada). *Canadian. J. Fisheries and Aquatic Sci.* 54 (8): pp. 1937-1952.

Hedge and Bharati (1984): Ecological studies in ponds and lakes of Dharwad. Trophic status. *Phykos.* 23 (1 and 2): pp. 71-74.

Hegde G. R. and Sujata T. (1997): Distribution of plankton algae in three fresh water lentic habits of Dharawad. *Phykos.* 36 (1 and 2): pp. 49-55.

Hosmani S. P. (1988): Seasonal changes in phytoplankton communities in a fresh water pond at Dharawad. Karnataka state. India. *Phykos*. 27: pp. 82-87.

Hosmani S. P. and Bharati S. G. (1980): Limnological studies in ponds and lakes of Dharwar. Comparative phytoplankton ecology of four water bodies. *phycos*. 19 (1): pp. 27-43.

Huden Christiane. (2000): Phytoplankton assemblages in the St. Lawrence river, downstream of its influence with the Ottawa river. Quebe. Canada. *Canadian. J. of Fisheries and Aquatic. Sci.* 57 (Suppl-1): pp. 16-30.

Hutchinson G. E. (1944): Limnological studies in connecticut. VII. A critical examination of the supposed relationship between phytoplankton periodicity and chemical changes in lake waters. *Ecology*. 25: pp. 3-26.

IImavirata K., Huttunen. P. and Merilainen J. (1984): Phytoplankton in 151 Eastern Finnish lakes: Species composition and its relation to the water chemistry. *Verh. Int. ver. Limnol.* 22: pp. 822-828.

Kalyani Y. and Charya Singara. M. A. (1999): Phytoplanktonic dynamics in Bhadrakali lake, Warangal. Andhra Pradesh. Edt. K. Vijaykumar. Fresh water ecosystems of India. *Daya pub.* New Delhi: pp. 208-225.

Kant S. and Raina A. K. (1985): Limnological studies of two ponds in Jammu-1. qualitative and quantitative distribution of phytoplankton. *Zoologica Orientalis*. 2: pp. 89-92.

Kilham S. S., Theirot E. C. and Fritz. S. C. (1996): Linking planktonic diatoms and climate change in the large lakes of the yellow stone ecosystem using resource theory. *Limnol and Oceanogr*. 41 (5): pp. 1052-1062.

Munawar M. (1970): Limnological studies on fresh water ponds of Hyderabad, India II. The Biocoenose, distribution of unicellular and colonial phytoplankton in polluted and unpolluted environments. *Hydrobiology*, 36: pp. 105-128.

Niemi A. (1972): Observation on phytoplankton in eutrophied and non-eutrophied archipelago waters of the southern coast of Finland. *Mem. Soc. Fauna. Flora. Fennica*. 48: pp. 63-74.

Noges-Peeter. and Laugaste-Reet. (1998): Seasonal and long-term changes in phytoplankton of lake Vartsjarv. *Limnologica*. 28 (1): pp. 21-28.

Padhi (1995): Algal environment of polluted and unpolluted fresh water ponds. Kargupta. A. N. and Siddiqui. E. N. (Eds). *Algal ecology*. An overview international book distributor. Dehradun. India.

Panth M. C., Sharma A. P. and Chaturvedi O. P. (1983): Phytoplankton population and diel variation in a subtropical lake. *J. Environ. Biol.* 4: pp. 15-25.

Pieturse A. J. H. and Roos. J. C. (1992): Diurnal variations in the Vaal, a turbid South African river: Physical, chemical and phytoplankton biomass characteristics. *WATER SA* (*Pretoria*) 18 (1): pp. 21-26.

Prasad K. V., Rajendra, Ghosh. Mita. and Gaur. J. P. (2000): A reconnaissance of species-environment relationships in pond phytoplankton at Varanasi (India). *Biologia. Bratislava.* 55 (1): pp. 35–42.

Prescott G. W. (1938): Objectionable algae and their control in lakes and reservoirs. Lousiana. Municipal. Rev. 1, 2 and 3.

Prescott G. W. (1951): Algae of the western great lakes area. Revised Edi. W. M. C. Brow Publ. Dubuque, Iowa.

Rajkumar R., Ramanibai R. and Niranjali D. (1994): Observation of species (Plankton) environmental relationships in urban aquatic ecosystems. *J. Envi. Biol.* 15 (3): 177-183.

Rao V. N. and Mahmood S. K. (1995): Nutrient status and biological characteristics of Hubsiguda pond. *J. Envi and poll.* 2 (1): pp. 31-34.

Ray P. and Rao N. G. S. (1964): Density of diatoms in relation to water condition Indian. J. Fish. 11: pp. 479-484.

Seenayya G. (1971): Ecological studies on the phytoplankton of certain fresh water ponds of Hyderabad. India. II. The phytoplankton. I. *ibid.* 13 (1): pp. 55-88.

Sharma K. P., Goel P. K and Gopal B. (1978): Limnological studies of polluted fresh waters I. Physico- chemical characteristics. *Int. J. Ecol. Environ. Sci.* 4: pp. 89-105.

Shastree N. K. (1992): Dynamics of phytoplanktonic fluctuations in a lentic water body. *Aquatic Environment*. 6: pp. 59-85.

Singh V. P. (1960): Phytoplankton ecology of the inland waters of Uttar Pradesh. *Proc. Symp. Algal.* ICAR, New Delhi. pp. 243-271.

Singh A. K. and Ahmed. S. H. (1990): A comparative study of the phytoplankton of the river Ganga and pond of Patna (Bihar), India. I. *Ind. Bot. Soc.* 69: pp. 153-158.

Trivedy R. K. and Goel P. K. (1986): Chemical and biological methods for water pollution studies. *Envirnomental publication.* Karad. India.

Verma J. P and Mohanty R. C (1995): Phytoplankton and its correlation with certain physico-chemcial parameters of Danmukundpur pond. *Poll. Res.* 14 (2): pp. 233-242.

Vijaykumar. K. and Paul R. (1994): Effect of nutrient enrichment on phytoplankton production. *Proc. Acad. Envi. And Biol.* 3 (2): pp. 171-175.

Vijaykumar, K. (1992): Limnological studies on perennial and seasonal water bodies of Gulbarga. Ph. D. Thesis, Gulbarga University Gulbarga.

Yoshimura. S. (1932): Seasonal Variation in the content of nitrogenous compounds and phosphate in the water of Takasuka pond. Saitma. Japan. *Arch. Hydrobiol.* 24 (1): pp. 155-176.

Zafar. A. R. (1964): On the ecology of algae in certain fishponds of Hyderabad. India I. Physico-chemical complexes. *Hydrobiologia.* 23: pp. 179-195.

Zafar A. R. (1967): On the ecology of algae in certain fishpond of Hyderabad, India. III The periodicity. *Hydrobiologia.* 30 (1): pp. 96-112.

**Aquatic Biodiversity in India: The Present Scenario, 2005** **123–145**
*Edited by*: D.R. Khanna, A.K. Chopra & G. Prasad
*Published by*: Daya Publishing House, New Delhi

# 8

# Intertidal Region and Biodiversity

**Amita Saxena**
*Department of Fisheries Biology, G.B. Pant University of Agriculture and Technology Pantnagar, Pantnagar – 263 145*

The earth is unique in the solar system for the enormous quantities of water, the world ocean cover approximately about the total surface area 361 milion (10 lakh) km$^2$ about (140 million mile$^2$) in the process occupying 71 per cent earth surface. In term of volume ocean contain 350 million cubic miles of water. The ocean on average contain 3.5 per cent of dissolved salts equivalent 165 million tonnes of salt per cubic mile. Thus amazingly the sea water is an unquestionably greatest georesources on planet earth, it constitutes the largest single mobile are "on the face of the earth".

1. The importance to oceans is staggering the oceans overwhelmingly influence nearly all the surface process on earth.
2. They regulate the water cycle flow of CO, major supply of $O_2$, to support life on earth. Meterologically it also regulates the climatic pattern on the earth distribution of heats and various gases.

3. It supports life within itself indirectly all life on planet earth.
4. Life undoubtedly began in ocean a few billion years ago.
5. Since that time onward dynamic growth and evolutionary process added vast diversified, life on the earth existing today from microorganism to whole in oceans.
6. The ocean also regulates geological process of weathering erosion. Being the greatest reservoir for minerals metallic deposits of Iron, Manganese, there are non metals such as Sulphur, Phosphates, Salt rocks. The oceanic rocks sediments hold much of the potassium, crude oil natural gas existing on earth.
7. Ocean provide a sink for billion of sediments carried by world rivers.
8. Ocean provides dynamic source of energy through its, current waves wind. Which can be converted into power and other energy sources express contend that desalinated water from ocean could increase terrestrial production of food.When therefore, the process of desalination of sea water is fully developed for commercial uses. Water shortage in many part of the world, particularly arid and semi arid regions, still plagued food shortages would certainly be alleviated. Because of higher organic matter then it's counterpart on land accurse a natural abundant supply of fertilizers.
9. Marine farming has been an attractive enterprise in sea bound lands. Ocean also contain millions of fish other related protein rich commercially important sea food which offer a great scope hope to solve the world food problems.
10. Due to their large volume rapid fluidity oceans are remarkably self-purifying and thus managed to survive uncontaminated for so long.
11. In recent years, however this has been increasing interest in the marine related field.
12. This interest apparently derived from growing awareness concerns all levels of society.
13. Recently, therefore, man's action have treated this unique self purifying capacity of the ocean.

14. These days oceans and sea have also come under the preview of war conflict and also as the ultimate dumping ground for the mounting burdens of human waste material. So causes pollution even in this persistent entity.
15. If the ocean are not plunged further to destruction, It could after man kind an important energy source, for they contain large quantities of heavy hydrogen (deuterium) is indispensable in the production of nuclear energy.
16. Ocean have provided man a powerful avenue of exploration and source of inspiration through out.
17. They are cheapest means of bulk transportation of the world's goods.
18. Since ocean going vessels can transport a ton of commodity commercial implication other means of transport.
19. Ocean shipping still is the main source of linkage for billions of people at our independent world.

## Importance of World Ocean-Conditions

The ocean environment is possible and adventurous to man and does not yields it secrets easily due to its impounding vastness this very alluring romance of sea has drawn many persons to study the aspects related to sea or ocean.

Marine biologist have studies vast diversity of floras and fauna found existing in the ocean today contacting their studies to find out details of life style behavior breeding pattern etc. within the sediments layers on sea floor are recorded.

## Biology of Intertidal Region

The intertidal zone lies at the junction of land and sea, subjected to tidal ebb and flow and it is the territory of sea bottom lying between high and low watermarks of the tide. Although it constitute by far the smallest area of all the realms of world oceans, extending as the narrow fringe from few cms to maximum of 15 meters, and inspite of very odd conditions, it is extremely rich in its biological resource and perhaps well studied also. The tides are of functional importance in shaping the intertidal environment. The physical conditions occurring in the intertidal zone are,therefore, quite dissimilar to those occurring elsewhere in the oceans, sea. Despite, its extremely restricted area it offers great variations in the environmental factors, tremendous diversity of life, still is the richest among all parts of ocean 'a' far as flora and fauna are concerned.

In the places where tide shows a widest range, its extensive area will be literally, expressed to arid condition or air or covered with water while remaining submerged. In addition to tides and waves as physical, factor, the fluctuations of other parameters like salinity, temperature, insulation and desiccation are associated with the cyclical change of time of exposure and submission. According to the nature of sediment and profile of substratum, the intertidal region is divisible into the 1) Rocky shore, 2) The sandy shore and the muddy shore.

## Physical Factor Influencing the Intertidal Region

### Waves

The one important physical feature which determine life activities of intertidal organism, and pattern of distribution is waves. The marine organisms have to face strong compounding effect of waves during high tide especially to the rocky shore are where the organism adhere or cling to rock,the strong wave action may dislodge these organisms although from their native habitat. While in sandy shore the continuous action of wave results in always shifting type of substratum for the organisms and does not allow them to settle and to have foot hold. In the muddy shore due to restricted area, however, the effect of wave action is of weakest kind.

Sometime the flora and fauna of specially top marginal belt of intertidal region is largely dependent on the frequency and intensity of waves, which supplement the wetting of that area. Therefore, it may prove as a boon for life to the organisms which struggle for their life in almost arid conditions of exposure from the top marginal belt of intertidal area, by their wash, splash and spray the waves are wetting these organisms.

### Fluctuations in Salinity

Intertidal waters continually receive the large quantities of fresh water from the adjacent land areas by streams and rivers, thereby reducing the salinity. The fluctuation in salinity from this region are often associated also with the seasons thus during the rainy season the salinity would by very low owing to heavy precipitation from land drainage, equally almost to that fresh water at the time of low tide. During the summer the salinity would be very high especially in rock pool areas as result of less depth and very highest rate of evaporation. The intertidal organisms are euryhaline it means they are able to sustain or tolerate very wider range of salinity fluctuations. Some of the organisms which are able to move avoid the fluctuations in salinity simply by moving up and down in tune with the tidal change as per their requirement, however the sedentary forms have to bear special adaptations to with stand salinity fluctuations.

## Insulation, Temperature and Dessication

Strong insulation and increase in temperature are determinant to organisms from this region during the low tide, when they are exposed to almost arid conditions, till the returns of high tide. The direct insulation from sun during low tide increases the body temperature of intertidal organisms considerably than the surrounding, this may lead to the drying and desiccation of organism if this conditions persist over long time especially at places receiving diurnal tide. The indirect effect of insulation and increase in temperature, coupled with the absence of watery medium results to the oxygen stress for the organisms leading to anaerobic situations and build of waste in the form of lactic acid in the body of organisms. In order to withstand such odds effectively almost all the intertidal organisms are thermal it means they able to tolerate widest range of temperature fluctuations. There are many organisms especially from higher latitudinal oceans or polar sea, which almost face process conditions when exposed to the low tide, still are able to survive at these low temperatures.

## Adaptations of Intertidal Organisms

### Tolerance of Fluctuations in Temperature Desiccation and Oxygen Stress

1. Tolerance to high temperature is one of the chief characteristic of intertidal organisms, which specially live in the upper area. Ex: Some molluscs of nudribranchi group can tolerate the high temperature of 40°-50°C.

2. Some resort or tolerate to very low as well as very high temperatures. Ex: *Modulus demises,* survive at temperature ranging from 22°C to 40°C.
3. The subtropical species of crab *Uca* can tolerate low temperature much effectively than tropical population of crab of the some species.
4. Many invertebrates of intertidal zone are able to sufficiently thermoregularise their body temperature specially during low tide to sustain higher rise in temperature Ex: *Barracks,* the fiddler crab *Uca pugnax* limpet (*Fissysella* species) and gastropod (*Nerita* species) therefore thermoregulase cooling.
5. Behavioural adaptations are shown by some organisms to avoid elevated temperatures. Ex: The Australian gastropod *Cerethrium* species decrease effect of environmental temperature by clustering together on falling tide. *Uca Pugnax* is attracted to light in morning but it avoids light during hottest part of the day by retrieving into its burrow during the low tide.
6. The sessile organisms at the higher tidal mark are able to avoid serious problem of desiccation for long period by conserving the water or by storing the water or moisture for more time in their body and lend to reduce the outside flow of water or body fluids during low tide. Ex.: Barnacles, oysters and mussels of rocky shore close themselves tightly into their shell and storing moisture and not allowing it to drain away.
7. Intertidal organisms experience periods of oxygen stress at the time of low tide specially in most burrowing animals which may suffer to low oxygen tension or even anaerobic condition during the low tide. Ex.: Barnacles, Oysters, Clams and mussels tightly close themselves during the low tide during the $O_2$ stress they can remain anaerobic for long period of time. *Uca pygnax* is capable to survive anoxia for 24 hours at 21°C by not utilizing oxygen to sun the metabolic process.

## Tolerance to Fluctuation in Salinity

8. Intertidal organisms tolerate extremely high hyper saline environment. Ex: *Uca sap ax* can survive on salt flats of 90 ppt salinity certain fishes in the tidal pool are found to live in salinities as high as 142.2 ppt. It is possible through osmotically expelling or rejecting salt from their body to the surrounding.
9. Many invertebrate organisms survive prolonged period of low salinity by active transport of salts in the body from the medium and also by reducing permeability of the body surface to the salts into the water and even by producing hypo osmotic urine.

## Adaptation to Cover All Parameters

10. Most of the intertidal sessile animals have pelagic larval stages. Through phototactic, hydrostatic and geotactic responses the larvae are guided to right time for settling. This is of great survival value for sessile intertidal forms: Ex: Edible oysters has the sea swimming larval period of 21 days. On 21$^{st}$ day it settles strongly over hard structure or any hard substratum.

## Adaption to Strong Wave Action

11. Most of the sessile intertidal organisms from the rocky shore area have several anatomical devices to adhere firmly to rocky habitat and avoid the organisms to be dislodged from the habitat from strong pounding actions of waves during high tide. Ex: Sea anemone have pedal disc, Barnacles, oysters secrete calcium carbonate for cementing themselves perfect to rock, Mussels and some clams have thread like structure called byssi's thread to perfectly cling to rocks.
12. Most of sessile organisms have strong of segmentation for loss of body parts.
13. Many sandy shore organisms avoid strong action of waves by moving up and down with the change in tune of the tide. Ex: Many small crabs many intertidal organisms from sandy shore are burrow deeply into the wet sand during

falling tide to accquire moisture and to avoid strong wave action Ex: Arinicolar tube worms, fodder crab.

## Tides

Tides occur due to interaction between the gravitational attraction of sun and moon on earth and centrifugal force generated by rotating earth and moon system. As a result of these forces the water in ocean basins is pulled in bulges. Gravitational attraction of one body for another is a function of mass of each body and their distance part. The earth and moon form an orbiting system that revolves around their common center of mass. Because of large size of earth relative to moon, this point is located inside the earth. The revolution of earth moon system creates a centrifugal force (acting outward) which is balanced overall by gravitational force acting between two bodies. However, gravitational force is much stronger than Centrifugal force on side of earth facing moon than on side opposite. As a result, the side facing the moon has the water pulled into a bulge (= high tide) on the opposite side of the earth the gravitational force of moon will be the least and the stronger centrifugal force will pull water into a buldge away from earth (another high tide) Fig. 8.1. Thus, we have two high tides. These then circle the earth following the position of moon as earth revolves on its axis once every 24 hours low tides are about half way between high tides means two high and two low tides of similar magnitude each day as earth is inclined 23.5° from vertical it is not possible. When two high and two low are not equal means mixed tide. At highest latitude (polar seas), only a singal tide would occur. Further changes in height of tides results from changes in moon relative to earth as moon move in its orbit around earth. Since orbit of moon moves in its orbit around earth. Since orbit of moon is elliptical there are times when moon is nearer to earth perigee and others when it is further away (apogee). Tides are greater at perigee and diminished at apogee when moon and seas are directly aligned and combine their forces and give greatest range of tide (both high and low) Spring tide. When sun and moon are at right angles to each other and thus counteract each other, tide showing minimum range-Neaptide. The well known extremely high tides of Bay of Fundy or Cook Inlet in Alaska are result of basic tide forces acting in geometry of basin (Nybakken, 1982).

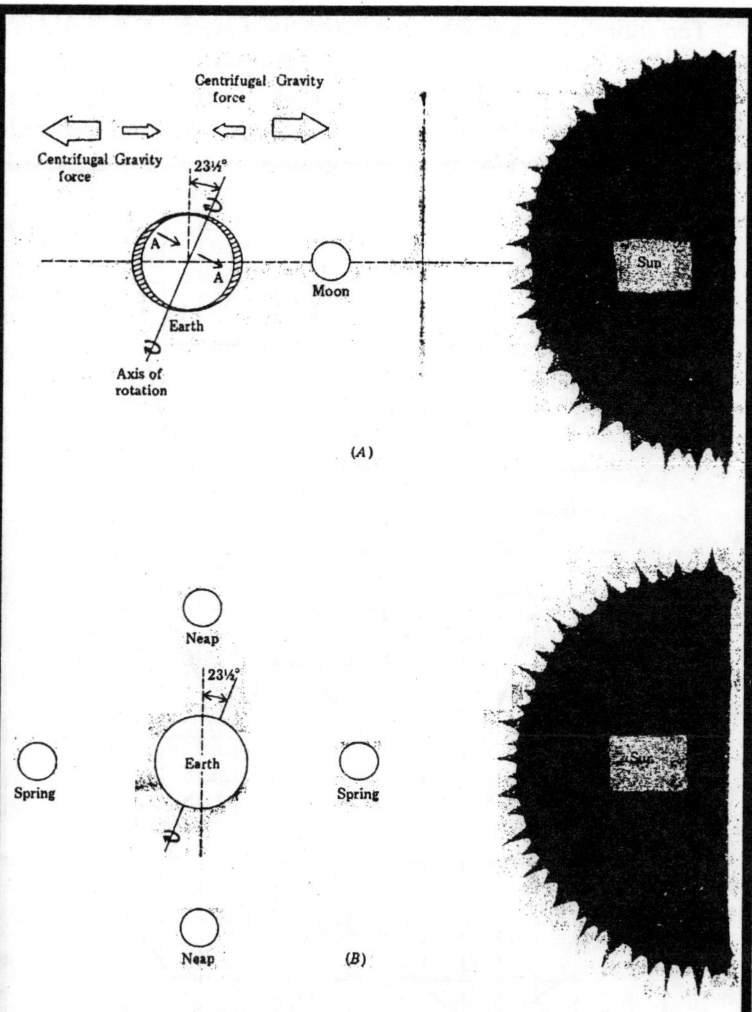


**Fig. 8.1: Origin of Tides**

**(A) Moon acts to raise a bulge on the side of earth nearest it due to greater gravitional force than counteracting centrifugal force. On the opposite side the centrifugal force is stronger and throws another bulge outward. Because of the inclination of the earth on its rotational axis, point A, as it rotates, will experience two high tides of different height (B) position of the moon and sun at neap and spring tides.**

The shoreline is a flexible boundary that marks the position of sea level between low tide and high tide. The region extending from the point of low tide to landward limit of effective wave action is called shore. Shore features change constantly because of variable

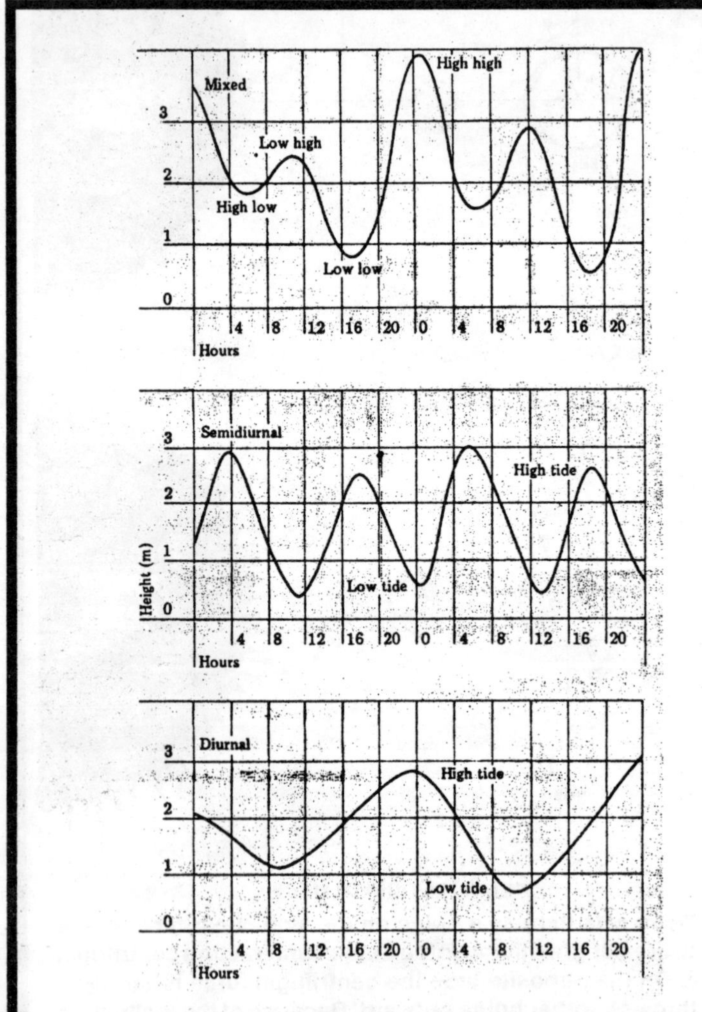


**Fig. 8.2: Semidiurnal, Diurnal and Mixed Tide Curves (Partly after D.A. Ross, Introduction to Oceanography, Prentice-Hall, 1970)**

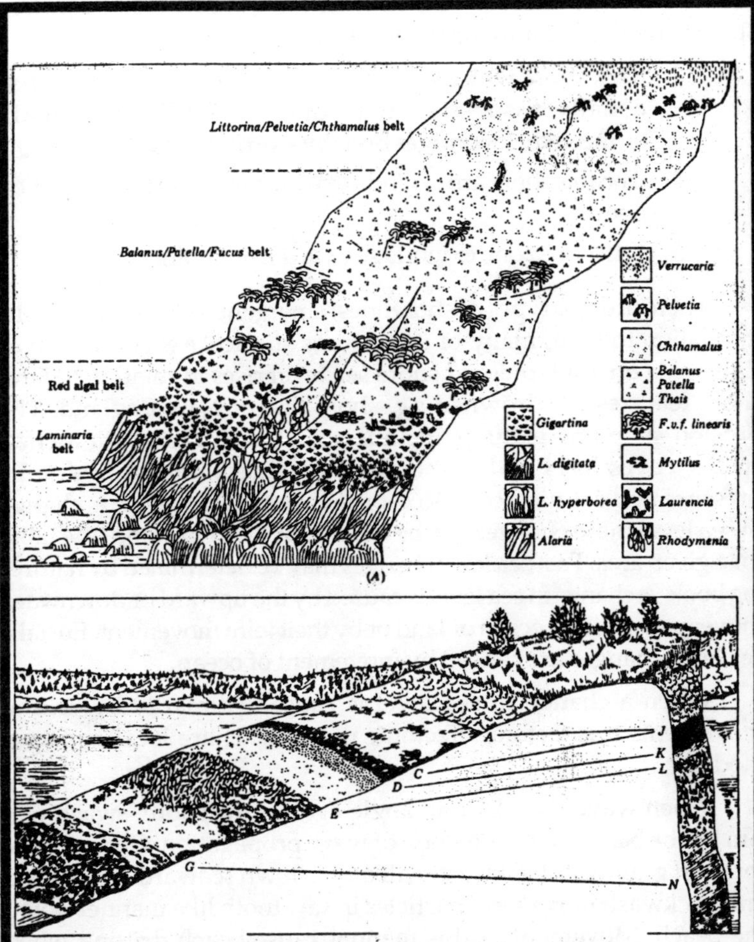


**Fig. 8.3: Zonation in the Rocy Intertidal**

**(A) An intertidal area in Scotland (B) An intertidal area in British Columbia. Note in (B) the difference in the extent of zones. In (B), line A-J indicates the lower limit of lichens; line C-J, the upper limit of the "black" zone; line D-K, the upper limit of barnacles' line E-L the upper limit of Fucus; line G-N, the lower limit of barnacles (A, form). R. Lewis, The ecology of rocky shores, 1964, Hodder and Stoughton Educational. B, from T.A. Stephenson and A. Stephenson. Life between the tidemarks on rocky shores, 1972, Freeman).**

nature of waves and currents so their topography is not constant Shores are divided into three zones:

1. Back shore which occupies the area in front of sea cliff is usually associated with one or two berms (small, terracelike, low ridges built by storm waves).
2. The foreshore are extends seaward from berms to low tide point
3. Offshore zone extends seaward from point of low tide

There are two types of shorelines, emergent shore sea level falls in relation to land, many of its features surface, *e.g.* shore area of California on a sub merged shore sea level rises in relation to land shore features are drowned *e.g.* shore line of New England. The Erosion shoreline features include sea cliff, terraces, sea arches, stacks, coves bays and bights includes direct impact hydraulic pressure and solution corrosion or abrasion. Daily there is change in sea level these changes can be determined by checking tide table for a given area. Past sea level changes may be determined by natures log book. A change in sea level is caused by the upward or downward movement of either ocean or land or by their joint movement. Eustalic sea level changes are caused by movement of ocean.

When a change in sea level is caused primarily by crustal movements it is tectonic. Tectonic movements are local and spas modic.

When waves break at an angle against the shore, swash will run up the beach in the direction of wave propagation. Owing to the force off gravity its backwash will move down seaward. The swash and backwash move sand particles in saw tooth like manner down the beach. Movement in this manner cause beach drifting when beach drifting and longshore currents move in the same direction they can efficiently transport both sand and gravel. However when their directions are opposite, the coarser sediment (gravel) will follow beach drifting and the finer sediment (sand) will travel with longshore currents the combined transportation of sand and gravel by beach drifting and longshore current is known as longshore drifting. Longshore drift can daily carry several hundreds to thousands kilograms of sand and gravel in suspension along many shores of world. When long shore drifting is obstructed by artificial features such as harbors and jetties, it fills up these sites with sand causing serious dredging and maintenance problems and erosion of beaches further down from the sites. To protect these sites ocean engineers have installed groins jetties and break waters. Groins are gentle walls extending from high tide areas seaward. Jetties unlikegroins are longer structures built at right angles to the shore Jetties are built to keep longshore drift sediments away from entrances to harbors or rivers.

Although there are many divisions of ocean study but in this chapter we deal with the biological aspects and environmental effect on biodiversity. Since the vastness of world ocean and its diversity here some lights are thrown on the intertidial zone organisms.

The littoral system can be divided into three divisions, from the essentially exposed supralittoral to the some times partially exposed eulittoral to deeper sublittoral.The supralittorial environment is extremely rugged. Animals living there are almost continuously exposed, being immersed only during periods of extremely high tides, and storms and by the spray from breaking waves. Animals of this environment are similar the world over, generally small gastropods baranacles and lichens (*Verrucaria* type) on rocks and crabs and amphipods on beaches. The eutittoral environment includes the region that is periodically exposed at low tide and extends out to depth of 40 to 60 m (131 to 197 ft.). The width of intertidial region depends upon the tidal range andslope of ocean bottom. Animal living in this environment also must withstand the effect of breaking of waves. Many animals accomplish this by burrowing into the bottom, thus also lessening some of harmful effects of exposure at low tide. The outer edge of eulittorial is near the depth limit at which most attached plants can grow on the bottom; they cannot grow in deep water because of the absence of sufficient light. Actually only a small portion of sea floor is available for growth of attached plants and even within this small area many parts cannot be utilized because the bottom may be muddy or otherwise unsuitable for plants. The outer part of littorial system is sublittorial system the edge of continental shelf, is an area exploited by commercial fisherman.

According to Stephenson term these three zones are supralittonal, Midilittorial and infralittorial. The high temperature promotes dessication and synergistic effect of these factors may be even more devasting than each acting alone. Sunlight itself may act adversely to limit organisms on shore sunlight includes wave lengths in ultraviolet region that are deleterious to living tissues water quickly absorbs these wave lengths and so protect marine animals. Intertidal animals have direct exposure to such rays at low tide. Higher the organisms greater exposure to these days.

Most intertidal fishes, because of turbulant environment are of small size. They are compressed and elongated (Blennidae, Pholidae) or depressed (Cottidae, Gobiesocidae). These fishes live in holes, tubes, crevices or depressions for protection against both desiccation and wave action. They are closely associated with substrates and lack air bladder and can tolerate great range of salinity and temperature. The fish *Periophthalmus* collectively called "mud-skippers" spend most of their time out of water, crawling around mud flat. They act like frogs or toads. They are having modified eyes the eyes set high on the head and are arranged so that they focus best in air, not in water, their gills are reduced they respire aerially by gill chamber, mouth cavity which are highly vascularised.

The fishes found in intertidal regions show almost similar life history, eggs are demersal and laid on stones, rocks or submerged vegetation. Eggs are guarded by male and after few week they hatch into planktonic larvae. Eggs hatch within two months. During this period larvae gradually acquire adult features and lastly settle down and become benthic. They live upto 2-10 years and in first or second year they attain sexuality. Among intertidal fishes some are migratory depends upon tides, season or diurnal.

Echinoderms sea stars, sea urchin and sea cucumber. These animals have a five sided pseudo symmetry and internal skeleton with water canal system. Class Holo thuroidea includes sea cucumbers which can live in all depths of oceans and of course intertidal also sea cucumbers have one peculiar ability. They can eviscerate in times of danger, they can discharge their internal organs leaving a meal for their tormentors and later regenerate another set of organs. Sea stars are member of class Asteroidea and are well represented world over. They frequently seen in littorial regions star fish feeds on oyster and clams and can cause considerable damage to these valuable food resources. In past seeing the loss fisher man used to cut the starfish and throw it into sea with the thought their desirable creatures are safe but they have regeneration power cutting the arms making them double. More recently a dramatic increase in population of starfish Acanthaster (crown of thorns) has been noted on some coral reefs in Pacific. This creature feed on corals and has been noted responsible for large scale destruction of reef in some places, such as Guam. 90 per cent of coral reefs are destroyed by starfish in a 2.5 year.

Chitons and gastropods have toothed radula and buccal mass from a complex and highly specialized means off rasping food from substratum.

Predatory molluscs have profound modifications of rhipidoglossan and docoglossan type of radula and jaws are developed in some forms such as Clathrus. In sponge eating molluscs Cerithiopsis and Triphora, the jaws are also prominent and used to separate off pieces of tissue which are then taken over to buccal cavity by radula. Members of Muricacea attack a variety of organisms including barnacles, gastropods and lamellibranchs. Thais (Nucella) Papillus *e.g.* feeds upon barnacles mussels and limpets whilst *Urosalpinx* and *Occnebra erinacea* eat young bivalves especially Venus and oysters. The penetration is achieved by a combination of chemical and mechanical activity in Urosalpinx and also in Thais, the prey is first gripped by foot and probosis is applied against the periostracum of foot and proboscis is applied against the shell for a short while but during this time makes little impression on shell. Then the animal creeps forward and applies the sucker like accessory boring organ to etached area. Other intertidal animals such as the regular echinoid *Psammechinus milidris* feeds upon barnacles cockles and occasionally adult oysters.

The pattern off distribution of shore molluscs is due to a complex set of circumstances, among which are chance settlement over a wide area with subsequent death except in favourable situations, an active choice of substratum or microhabitat, migration to situation; favouring survival, behavioural activities such as homing of various gastropods as well as less well defined kineses and taxes in response to environmental cues, and maintenance of position by boring, burrowing or adhesion.

A sequence off seasonal changes in respiratory rate was shown to occur in the limpet *Patella vulgata* and *Patella aspera* Davis (1967) superimposed upon such changes are the effects of tidal level.

The adults of the small prosobranch *Periugia* (Hydrobia) *ulvae*, occupy a mainly mid tide level zone on mud flats at low tide, crawing and browsing on the surface. Then they burrow for a time and resurface just before the incoming tide reaches them. After that they float by means off mucous raft which also serves to entrap the plankter that is engulfed by means of radula. The cycle of tidal activities is completed by the animals sinking to the bottom again when the receding tide reaches the zone occupied by animals at low water. This is an example of an intertidial gastropods with unusual range of movement, but most gastropods (some bivalves) retain their motility although some of them have remarkable powers of adhering to rocks. Patella an extreme example can resist dislodgment by a force equal to 15 kg, or a straight pull equal to a force of 70lb for a limpet having a base area of little less than 1 square inch. Other *e.g.* Chiton trochids and litorinids although able to cling to surfaces avoid wave action or desiccation by the less exposed places, often congregating in crevices or beneath stones or algae. Most of them crawl about and feed for part of tidal cycle. If their movements were disoriented it is obvious that after a time the species would be randomly dispersed. They orientate to environmental cues such as light, gravity and kineses. Eltringham (1971) reports that on Ghana coast *Littorina punctata* orientates not only to sun and to gravity but also to configuration of shore line which implies that in this animal there is, even if some what limited, a degree of form vision recalling that postulated for *Talitrus saltator* by Williamson (1951) a sign of response to a directional stimulus reverses after a time so that animal retraces it path arrives back at roughly the place from which it started (Brafield,1978). The area of intertidal claim fishery community management board for west coast Vancouver island clam fishery was formely established in 1998 for expanison of selfish aqua culture. (Dunlop,2000) *Caprella* species, mussels and neogas tropods have been mainly used for monitoring butylin in shallow water ecosystem.These invertebrates mainly used for monitoring butylin in shallow water ecosystem inhabit the intertidal zone where butylin levels vary widely depending on immersion period and exposure to sea surface microlayer. Monitoring using neogastropods may also over estimate exposures after restrictions on tribuly since neogastropods show an irreversible response to residue changes owing to their long life spans. Thus, it is proposed usage of *Caprella* spp. to monitor temporal and spatial changes in baseline concentrations of butyltins (Taheuchi, 2001).

The Shore crab *Carcinus maenas*, the sand hopper *Tatitrus saltator*, *Nassarius whelk*, *Buccinum undatum* although normally sub-littorial, is occasionally to be found at extreme low water, and the isopod *Idotea* which is often abundant amongst intertidal seaweed. *Idotea* is a good example of an omnivorous scavenger.

The physical and biological factors cause the large scale vertical distribution patterns of algae an invertebrates.

In case of *Chthamalus stellatus* and *Balanus glandula* without a refuge in highest intertidial (physical factor) both would be eliminated from intertidial the former by competition and the latter by predation (biological factor). Similarly, persistence of *B. cariosus* in mid intertidal of pacific coast depends upon the predation of *Pisaster ochraceus* for, without this predator to remove *Mytilus* and *Thais*, *B. cariosus* would be out competed by *Mytilus* or completely consumed by *Thias*.

Some species of algae or invertebrates cannot tolerate wave action and are absent from wave swept areas. They are delicate or with poor means of holding on to substrate. Certain species cannot tolerate quiet water areas they unable to compete, lack of food or presence of predator that are not present in wave swept areas. This *Mytilus edulis* out competes *Balanus balanoides* and *Chondrus crispus* in exposed area of New England because no predators can tolerate the wave action; in protected areas predators reduce competition allowing the barnacle and algae to exist and thus altering the appearance of the zones (Nybakken, 1982).

In many rocky shores presence of tidepools of various sizes, depth and locations how level pools often are little different from surrounding sea and are essentially extension of sub-littorial. These tide pool are the escape places from hardship during exposure to air, temperature salinity tide pools are the refuge from desiccation for intertidial organisms.

Sand beaches are recreational places in all over the world they devoid of macroscopic life. The organisms existing in tiny spaces between sand grains. The importance of the particle size to organism distribution and abundance rests with its effect on water retention and its suitability for burrowing.

Since the organisms inhabiting the intertidal are aquatic they are well protected against desiccation in fine sand beach but subject to desiccation. With each passing wave the substrate particles are picked up, churmed in water and redeposited. Particles are thus being continually moved and sorted. The reason that fine sand beaches occur only where wave action is light (effecting small depth) and coarse ones coarse ones where it is heavy is that in heavy wave action (affecting greater depth). Smaller particles remain in suspension so long they are carried away from beach leaving only coarse.

Because of ceaseless movement of surface layers of sediment few large organisms have capability of permanently occupying surface of sand or gravel beaches appeared barren. Fine sand beaches has reduced oxygen than coarse sediment. Rocky beaches lack the great topographic diversity no Crevices, no slopes facing different direction and offering different moisture conditions nor overhangs or tide pool. Environmental factors act uniformly. The presence of an impermeable shell cuts down the evaporative water loss. Limpets of the genera Patella, Acmaea, Collisella are also dominant animals of rocky intertidal. Certain species of limpets have a home scar into which their shells exactly fits. At low tide they return to these "homes" and by fitting into these grooves greatly reduce the water loss. Other limpets have no sar simply clamp down tightly against the rock so no tissue is exposed, only shell.

They also reduce heat gain and increasing heat loss. Heat loss can be attributed by greater elaboration of ridges and sculpturing on light coloured shell. These acts as radiator fins indicator tuns and help in losing the heat *e.g. Tectarius muricata* and *Nodolittorina tuberculata.*

To maintain position in various wave action barnacles, oysters and serpulid polychaetes worms fixed themselves to a substratum. But attachment is not permanent.

Shore molluscs mainly or partly reply on cutaneous respiration and many prosobranchs which inhabit pools have no ctenidia and only small mantle cavities *e.g.* members of Omalogyridae. Rissoellidae and Pyramidellidae. There are some evidence that to a varying degree bivalves can respire anaerobiosis has great survival value to intertidial animals.

Most probably all, shore proso branchs have hemocyanine as a respiratory pigment but at best this serves to increase the oxygen carrying capacity of blood only about 3 per cent many of them also have haemoglobin in certain muscles which may act as an oxygen store. The blood of stenoglossan *Busycon canaliculatum* will combine with more carbon dioxide than will reduced blood under comparable conditions and the presence of carbon dioxide increases the affinity of haemocyanin for oxygen. This is the reverse of what happens in bloods containing haemoglobin or indeed, in the haemocynin containing blood of cephalopods in *Busycon* animal can live in oxygen deficiency "even down to 10-20 mm tension. When most of the molluses are away from tides they become relatively inactive and their oxygen demand become less. Molluscan aquatic species are amoniotelic or Ureotelic. In all molluscs a high but variable proportion of non protein nitrogen is excreted as amino acids, purines and other compounds as well as ammonia. It is also noted that the ratio of body weight/weight is greater is higher latitude populat Namejko (1951) report that American Oyster (*Crassostrea virginica*) are divisible into different geographical races with the more northern ones breeding at lower temperatures than more southern ones of comparative studies on *Mytilus edulis*. Schlieper (1957) reports the results which seems to be poikilosmotic, *i.e.* incapable of active regulation acclimated to north sea (30 per cent Salinity) and Western Baltic (15 per cent salinity). The brackish water mussels have less active gill cilia, a slower heart beat and less resistance to heat, but their gills have a higher oxygen consumption. The size of the gonads also varies with tidal level, and low level animals have a larger gonad in the winter than higher level animals. The two groups spawn out of phase and high level animals do not contribute to the breeding populations *Lasaea rubra* specimen observed by Morton (1958) from high up the shores as a rule are smaller in absolute in absolute size but have a higher pumping rate than those from lower down the shore can tolerate a greater ranger of salinity and more quickly become active in response to splashing by waves, all of which are obviously adaptive compensation (Morton, 1958). In a study of a population of *Littorina saxatilis* (=rudis) on a stony beach has shown that animals high up shore grow faster, mature quicker, and produce more young than those at low level. Their breeding cycle is closely synchronized with tides. *Mytilas edulis* in which gonads mature during periods of spring tides and spawning in neap tide.

These areas have very little primary productivity. Diatoms population is restricted to very surface layers. Most of the organisms in beach are either suspension filter *e.g. Siliqua, Ensis, Tivela spisula, Domax,* detritus feeders *e.g.* Sand crabs family Hippidae such as *Emerita analoga* and *Blepharipoda accidentalis, Olivella columeltaris,* Polychaetes, sanddollars and molluscs of pacific coast of America (sand beaches) *Dendraster excentricus* Polychaetes *Nepthys* and *Glycera*. At low tides birds invaited.

High Tide Fishes: The role of competition, predation is also important for the distribution of organism of their zonal pattern.

Dayton (1975) was directed at larger kelps and found that the dominant kelps *Hedophyllum sessile, Laminaria setchelli* and *Lessionopsis littoralis* all outgrew and outcompeted certain smaller species in lower intertidal. These smaller species are fast growing species which quickly colonized open areas. They are opportunistic or fugitive species. *Hedophyllum* was outcompeted by other two such that they came to dominate that area.

*Pisaster ochraceus* can consume *Thais* and barnacles of any size and is primary predator of small and medium sized *Mytilus* it is predator keystone species. They have profound effect not only on their population but also on entire community of which they are part. The removal or depletion of numbers of predator has the effect of causing great changes in the presence and abundance of entire community, most of which are not prey of predator change occurred in community structure.

Lub chenko and Menge (1978) have also demonstrated that predation is important in setting zonal patterns in the intertidal of North Atlantic Ocean. In low intertidal of New England the competitive dominant species in *Mytilus edulis*. It is able to out compete and eliminate the barnacle *Balanus balanoides* and alga *Chodrus crispus*. It is prevented from doing this by starfish. *Asterias forbesi* and *Asterias vulgaris* and the snail *Thais lapillus*, all of which prey upon *Mytilus edulis*. In the area where these predators are absent, namely the most exposed wave beaten areas *Mytilus edulis* eliminates *Balanus* and *Chondrus*.

As shore become more protected from wave action they tend to become finer grained and accumulate more organic matter, thus the muddy areas have finest grain size. Most of the muddy shores of world associated with estuaries and similar embakments.

Proches and Marshall (2001) studied taxonomic ecological biogeographical trends of non hala carid marine mites to study their origins in littorial habitats including rocky shores, boulder beaches salt marshes and mangrove forest floors. Only 162 species of mites were recorded in literature.

*Himasthia elongata* larvae were also recorded from intertidial whelh *Littorina saxatilis* in Kandalaksha bay white sea studies revealed blue mussel *Mytilus edulis* as second intermediate host and seagul *Larus earnus* as final host by Ishkulor (2001). Cordaux *et al.* (2001) detected two distant Wolbachia strains in two amphipod (intertidil region) these strains were closely related to different coastal isopod symbionts, suggesting wolbachia transmission may occur between distantly related crustacean hosts living under the same ecological conditions.

It means that intertidal zone is quite rich in fauna may be leading independent life or in parasitic form.

Parasitism and diseases are occurred for biological control of a population. According to McCurdy *et al.* (2001) stated that life history theory predicts that hosts should reproduce when first infected by parasites if hosts are capable and if parasites have a lower cost on current than on future reproduction of host.

## References

Behenke, J. (1978). Plant species diversity in a marine intertidal community, Importance herbivorous food preference and algal competitive abilities, Amar, Nath 112 (1983) 23-32.

Bhatt, J. J. (1978). Oceanography, Exploring the Planet ocean: D. Van Nostrand Company, New York.

Brafield, A. E. (1978). Life in sandy shores, studies in biology. No. 89 Educard Arnold: London.

Cammaughey, B. N. (1974). Introduction to marine Biology II St. Louis.

Cordaux, R., Michel-Salzat. A, Bouchom, D. (2001). Journal of Evolutionary Biology 14: 2, 237-243.

Davies, P. S (1967) Physiological ecology of *Patella* I. The effect of environmental accumulation on metabolic rate J. mar. boil. ass. UK. 47, 61-74.

Dunlop, R. (2000). Area F. clan management Board, Nuu chah muth. Tribal council. Bulletin of the Aqua culture Assoc. of Canada No. 2, 20-26.

Eltringham, S. K. (1971). Life in mud and sand Russak, New York.

Fell, H. B. (1975). Introduction to Marine biology. Harper and Row, Greenwood pH New York. A history of marine fishes-IIIrd, New York, Inlay.

Imman, D. L. and B. M. Brush (1973). The coastal challenges science 181, 20-32.

Ishkulav, DG (2001). On the taxonomy status of trematode larvae of genus Himasthia (Trematoda) *Ichinos tomatidae* from intertidal whelk *Littorina saratilic* living in Kandalaksha Bay of white sea. Parasitologiya: 351, 81-85.

Lewis JR. (1964). The ecology of Rocky shores, The English Universities Press Ltd. London.

Mc Curdy D, Boates, J. S. and Forbes M. R (2001). An empirical model of optimal timing of reproduction for female, amphipodes infected by trematodes, J. Parasitology 87, 124-130.

Morton, J. E. (1958). Molluscs, Hutchinson Press London.

Newell, R. C. (1970). Biology of Intertidal animals. Elsevier Publication, N. Y.

Newell, R. C. (1979). Biology of intertidal animals, Marine ecological surveys Ltd. Faversham, Kent UK.

Nybakken W. James (1982). Marine Biology and an Ecological Approach Harper and Row, Publishers, New York.

Parson, T. Rand M. Takahasi (1973). Biological oceanographic processes. Oxford, Pergamon.

Ricketts, E and J. Calvin (1968). Between pacific tides 4$^{th}$ ed. Rev. by J. Hedgpeth, Stanford University Press, Stanford.

Sanith, J. E. and Newell G. E. (1955). The dynamics of zonation of common periwinku (*Littorina littorea* L). on a stony beach. J. Animal Ecol. 24, 35-56.

Schlieper, C. and K. O. Walskir (1957). Weitere Beobach tunger Zor okologistun Physiology der Micsmuschel: *Mystilus edulis* L. kiel Meeres for Sch; 13, 3-10.

Segal, E., Rovok, P. and James, T. W. (1953). Rate of activity as a function of intertidal light within population of some littoral molluscs, Nature: 172, 1108-1111.

Simkiss, K. (1960). Some properties of organic matrix of the shell the cockle (*Cardium edule*) proc. Malacol. Soc. London 34, 88-95.

Southward, A. J. (1958). The zonation of plants and animals on Rocky shores, Biol Revs. Cambridge Phil Soc. 13, 137-177.

Southward, A. J. (1965). Life on seashore. Havard UniversityPress Cambridge.

Strauber, L. A. (1950). The problem of Physiological species with special reference to oysters and krill. Ecology 31, 108-118.

Takeuchi, I., Takahashi, S. and Miyazaki N. (2001) Caprela watch: a new approach for monitoring butylin residues in ocean. Marine Environmental Research 52, 97-113.

Thompson T. E. (1960b). Defensive acid secretion in marine gastropods J. Marine Biol Assoc. U. K. 39, 115-122.

Thorpe, W. H. (1956). Learning and instinct in animals methenen London.

Thorson, G. (1950). Reproductive and Larval Ecology of Marine bottom invertebrates. Biol. Rervs. Cambridge Phil Soc. 25, 1-45.

Williamson, D. I. (1951). Studies in the biology of Taletridae (Crustacea, Amphipoda). Visual orientation in *Talitrus saltator* J. Marine Biol. Assoc. U. K. 30, 91-99.

Yonge, CM (1960). Mantle cavity, habits and habitats in blind lampet, *Lepeta concentrica*, Middendorff. Proc. Calif Acad. Soc., 31, 103-110.

Zeuthen, E. (1947). Body size and metabolic rate in animal Kingdom with special regard to marine microfauna compt. Rend. Frav. Lab. Carisberg. Ser Chim. 26, 17-161.

**Aquatic Biodiversity in India: The Present Scenario, 2005** **146–153**
*Edited by*: D.R. Khanna, A.K. Chopra & G. Prasad
*Published by*: Daya Publishing House, New Delhi

# 9

# Fish Fauna of the District Udham Singh Nagar with Some Recommendations for Improvement of Fisheries

**S.P. Badola, Smita Badola and D.R. Khanna***
*Smriti Bhawan, Upper Kalabarh, Kotdwar – 246 149, Distt-Pauri Garhwal (U.A.)*
** Deptt. of Zoology and Environmental Sciences, Gurukul Kangri University, Hardwar*

Fish fauna of Kumaon hills, including the streams and lakes of Nainital hills, have been reported by (Chaudhury and Khandelwal, 1960; Hora, 1937; Menon, 1949, 1951, 1962, 1971 and Walker, 1888 and Pant, 1970) has described the fish fauna of the kumaon hills (Almora and Pithoragarh Districts).Badola and Pant (1973); Badola (1975) and Badola and Singh (1977) have reported the fish fauna of the neighbouring Garhwal Himalaya. Hora and Mukherji (1936); Lal and Chatterjee (1962) and Singh (1964) have reported the fish fauna of Dehradun district. Mahajan (1961); Motwani and Saigal (1974); Hussain (1975); Grover and Gupta (1977); Singh (1974) and

Sinha and Shiromani (1953) and Srivastva (1968) have described the fish fauna from different parts of Uttar Pradesh.

But our knowledge on the fish fauna of the U.S. Nagar is still scanty, which harbour a large number of economically important food fishes.

## Material and Methods

Fishes were collected form different rivers, ponds, reservoirs, ditches, and paddy fields. For the taxonomic study the fishes were examined in fresh condition. Some specimens were fixed in 10 per cent formaline and then examined for their morphological characters. Fishes were identified with the help of Day's (1878).

## Physiogeographical Features

The district Udham Singh Nagar (Uttaranchal) was formerly a Southern plain part of Nainital district, which was known as Nainital Tarai. This district is the Southern part of the Uttaranchal and located in North of the Uttar Pradesh. It is surrounded by Nainital, Champawat and Pauri districts of Uttaranchal. Similarly Bijnor, Moradabad, Rampur, Bareilly and Pilibhit districts of Uttar Pradesh. Its eastern part adjoins with Nepal border.

The Udham Singh Nagar district bears a network of rivers, rivulets, ponds, lakes and some reservoirs which contain a colourful fish fauna of the plains, although there are some hill stream fishes. The important rivers of U.S. Nagar are Feeka, Dhela, Kosi, Dhbka, Bore, Jhakra, Dimari, Gaula, Baigul, Dehawa, Khakra, Chaka and Sarda river in Indo-Nepal Border.

In this district for the purpose of irrigation and fisheries, Tumaria, Gularbhoj, Baigul, Dhora and Nanak Sagar reservoirs are constructed. These reservoirs are a good source of Government revenue.

## Recommendations for Improvement of Fisheries

During 1945-52 the Nainital Tarai (U.S. Nagar) has been colonized by the Government but earlier this region was covered with dense forests. For the purpose of colonization the forests were cut and people from different parts of the country were invited for rehabilitation. During that time Tarai region was rich in fish fauna but within last 30 years the fish fauna of Tarai has suffered very much. Now-a-days this region has become thickly populated and is playing a good role in green revolution. Due to over–exploitation the fish population has been reduced. Another main reason for decline of fish population is the enormous use of pesticides, insecticides, weedicides and herbicides. These chemicals are used in agricultural fields to protect the crops from pests and to eradicate the weeds. The Toxic effects of these chemicals kill large number of fishes in all stages of their life cycle. A large number of dead fishes were observed in paddy fields, ditches, ponds and rivers. Hence the enormous use of pesticides and weedicides should be stopped to save the decreasing fish population. The U.S. Nagar is still rich in fish fauna and holds bright prospects of fish culture. Like the bumper agricultural production, the fish production may also be done on scientific lines and will provide supplementary jobs to the farmers of this region. The vast water resources of tarai can be utilized for Pisciculture practices.

**Table 9.1**

| | | *Name of Species* | *Local Name* |
|---|---|---|---|
| Series | : | Pisces | |
| Class | : | Teleostomi | |
| Sub class | : | Actinopterygii | |
| Order | : | Clupeiformes | |
| Family | : | Clupeidae | |
| | | 1. *Gadusia chapra* (Ham.) | Suhia |
| | | 2. *Gadusia godanahia* (Srivastva) | Suhia |
| Family | : | Notopteridae | |
| | | 3. *Notopterus notopterus* (Pallas) | Patra |
| | | 4. *Notopterus chitala* (Ham.) | Chital |
| Order | : | Cypriniformes | |
| Division | : | Cyprini | |
| Family | : | Cyprinidae | |
| | | 5. *Catla catla* (Ham.) | Bhakur |
| | | 6. *Labeo rohita* (Ham.) | Rohu |
| | | 7. *Labeo calbesu* (Ham.) | Kalbasu |
| | | 8. *Labeo dero* (Ham.)* | Aragi |
| | | 9. *Labeo dyocheilus* (Mo. Clell.)* | — |

*Contd...*

**Table 9.1–Contd...**

| Name of Species | Local Name |
|---|---|
| 10. *Labeo gonius* (Ham.) | Bata |
| 11. *Labeo angra* (Ham.) | Raiya |
| 12. *Labeo boga* (Ham.) | — |
| 13. *Cirrhinus mrigala* (Ham.) | Nain |
| 14. *Cirrhinus reba* (Ham.) | Reba |
| 15. *Puntius sarna* (Ham.) | Darahi |
| 16. *Puntius ticto* (Ham.) | Shidhari |
| 17. *Puntius chola* (Ham.) | Shidhari |
| 18. *Puntius sophore* (Ham.) | Shidhari |
| 19. *Puntius conchonius* (Ham.) | Darahi |
| 20. *Puntius phutunio* (Ham.) | — |
| 21. *Tor tor* (Ham.)* | Mahseer |
| 22. *Tor putitora* (Ham.)* | Mahseer |
| 23. *Cyprinus Carpio* (Linn.) | Carp |
| (i) *Cyprinus Carpio* var *communis* (Linn.) | |
| (ii) *Cyprinus Carpio* var *specularis* (Linn.) | |
| 24. *Chagunius chagunio* (Ham.) | — |
| 25. *Rasbora daniconius* (Ham.) | Dandwa |
| 26. *Rasbora rasbora* (Ham.) | Dandwa |
| 27. *Barilius bola* (Ham.) | Bola |
| 28. *Barilius bendelisis* (Ham.)* | — |
| 29. *Oxygaster bacaila* (Ham.) | Calwa |
| 30. *Oxygaster gora* (Ham.) | Calwa |
| 31. *Bsomus danricus* (Ham.) | Dandwa |
| 32. *Danio devario* (Ham.) | Fatukari |
| 33. *Brachydanio reño* (Ham.) | — |
| 34. *Osteobrama cotio* (Ham.) | Khurahi |
| 35. *Chela laubuca* (Ham.) | Dendul |
| 36. *Crossocheils latius latius* (Ham.)* | — |
| 37. *Garra gotyla gotyla* (gray)* | — |

*Contd...*

**Table 9.1–Contd...**

| | | Name of Species | Local Name |
|---|---|---|---|
| Family | : | Cobitidae | |
| | | 38. *Botia dario* (Ham.)* | — |
| | | 39. *Botia lohachata chaudhuri* | — |
| | | 40. *Lepidocephalus guntea* (Ham.) | Hakati |
| | | 41. *Somileptes gongota* (Ham.) | — |
| Division | : | Siluri | |
| Family | : | Siluridae | |
| | | 42. *Ompok bimaculatus* (Bl.) | Pabda |
| | | 43. *Wallago attu* (Bl. and Schn.) | Lanchi |
| | | 44. *Chaca chaca* (Ham.) | Chakwa |
| Family | : | Bagridae | |
| | | 45. *Mystus seenghala* (Sykes) | Tengar |
| | | 46. *Mystus bleekeri* (Day) | Tengar |
| | | 47. *Mystus cavasius* (Ham.) | Tengar |
| | | 48. *Mystus tengara* (Ham.) | Tengar |
| | | 49. *Mystus vittatus* (Bl.) | Tengar |
| | | 50. *Mystus aor* (Ham.) | Tengar |
| | | 51. *Rita rita* (Ham.) | Reeta |
| Family | : | Sisoridae | |
| | | 52. *Bagarius bagarius* (Ham.) | Goonch |
| | | 53. *Gagata cenia* (Ham.) | — |
| Family | : | Schilbeidae | |
| | | 54. *Ailia coila* (Ham.) | Patsi |
| | | 55. *Eutropiichthys vacha* (Ham.) | — |
| | | 56. *Pseudeutropius atherinoides* (Bl.) | Barusa |
| | | 57. *Clupisoma garua* (Ham.) | Karahi |
| | | 58. *Silonia silondia* (Ham.) | Silund |
| Family | : | Saccobranchidae | |
| | | 59. *Heteropneustes fossilis* (Bl.) | Singhi |
| Family | : | Clariidae | |
| | | 60. *Clarias batrachus* (Linn.) | Mangur |

*Contd...*

**Table 9.1–Contd...**

| | | Name of Species | Local Name |
|---|---|---|---|
| Order | : | Mugiliformes | |
| Family | : | Mugilidae | |
| | | 61. *Rhinomugil corsula* (Ham.) | Mugil |
| | | 62. *Sicamugil cascasia* (Ham.) | Lorhia |
| Order | : | Beloniformes | |
| Family | : | Belonidae | |
| | | 63. *Xenentodon cancila* (Ham.) | Kauwa |
| Order | : | Ophiocephaliformes | |
| family | : | Channidae | |
| | | 64. *Channa marulius* (Ham.) | Saur |
| | | 65. *Channa striatus* (Bl.) | Saur |
| | | 66. *Channa gachua* (Ham.) | Girai |
| Order | : | Symbranchiformes | |
| Family | : | Amphipnoidae | |
| | | 67. *Amphipnous cuchia* (Ham.) | Baam |
| Order | : | Perciformes | |
| Family | : | Centropomidae | |
| | | 68. *Chanda nama* Ham. | Chanari |
| | | 69. *Chanda ranga* Ham. | Chanari |
| | | 70. *Chanda baculis* Ham. | Chanari |
| Family | : | Nandidae | |
| | | 71. *Badis badis* (Ham.) | Sumha |
| | | 72. *Nandus nandus* (Ham.) | Dhebari |
| Family | : | Anabantidae | |
| | | 73. *Anabas testudineus* (Bl.) | Kawai |
| | | 74. *Colisa fasciatus* (Bl. and Schn.) | Khosti |
| Family | : | Gobiidae | |
| | | 75. *Glossogobius* giuris (Ham.) | Garua |
| Order | : | Mastacembeliformes | |
| Family | : | Mastacembelidae | |
| | | 76. *Macrognathus aculeatus* (Bloch) | Patya |
| | | 77. *Mastacembelus armatus* (Lac.) | Baam |
| | | 78. *Mastacembelus pancalus* (Ham.) | Patya |

* Hill Stream Fishes

## References

Badola, S. P. (1975). Fish Fauna of the Garhwal hills Part II (Pauri Garhwal). *Ind. J. Zool.* 16 (1): 57-70.

Badola, S. P. and Pant, M.C. (1973). Fish Fauna of the Garhwal hills Part I (Uttarkashi district). *Ind. J. Zool.* 14 (1): 37-44.

Badola, S. P. and Singh H. R. (1977). Fish Fauna of the Garhwal hills Part IV (Tehri district). *Ind. J. Zool.* XVIII (2), 115-118.

Badola, S. P. and Singh, H. R. (1977). Fish Fauna of the Garhwal hills Part III (Chamoli district) *Ind. J. Zool.* XVIII (2), 119-122.

Chaudhary, H. S. and Khandelwal, O. P. (1960). The Fish survey of Nainital district, *Vigyan Parishad Anusandhan Patrika* III: 139-145.

Day, F. (1878). The fish of India being a Natural History of the fishes known to inhabit the sea and fresh water of India, Barma and Ceylon, Vols. I and II: XXX 778, pls CXCV.

Grover, S. P. and Gupta, S. K. (1977). Fish and Fisheries of Banda District (U. P). *Proc. Nat. Acad. Sci. India* 473 (IV): 204-218

Hora, S. L. and Mukerji, D. D. (1936). Fishes of Eastern Doon, United Provinces with Introduction and Remarks on Mahseer Fisheries *Rec. Indian. Mus.* 38: 317-331.

Hora, S. L. (1937). Notes on Fishes in the Indian Museum On A Collection of Fish from Kumaon Himalaya. *Rec. Indian. Mus.* 39: 338-341.

Husain, A. (1975). Fish Fauna of Corbett National Park Uttar Pradesh *Cheetal* 17 (2): 39-42.

Lal, M. B. And Chatterjee, P. (1962). Survey Of Eastern Doon Fishes with Certain Notes on their Biology. *J. Zool. Soc. India.* 14 (1-2): 229-243.

Mahajan, C. L. (1961). Fish Fauna of Muzaffarnagar District, Uttar Pradesh *J. Bombay Nat. Hist. Soc.* 62 (3): 440-454.

Menon, A. G. K. (1949). Fishes of Kumaon Himalaya, *J. Bomay Nat. Hist. Soc.* 48 (93): 535-542

Menon, A. G. K. (1951). Fishes from the Kosi Himalaya, Nepal *Rec. Indian. Mus.*, XLVII (3-4): 231-237.

Menon, A. G. K. (1962). A Distributional List of Fishes of the Himalaya, *J. Zool. Soc. India*, 14 (1-2): 23-32.

Menon, A. G. K. (1971). Taxonomy of Fishes of the Genus *Schizothorax* Heckel with The description of a New Species from Kumaon Himalaya. *Rec. Zool. Surv. India* 63 (1-4): 195-207.

Motwani, M. P. and Saigal, B. N. (1974). Fish Fauna of Sarda Sagar Reservoir in Pilibhit (U. P). and Some Recommendations for Development of Reservoir Fisheries, *India J. Fish.* 21 (1): 109-119.

Pant, M. C. (1970). Fish Fauna of the Kumaon Hills *Rec. Zool. Surv. India*. 64 (1-4): 85-96.

Singh, P. P. (1964). Fishes of The Doon Valley *Ichthyologica* II(1-2): 86-92.

Singh, S. P. (1974). Fishes of Mordabad (U. P). *Ind. J. Zool.*: XV (2): 75-78.

Sinha, B. M. and Shiromani, A. A. (1953). The Fishes of Meerut, *Rec. Indian Mus.* Li: 61-65.

Srivastava, G. (1968). Fishes of Eastern Uttar Pradesh, Vishwa Vidyalaya Prakashan Varanasi: 1-163.

Walker, W. (1888). Angling in Kumaon Lakes, Calcutta, 105 pp.

**Aquatic Biodiversity in India: The Present Scenario, 2005** **154–183**
*Edited by*: D.R. Khanna, A.K. Chopra & G. Prasad
*Published by*: Daya Publishing House, New Delhi

# 10

# Role of Water Hyacinth and Other Aquatic Weeds in Wastewater Treatment: An Eco-friendly Approach Having Vast Potential

**V. Singhal and A. Kumar**
*Department of Environmental Sciences,*
*G.B. Pant University of Agriculture and Technology, Pant Nagar*

Water pollution is an age-old serious environmental problem. From the very beginning, people used to filter water through cloth, which is now replaced by gravel and sand filter. About pure water the very old proverb 'water water everywhere but no water to drink' implies correct today also. More than 97 per cent of earth's water supply is in the ocean, which is unfit for human consumption and other uses because of its high salt content. Of the remaining 3 per cent, 2 per cent is locked in the polar ice caps and only 1 per cent is available as fresh water in rivers, lakes, streams, reservoirs and ground water, which is suitable for human consumption. These fresh water resources continue to be contaminated with run-off water from agricultural fields, containing pesticides, fertilizers, soil particles, waste chemicals from industries, sewage and household detergents from cities and rural areas. This signifies the need and importance of developing an efficient, environmental-friendly and potential method of wastewater treatment.

## Need for Effluent Treatment

Effluent treatment is needed to achieve one of the following (Mahabal, 1993):

1. To satisfy legal requirements, laid down by pollution control authorities or local municipal authorities. These authorities prescribe limits (upper limits for most of the parameters except dissolved oxygen content and lower and upper limits for pH) for parameters of treated effluent relevant to the final mode of disposal or endure. The final disposal permitted may be on land or into nala, river or sea.
2. To recycle the same in industry or factory in the same process.
3. To reuse the same in industry/factory or outside the factory for a different purpose.

According to Vesilind *et al.* (1994) the objective of wastewater treatment is to reduce the concentrations of specific pollutants to the levels at which the discharge of the effluent will not adversely affect the environment or pose a health threat. Moreover reductions of these constituents need only to be on some required level. Although water can technically be completely purified by distillation and deionization, this is unnecessary and may actually be detrimental to the receiving water. Fish and other organisms can not survive in deionized or distilled water.

Quantity and quality of raw effluents change not only for industry to industry but for the same type of industry also. Even for the same industry, the quantity and quality fluctuations take place from day to day and even from hour to hour on the same day. Ecological modernization of industry aims at raising the levels of both ecological and economic efficiency by increasing material and energy effectively in production and consumption process in order to minimize the expense on environmental protection while keeping the cost of natural resource exploitation within acceptable limits. In fact, ecological modernization aims at restructuring of economy based on ecotechnological principles. Ecotechnology encompasses a paradigm shift from a growth model based entirely on economic criteria to one based on concurrent attention to the principles of ecology, equity and employment. Besides, it designs the human society with its natural environment for the benefit of both, helps to combine traditional wisdom and techniques with modern science and frontier technologies (Dash, 1999). This particular approach compels us towards phytoremediation technology *i.e.*, use of green plants in wastewater treatment.

## Phytoremediation

Phytoremediation (Phyto-Greek word meaning plant, and remediation-the process of correcting a problem or to put back in proper condition) is probably the most descriptive terminology. Phytoremediation is defined as the use of green plant to remove pollutants from the environment or to render them harmless. The main strategies are either storage and disposal or conversion of the contaminants by plant enzymes and microbial activity to nontoxic substances. With a growing market, research and fieldwork are being done to explore phytoremediation efficiency in the treatment of soils and groundwater contaminated with metals, radionuclides, and organics, and in wastewater and landfill applications.

Several comprehensive reviews have been written on this subject, summarizing many important aspects of this novel plant-based technology Cunningham *et al.* (1996); Cunningham and Ow (1996); Raskin *et al.* (1997); Salt *et al.* (1995); Salt *et al.* (1998) and Srivastava and Purnima (1998). The basic idea that plants can be used for environmental remediation is very old and can not be traced to any particular source. However, a series of fascinating scientific discovers combined with an interdisciplinary research approach have allowed the development of this idea into a promising, cost-effective and environment friendly technology. Phytoremediation can be applied to both organic and inorganic pollutants, present in solid substrates (*e.g.* soil), liquid substrates (*e.g.* water) and the air.

Cunningham *et al.* (1997) defined phytoremediation as the use of green plant-based systems to remediate contaminated soils, sediments and water. Relative to many traditional techniques, phytoremediation is a fledgling technology intended to address a wide variety of surficial contaminants. Phytoremediation targets currently include contaminating metals, metalloids, petroleum hydrocarbons, pesticides, explosives, chlorinated solvents and industrial by-products. On the other hand, Watanabe (1997) defined phytoremediation on the brink of commercialization. Although phytoremediation has been tested on sites contaminated with petroleum products, heavy metals, munitions and radionuclides and at abandoned mines, wood treatment sites, and sewage treatment sites, it remains unclear how large the phytoremediation market will be. Phytoromediation experts opine that the growth of interest in the field is driven by its relative cost-efficiency compared to standard remediation methods for government-mandated site clean up.

According to Salt *et al.* (1998) phytoremediation could be currently divided into the following areas:

## Phytoextraction

The use of pollutant accumulating plants to remove metals or organics from soil by concentrating them in the harvestable parts. High biomass metal-accumulating plant and appropriate soil amendments are used to transport and concentrate metals from the soil into the above-ground shoots.

## Phytodegradation

The use of plants and associated microorganisms to degrade organic pollutants. Phytodegradation is the metabolism of contaminants within plant tissues. It operates by the same principle as selective herbicides: desirable plants metabolize the herbicide into a nontoxic compound, while undesirable weeds do not and are killed. In Phytodegradation, certain plants are used to convert harmful contaminants into non-toxic substances.

## Rhizofiltration

The use of plant root to absorb and adsorb pollutants, mainly metals from water and aqueous waste streams. Rhizofiltration is a type of phytoextraction where plants are placed along aqueous waste streams. Their roots absorb the metal contaminants, and when they become saturated, the plants are harvested and replaced if further treatment is required.

## Phytostabilization

The use of plants to reduce the bioavailability of pollutants in the environment. Phytostabilization is the use of plants to immobilize contaminants in soil, preventing their entry into groundwater and food chains. Organics and inorganics are tightly bound to the plant root tissue and stabilized.

## Phytovolatilization

The use of plants to volatilize pollutants termed as phytovolatilization. Phytovolatilization helps with the disposal in some instances, as the extracted contaminants pass through the leaves and volatilize and disperse into the atmosphere.

## The Use of Plants to Remove Pollutants from Air

Most of the early research and fieldwork in phytoremediation has focused on phytoextraction, the ability of plants to absorb metals and radionuclides from the ground and store them in their tissues. Often, plants can not distinguish between heavy metals such as cadmium and metals such as zinc or copper, which are needed for growth. Some plants, known as hyperaccumulators, can take up exceptionally high levels of contaminants from soil. Their tissues can contain from 1,000 to 10,000 parts per million of certain heavy metals (Black, 1995). Several researchers have worked on metal sequestration studies (Table 10.1) in aquatic weeds but very little literature is available on total phytoremediation of industrial effluents.

## Selection of Plant Species

The majority of current research in the phytoremediation field revolves around determining which plant works most efficiently in a given application. It is not possible that all the plant species will metabolize, volatilize, and/or accumulate pollutants in the same manner. Hence, plant species are selected to find out that which plants are most effective for remediation of a given pollutant or chemical (Table 10.2).

The following criteria was suggested by Abbasi and Ramasami (1999) over the years for selecting a plant species or combination of species as the main bioagent (s) in water treatment systems:

**Table 10.1: Metal Sequestration Studies Performed by Various Workers**

| *Sl.No.* | *Metal Sequestration Studies* | *Reference* |
|---|---|---|
| 1. | Removal of lead by water hyacinth | Akcin *et al.* (1994) |
| 2. | Heavy metal absorption by aquatic weeds | Moenandir and Murgito (1994) |
| 3. | Removal of chromium, copper and nickel by water hyacinth | Saltabas and Akcin (1994) |
| 4. | Bioaccumulation and toxicity of Cu and Cd in *Vallisneria spriralis* (L.) | Sinha *et al.* (1994) |
| 5. | Metal-scavenging plants to cleanse the soil | Comis (1996) |
| 6. | Bioremediation of Cr from tannery effluent by aquatic macrophytes | Vajpayee *et al.* (1995) |
| 7. | Treatment of tannery wastewater by water hyacinth application | Gupta and Sujatha (1996) |
| 8. | Bioremediation of chromium from water and soil by vascular aquatic plants | Chandra *et al.* (1997) |
| 9. | Phytoextraction of cadmium and zinc from a contaminated soil | Ebbs *et al.* (1997) |
| 10. | Phytoextraction of lead from contaminated soils | Huang *et al.* (1997) |
| 11. | Phytoremediation of Selenium | Terry and Zayed (1998) |
| 12. | Phytoaccumulation of trace elements by wetland plants I. Duckweed | Zayed *et al.* (1998) |
| 13. | Phytoaccumulation of trace elements by wetland plants III uptake and accumulation of ten trace elements by twelve plant species | Jin Hong *et al.* (1999) |
| 14. | Phytoaccumulation of trace elements by wetland plants II water hyacinth | Zhu *et al.* (1999) |
| 15. | Spectroscopic studies of interaction of Eu (III) with roots of water hyacinth | Kelley *et al.* (2000) |
| 16. | Phytoremediation of toxic metals | Raskin and Ensley (2000) |

1. Adaptability to local climate,
2. Ease of management
3. High oxygen transport capability,
4. High photosynthetic rates/high growth rate,

5. High pollutant-uptake efficiency,
6. Resistance to pests and diseases,
7. Tolerance to adverse climate conditions, and
8. Tolerance to adverse concentration of pollutants

**Table 10.2: Partial Listing of Plants and the Chemicals Remediated by them**

| *Sl.No.* | *Plant* | *Chemicals* |
|---|---|---|
| 1. | Arabidopsis | Mercury |
| 2. | Bladder campoin | Zinc, Copper |
| 3. | Brassica family (Indian Mustard and Broccoli) | Selenium, Sulphur, Lead, Cadmium, Chromium, Nickel, Zinc, Copper, Cesium, Strontium |
| 4. | Buxaceae (Boxwood) | Nickel |
| 5. | Compositae family | Cesium, Strontium |
| 6. | Euphorbiaceae | Nickel |
| 7. | Genus *Lemna* (Duckweed) | Explosives wastes |
| 8. | Parrot feather | Explosives wastes |
| 9. | Pennycress | Zinc, Cadmium |
| 10. | Perennial Rye grass | Polychlorinatedphenyls (PCP's), Polyaromatichydrocarbons (PAH's) |
| 11. | Pondweed, Arrowroot, Coontail | 2,4,6-trinitrotoluene (TNT), hexahydro-1,3,5-trinitro-1,3,5-triazine (RDX) |
| 12. | Sunflower | Cesium, Strontium, Uranium |
| 13. | Tomato plant | Lead, Zinc, Copper |
| 14. | Trees in the *Populus* genus (Popular, Cottonwood) | Pesticides, Atrazine, Trichloroethylene (TCE). Carbon tetrachloride, nitrogen compounds, TNT, RDX |

## Floating Aquatic Plant Treatment Systems

Aquatic weeds may be used in several types of floating aquatic plant treatment systems. According to Metcalf and Eddy (1995) the principle types of floating aquatic plant treatment systems used for wastewater treatment are those employing water hyacinth and duckweed, which are described here below:

## Water Hyacinth Systems

Majority of constructed aquatic plant system consist of water hyacinth. On the basis of dissolved oxygen content and method of aerating the pond three types of water hyacinth systems can be described:

1. Aerobic nonaerated
2. Aerobic aerated, and
3. Facultative anaerobic

### Non-aerated Aerobic Water Hyacinth Systems

These systems will produce secondary treatment or nutrient (nitrogen) removal depending on the organic-loading rate. This is most commonly used in water hyacinth systems. Excellent performance with few mosquitoes or odours is the main advantage of this type of system.

### Aerated Aerobic Water Hyacinth Systems

These systems are constructed at those plant locations, where no mosquitoes or odours can be tolerated. Besides these systems also provide an added advantage that higher organic-loading are possible with aeration and reduced land area is required.

### Facultative Anaerobic Water Hyacinth Systems

These systems are also operated at very high organic-loading rates. Increased mosquito populations and odours are the main disadvantages of these systems and due to these problems seldom these system are used.

## Duckweed Systems

The primary objective behind the use of duckweed and pennywort is to improve the effluent quality from facultative lagoons or stabilization ponds reducing the algae concentration. Conventional lagoon design may be used for the application of these type of systems, except for the need to control the effect of wind. Unless controls, duckweed will be blown to the downwind side of the pond resulting in exposure of large surface areas and defeating the purpose of duckweed cover. In this type of systems floating baffles can be used to form cells of limited size to minimize the amount of open surface area exposed to wind action.

## Mechanisms Responsible for Phytoremediation

Anderson and Coats (1995) showed an overview of microbial degradation in the rhizosphere and its implications for bioremediation. They showed that the complex nature of most synthetic chemicals encountered by microorganisms can require interaction of microbial communities to achieve transformation. The plant root zone fosters these types of interactions. For chemicals that are easily degraded, the presence of 100 fold more microorganisms in the rhizosphere compared with non-vegetated soil leads to increased rates of chemical transformation. By providing a niche suitable to a diverse population of microorganisms, vegetation may enhance microbial degradation because of the presence of a key group of organisms involved in the metabolism of the contaminant. Root exudates may serve as structural analogs to contaminants as well as enhance co-metabolism of contaminants.

## Factors Responsible for Phytoremediation of Contaminants

The key factors for phytoremediation of contaminants were described in details by (Anderson, 1996) as follows:

1. Enhancing microbial degradation in the root zone requires the presence of root surface area for microbial colonization as well as root exudates for microbial growth. While plant roots can grow into the subsurface, maximum root density occurs near the soil surface.
2. High concentrations of organic contaminants can prohibit the growth of vegetation at site as well as be toxic to microorganisms. This is especially true with pesticides contamination where mixtures of herbicides may be present.
3. In addition to high concentrations of chemical contaminants, environmental conditions often prohibit plant growth or dictate which type of plants may survive.

## Water Hyacinth as Phytoremediator

Owing to the suitability of phytoremediation, *E. crassipes* and other aquatic weeds are being used to phytoremediate sewage and industrial effluents. Fate of nitrogen and phosphorous in a wastewater reservoir containing *E. crassipes* and others aquatic macrophytes was evaluated by (Reddy, 1983). Results showed that 34 to 40 per cent of the added inorganic $^{15}N$ ($^{15}NH_4^+$ + $^{15}NO_3^-$) was removed through plant uptake, while 45 to 52 per cent of the added $^{15}N$ was unaccounted for, presumably lost through $NH_3$ volatilization and nitrification denitrification processes.

The potential use of water hyacinth for biomass production and for nutrient removal from wastewaters was suggested by Reddy and Sutton (1984) especially in warm climates. Reddy and Debusk (1985) reported that nitrogen removal by aquatic macrophyte systems was in the order of water hyacinth > water lettuce > pennywort > *Lemna* > *Salvinia* >*Spirodela* > Egeria during the summer season, while pennywort ranked first during the winter followed by water hyacinth, *Lemna*, Water lettuce, *Spirodela, Salvinia* and Egeria. Phosphorous removal in summer was highest by water hyacinth and Egeria systems, while pennywort and *Lemna* showed high P removal rates during the winter compared to other plants. Debusk *et al.* (1989) worked on effectiveness of mechanical aeration in floating aquatic macrophyte-based wastewater treatment systems. In continuous-flow raceways fed primary domestic effluent, vigorous aeration (6.1 $Lm^{-2} min^{-1}$) improved effluent quality with contaminant removal rates averaging 77 per cent ($BOD_5$) and 76 per cent (suspended solids) in non-aerated raceways, and 94 per cent ($BOD_5$) and 89 per cent (SS) in aerated raceways. Results of this study suggested that the high aeration requirement for enhanced contaminant removal in floating aquatic macrophyte systems (FAMS) was due not only to poor $O_2$ transfer efficiencies (1.6-4.0 per cent), but also to the inefficient utilization of $O_2$ for $BOD_5$ removal.

Casabianca *et al.* (1991) treated the paper industry effluents with *Eichhornia crassipes*. The percentage of removal was doubled with a detention time of 2-days. COD and $BOD_5$ removal, which is related to both physical setting and plant absorption, was improved with a 2-d detention time, but the difference was not significant, and the removal of nitrogen and phosphorous by biomass production (0.43-1.94 g $N/m^2d$ and 0.057-0.23 g $P/m^2d$, with a production of 14-65 of g D.W./$m^2$ d) was correlated with factors favouring this production. Granato (1993) studied cyanide degradation by water hyacinths. In batch tests, *E. crassipes* removed free cyanide in the first 8 hours. When synthetic gold mill effluent containing free cyanide (9-20 mg), thiocyanate (14-23 mg) and metallocyanides (iron, copper and zinc) was fed to *E. crassipes* at 6 litres/h, the weed degraded free cyanide (removing 46-56 per cent after 8 d compared to no-plant control values of 22-27 per cent) and removed zinc (26-62 per cent compared to control values of 19-29 per cent) and small amounts of iron (18-31 per cent compared to control values of 8-22 per cent). However, *E. crassipes* did not degrade copper and thiocyanate.

Eyini *et al.* (1993) reported that the epicuticular wax deposits showed clear disorganization in the plants grown in distillery effluent. The epicuticular waxes of both control and test plants showed the presence of fatty acids, OH-beta diketones, hydrocarbons, esters and ketones. The content of the leaf epicuticular wax was significantly lower in the effluent grown plants than in the control plants. (Saltabas and Akcin, 1994) grown water hyacinth plants in solutions containing 10 ppm chromium, 35 ppm copper and 14 ppm nickel for 24 hours and found that these plants can be used for removing heavy metal from industrial effluent under Turkish conditions. Singaram (1994) worked on removal of chromium from tannery effluent in India using three water weeds, water hyacinth, pseudo water hyacinth and *Lemna* species; water hyacinth accumulated the most chromium (38 ppm) followed by *Lemna* and pseudo water hyacinth. In all three species the root accumulated higher amounts of chromium than the foliage. Saini (1995) exploited the potential of water hyacinth as a natural de-pollutant employing a short detention period of 48 hours. The effluent obtained was colourless and odourless with a DO content of 0.5-1.0 mg/l. The pH of the effluent was in the range of 6.8-7.2, turbidity was less than 5.0 mg/l and the average percentage removal of BOD, COD, total solids and suspended solids were 87, 90, 77.6 and 76.6 per cent, respectively. Similar investigations were made by (Gupta and Sujatha (1996) and Jebanesan (1997) with tannery and dairy wastewater, respectively.

Montien-art *et al.* (1998) utilized some aquatic plants for water turbidity treatment and animal feed was investigated at the Kumamoto Zoological Park, Japan. When *E. crassipes, Pontederia cordata* and *Zizania latifolia* were cultivated in a basin with non-toxic eutrophic wastewater, the concentration of suspended solid in the water significantly decreased with the increment of plant biomass. The effect of a fertilizer factory effluent on total chlorophyll content and biomass of some aquatic macrophytes was studied by Srivastava and Pandey (1999a). The total chlorophyll content and biomass of *Eichhornia crassipes, Pistia stratiotes* and *Hydrilla verticillata* were reduced significantly with an increase in treatment dilutions from 7, 4 and 2 days of expose in different fertilizer effluent concentrations. Srivastava and Pandey (1999b) determined toxicity of paper mill effluent on chlorophyll-a content and biomass of *E. crassipes* and *Spirodela polyrrhiza* at different intervals. Chlorophyll content and biomass of both the macrophytes decreased when exposed to the paper mill effluent. Extent of reduction was in proportion to the concentration of effluent and duration of exposure.

Shoyakubov and Aitmetova (1999) examined the chemical composition of *E. crassipes* and *P. stratiotes* grown in effluent from the industrial production of nitrogen fertilizers, following thermal treatment of the biomass. Protein and cellulose were the major component of *E. crassipes*, while *P. stratiotes* contained more fat and ash. Carotene contents of 332 and 34 mg/kg were observed in *E. crassipes* and *P. stratiotes*, respectively. The ability of water hyacinth to take up and translocate As (V), Cd (VI), Cr (VI), Cu (II), Ni (II), and Sc (VI) under controlled conditions has been studied by (Zhu *et al.* (1999). According to Jain (2000) the water hyacinth had ideal characteristics for water purification and pollution control. The production of high quality vegetable protein, vitamins, minerals, fertilizer, chemicals and energy (in the form of biogas) from water hyacinth had reduced it nuisance value and made it a potential provider.

Narayana and Parvez (2000) worked on treatment of paper mill effluent using water hyacinth. The results of water hyacinth treatment, favoured pH from 7.7-6.96. However DO content declined from 6.96-2.4. Similarly, there was major decline in BOD and COD and the results recorded before and after treatment. The values showed 800-67 and 400-132 mg/l respectively. Recently, the roots of water hyacinth removing large quantities of Eu (III) from water has been reported by (Kelley *et al.,* 2000). They further showed that carboxylic acids are responsible for binding the intracellular proportion of Eu (III) in the roots of water hyacinth using the techniques of Nuclear Magnetic Resonance (NMR) and Infra red (IR) spectroscopies. Mehra *et al.* (2000) conducted a study of

*E. crassipes* grown in the over bank and flood plain soils of the river Yamuna in Delhi, India. They have investigated the impact of wastewater discharges from four major drains (Najafgarh, Power House, Barapula, Kalkaji) on the overbanks, flood plains and *Eichhornia* in river Yamuna in Delhi. They concluded that the wastewater discharges from the drains, with the exception of Barapula drain, generally increase the elemental concentrations of overbank soils downstream of the discharges. *Eichhornia* plants growing along the banks receiving wastewaters from the Najafgarh and Barapula drains are unhealthy and reduced in population, which can be attributed to a combination of alkaline pH of growth medium, metal toxicity and high BOD at the site receiving effluents from the Najafgarh drain, and alkaline pH, metal toxicity and the turbid conditions of water with fly ash particles deposition on the plant surfaces at the site receiving effluents from the Barapula drain. Singhal and Rai (2003) found that maximum reduction by water hyacinth plants was in colour unit (80.93 and 80.50 per cent) and minimum reduction in sodium (38.34 and 45.84 per cent) amongst all the studied parameters at 20 per cent concentration of pulp and paper mill and distillery effluents, respectively.

The above mentioned studies clearly depict the popular use of *E. crassipes* and other aquatic weeds in phytoremediation of variety of sewage and industrial effluents. However, the huge growth of these plants adhering enormous amount of energy may pose serious disposal problem, unless it is tapped logically. To achieve this, several efforts have been made by different workers, which will be discussed in later part of this chapter.

## Research Work on Other Aquatic Weeds

The effectiveness of mechanical aeration in floating aquatic macrophyte based wastewater treatment system was studied by Debusk *et al.* (1989). Light aeration (0.003 and 0.021 $Lm^{-2} min^{-1}$) had no effect on the treatment of primary domestic effluent in the batch-fed water hyacinth tanks. Heavy aeration (1.03 and 3.53 $Lm^{-2} min^{-1}$) raised wastewater D.O. concentrations but did not improve Biochemical Oxygen Demand ($BOD_5$) removal efficiency or increase plant growth rates during 21-days experiments. Debusk *et al.* (1991) reviewed wastewater treatment methods based on floating aquatic macrophytes in pond systems. Pennywort was the most efficient plant to provide secondary wastewater treatment, followed by water hyacinth and common duckweed. Similar study has been reported by Walsh *et al.* (1991) with *Echinochloa crusgalli* and *Sesbania macrocarpa*.

The phytoremediation experiments have been popular in several academic laboratories and industries (Brown, 1995). At an April conference hosted by the University of Missouri, more than 250 researchers from around the world gathered to hear 35 phytoremediation presentations. The stars of phytoremediation could be plants called hyper-accumulators. They can be herbs, shrubs, and even trees. Hyper-accumulators concentrate trace elements, heavy metals or radio nuclides at level 100-fold or greater than normal.

Comis (1996) also described eco-friendly approach for remediation of soil *i.e.* used plants to clean the soil. Cunningham and Ow (1996) described in details the promises and prospects of phytoremediation. According to Cunningham *et al.* (1999) phytoremediation is an affordable technology that is most useful when contaminants are within the root zone of plants (top 3-6 ft [1-2 m]). He also reported various plants for the treatment of wastewater *viz., Lemna, Pistia,* Alfa alfa water hyacinth etc. Further popularization of phytoremediation technology from laboratory to the market place has been attempted by Dushenkov *et al.* (1997); Nyer and Gatliff (1996); Boyajian and Carreira (1997) and Bishop (1997). Plant's ability to absorb, translocate and concentrate metals has been studied by Chandra *et al.* (1997). The potential to accumulate chromium by *Scirpus lacustris, Phragmites karka* and *Bacopa monnicri* was assessed by subjecting them to different chromium concentrations under laboratory condition. The plants caused significant reduction in chromium concentrations. While there was an increase in biomass, no visible phytotoxic symptoms were shown by treated plants. Similar reports were made by Satyakala and Jamil (1997) with *Pistia stratiotes.*

A hypothesis to test that herbicide-tolerant aquatic plants could remediate herbicide-contaminated water was conducted by Rice *et al.* (1997). The addition of *Ceratophyllum demersum* (Coontail, hornwort), *Elodea Canadensis* (American elodea, Canadian pondweed) or *Lemna minor* (common duckweed) significantly (p 0.01) reduced the concentration of [$^{14}$C] metolachlor (MET) remaining in the treated water. Sarkar (1997) studied seasonal influence of the plant *Lemna major* in the treatment of eutrophicated ponds. The concentrations of total ammonical nitrogen (TAN), nitrate-nitrogen (NN) and phosphate of water in different seasons decreased in ponds exposed to aquatic plants and the mean percentages of nutrients in ponds with plants exhibited different trends and in the following orders: monsoon > winter > summer for TAN, monsoon > summer > winter for NN and summer > winter > monsoon for phosphate. Using water lettuce plants, heavy metals can be effectively removed when they are present at a concentration of 10 mg/l or less Selvapathy *et al.* (1998). Siciliano and Germida (1998) described the mechanisms of phytoremediation. They further specified the biochemical and ecological interactions between plants and bacteria. Zayed *et al.* (1998) reported phytoaccumulation of trace elements by wetland plants *viz.*, duckweed.

Relationship between phytoremediation and on-site treatment of septic effluents in sub surface flow constructed wetlands was studied by Neralla *et al.* (1999). Phytoremediation generally reduced B.O.D., $NH_4^+$-N, P, turbidity and volatile suspended solids. Populations of faecal coliforms were consistently reduced by 90 to 99 per cent by 2-days detention in microcosms with and without plants. Trivedy and Nakate (1999) studied aquatic weeds based waste water treatment plants in India and reported a very high degree of reduction in suspended solids, BOD and COD, nitrogen, phosphorous and oil and grease. Rose (2000) inferred that *Lemna minor* is efficient in removing BOD, solids and nutrients from the wastewater and has high potential for treating organically rich wastewater and reuse possibilities. Trivedy and Nakate (2000) studied treatment of diluted distillery waste by using constructed wetland on laboratory scale using *Typha latifolia*. The reduction in BOD and COD was 47.59 per cent and 78.77 per cent in 10 days.

On the basis of above reports Central Pollution Control Board (CPCB, 2001) emphasized that phytoremediation is a natural biological treatment of wastewater. Wetlands are inundated land areas with water depths typically less than 2 ft that can support the growth of emergent plants. Floating aquatic plants system contains the floating species such as water hyacinth and duck weed, where an average depth of water ranges from 1.6–6.0 ft. Supplementary aeration has been used with floating plant system for improved treatment efficiency and to maintain an aerobic condition for the biological control of mosquitoes. So far the studies of phytoremediation have been done on sewage and municipal waste treatment. Very few workers have investigated on the phytoremediation using industrial effluent. Higher plants (common duckweed, lettuce and rice) were used by Wang and Wang (1991) for effluent toxicity assessment. Whilst the effect of paper mill effluent on the water quality of receiving wetlands was studied by (Baruah *et al.*, 1996). The different parameters of effluent studied included pH, temperature, DO, BOD, COD, chloride, sulphate, alkalinity, hardness, calcium, magnesium, sodium, potassium, residual, chlorine, mercury, colour and lignin and most of them revealed elevated values indicating toxic nature of effluent. Before confluence with effluent the water quality showed its suitability for sustenance of life processes but after confluence the water quality degraded in several parameters making it unsuitable for survival of life processes. A similar observation has been made by Chen *et al.* (1995) with dairy lagoon effluent.

The role of glutathione and phytochelatin in *Hydrilla verticillata* and *Vallisneria spiralis* under mercury stress was studied by (Gupta *et al.*, 1998). Both free (*Hydrilla*) and rooted (*Vallisneria*) submerged plants showed high potential to accumulate mercury, the greatest amount being in the roots of *V. spiralis*. During mercury stress, these plants synthesized different species of phytochelatins (PCs), which bound with the accumulated mercury, and showed high levels of cysteine and non-protein thiols. It was concluded that phytochelatins were synthesized in these plants and played a role in mercury detoxification. Welsch and Denny (1979) reported that *Vallisneria spiralis* extracted nutrients mostly from the sediment, which is translocated to the upper part. Gupta and Chandra (1994) observed lead accumulation and toxicity in *Vallisneria spiralis* and *Hydrilla verticillata*. They determined the lead uptake potential of two aquatic macrophytes *Vallisneria spiralis* and *Hydrilla verticillata* in solution cultures under the laboratory conditions. Both the plants showed substantial accumulation of Pb, though it was more in *H. verticillata*.

Bioaccumulation and toxicity of Cu and Cd in *Vallisneria spiralis* (L) under laboratory conditions was studied by (Sinha *et al.*, 1994). Plants showed ability to reduce 5 $\mu g ml^{-1}$ Cu background concentration to below 0.05 $\mu g ml^{-1}$ within 48 hours. (St-Cyr and

Campbell, 1994) observed trace metals in submerged plants of the St. Lawrence river. *Vallisneria americana* showed potential as a bioindicator species for trace metals in its environment. The Cd, Cr, Cu, Ni, Pb and Zn concentration in *V. americana* leaves reflected spatial variation in environmental contamination and represented bioavailable trace metals potentially transferable through the food chain to higher organisms. With the exception of Pb, *V. americana* concentrated metals to higher levels in its above ground tissues than did *Potamogeton richardoonil*, another abundant species in the lake studied. Similar study was made by Rai *et al.* (1995) with channel grass, coontail, duckweed and bacopa. Singhal *et al.* (2003) studied that treatment by channel grass resulted into maximum reduction in COD of pulp and paper mill effluent (74.66 per cent) and in colour unit of distillery effluent (75.19 per cent), whilst the minimum reduction was observed in sodium content of the two effluents at 20 per cent concentration (31.84 and 38.77 per cent).

## Phytoremediated Plant Biomass a Source of Energy

In order to effectively remove nutrients and other chemicals from waste effluents, water hyacinths must be harvested at interval that allow for maximum biomass production. One acre (0.40 hectare) of water hyacinths has the potential of producing over 70 tonnes (63, 640 kg) of dry plant material annually when grown in a desirable nutrient media such as domestic sewage effluent under proper climatic conditions. This large volume of biomass has the potential of producing about 30,000 $m^3$ of biogas through anaerobic decomposition with 70 tonnes (63.5 metric tonnes) of residual high-grade fertilizer being produced as a by-product. Table 2.8 clearly displays some of the processing alternative and products that may be derived from the harvested biomass (Chawla, 1986).

Samples of *Typha angustata, Marsilea quadriflora, Ipomoea aquatica, Eichhornia crassipes, Limnanthemum oristatum, Azolla pinnata, Vallisneria spiralis, Hydrilla verticillata* and *Chara* sp. from ponds in Gujarat were analyzed by Rana and Kumar (1988), who observed great variation in ash mineral content. Sodium content was highest in *T. augustata* (20.1 per cent), magnesium in *H. verticillata* (11.9 per cent), aluminium, silica and calcium in *Chara* sp. (9.6, 16.0 and 32.3 per cent, respectively), phosphorous in *A. pinnata* (16.7 per cent), sulphur in *T. aquatica* (9.0 per cent) and potassium and chloride in *E. crassipes* (33.4 and 37.0 per cent, respectively). Water hyacinth was described as a potentially vast resource available to many tropical and subtropical countries. Some of the major possibilities for use were analyzed *e.g.*, incineration, briquetting, gasification, pyrolysis, anaerobic digestion (Thomas and Eden, 1990). Studies were conducted by Delgade *et al.* (1992) to determine the production of water hyacinth in Madrid. To evaluate the nutritive value of the water hyacinth the biomass was fractionated to obtain a liquid fraction with 23.78 per cent protein and a solid fraction with 5.61 per cent protein. For animal food, the both fractions should be used *i.e.* solid and liquid. The protein content was 21.14 per cent (dry weight), which was higher or at least equal to normally used green fodder.

The effects of harvesting frequency on productivity, nutrient storage and uptake, and detritus accumulation by water hyacinth were described by Reddy and D'Angelo (1990). Significant differences in hyacinth standing crop and productivity were measured with harvesting regimes of 1, 3 (harvest at maximum density) and 21 harvests over a 13-months period. The average plant standing crop decreased from 55 to 20 kg (fresh weight)/m$^2$ for systems with 1 and 21 harvests, respectively. Chin and Kee (1989) also recovered energy from waste and biomass from water hyacinth grown in palm oil mill effluent, refuse in landfills, piggery effluent. Amongst the plant energy sources, biogas generation could be visualized as major source, where aquatic macrophytes may serve better. Sen (1930) and Sen and Chatterjee (1931) were the first demonstrate the possibility of using water hyacinth for the generation of power alcohol and fuel gas by virtue of the presence of 11.3 per cent lignin, 13.3 per cent pentosans, and 21.9 per cent cellulose but only 0.018 per cent starch in plant body. They also described the following three methods of utilizing the plant: a) saccharification and subsequent fermentation, b) gasification and c) bacterial fermentation.

Biogas production from crop residues and aquatic weeds was observed by El-Shinnawi *et al.* (1989). Maize stalks gave the highest value of cumulative yield of biogas, whereas rice straw, water hyacinth, and cotton stalks produced progressively less biogas. Such a sequence was paralled by the rate of loss of volatile solids. Peaks of biogas generation apapeared after 17-21 days of fermentations according to the type of feedstock. Bioenergy potential of eight common aquatic weeds *Salvinia molesta, Hydrilla verticillata, Nymphaea stellata, Azolla pinnata, Ceratopteris* sp., *Scirpus* sp., *Cyperus* sp. and *Utricularia reticulata* was assessed by Abbasi *et al.* (1990). Natural stands of *Salvinia* such as the one employed in study, would yield energy (methane) of the order of $10^8$ Kcal $ha^{-1}$ $year^{-1}$, while *Azolla, Scirpus, Hydrilla* and *Nymphaea* had energy potentials of the order of 10 Kcal $ha^{-1}$ $year^{-1}$. Anaerobic digestion was used by Chynoweth and Reddy (1990) to digest a mixture of harvested aquatic plant biomass and primary sludge for biogas generation. The production of biogas from aquatic biomass with that of terrestial biomass was compared by (Jain *et al.*, 1990). Water hyacinth showed promise and gave maximum biogas yield followed by water velvet and duckweed.

Various conditions such as temperature, total solid content, water hyacinth-cattle dung ratio, retention time, pH and stirring to improve anaerobic digestion of water hyacinth-cattle dung were optimized by (Madamwar *et al.*, 1990a). Maximum methane production (0.64 lit/lit of digester per day) was with a retention time between 7 and 9 days at 35°C and a 7 to 9 per cent (w/v) total solid content of water hyacinth-cattle dung (7: 3 w/w). (Wolverton *et al.*, 1975) noted that the rate of methane production was higher (81.8 ml/day) in plants contaminated with nickel and cadmium than in uncontaminated plants (51.8 ml/day). The methane content was also higher (91.1 per cent) in contaminated plants than in the normal ones (69.2 per cent). Temperature has a marked effect on biogas production. The biogas was produced quickly and had higher methane content (69.2 per cent) at 36°C than at 25±5°C (59.9 per cent). The effect of various residues, such as *Polyalthia longifolia* leaves, *Azadirachta indica* leaves, eucalyptus leaves, sugarcane baggasse, banana stem, poultry waste, cheese whey, algal powder (*Enteromorpha* sp.) and sugarcane filter cake was studied by Madamwar *et al.* (1990b) to help improve the anaerobic digestion of water hyacinth-cattle dung. Among the residues tested, sugarcane bagasse, banana stem, poultry waste, cheese whey and algal powder showed more than 100 per cent increase in gas production with 5-10 per cent higher methane content.

Mallik *et al.* (1990) assessed the potentialities of *Cannabis sativa*, water hyacinth and crop wastes mixed with dung and poultry litter for biogas production. Some wastes could substitute for cow dung when dung was in short supply. Misra *et al.* (1992) and Patel *et al.* (1992) developed a new temperature controlled digester for anaerobic digestion for biogas production. Using six aquatic weeds, *viz.*, *E. crassipes*, *Najas indica*, *Pistia stratiotes*, *Salvinia molesta*, *Ottelia alismoides*, and *Ipomea carnia* (Sasmal, 1992) evaluated suitability for their biogas production. The stimulatory effect of the aquatic weeds in biogas production was in the order of *Najas*> *Pistia*> *Eichhornia*> *Ottelia*> *Salvinia*> cowdung.

Effect of particle size, plant nitrogen content and inoculum volume on batch anaerobic digestion of water hyacinth was evaluated by (Moorhead and Nordstedt, 1993). Cumulative biogas production was maximum at 15 days. Cumulative biogas production was highest for a plant particle size of 6.4 mm. Cumulative biogas production at 15 days increased with increasing inoculum volume for plants with a high N content but not for plants with a low N content. (Patel *et al.*, 1993) examined the effect of several salts $FeCl_3$, $NiCl_2$, $CoCl_2$, $CuCl_2$ and $ZnCl_2$, on anaerobic digestion of water hyacinth-cattle dung. Among the salt studied, $FeCl_3$ caused a more than 60 per cent increase in gas production with high methane content. Profitability of biogas from a Chinese type digester was evaluated by (Versters, 1994). A 20 $m^3$ digester of Chinese type was fed with water hyacinth and pig manure. Enough gas was produced to cook 3 meals in a day for 50 students; wood was no longer required.

Control of the diurnal pattern of methane emission from emergent aquatic macrophytes was reported by Whiting *et al.* (1996) by gas transport mechanisms. Experimental manipulation of elevated and reduced $CO_2$ levels in the atmosphere surrounding the plants and of light/dark periods suggested that stomatal aperture had little or no control of methane emissions from *Typha latifolia*. Abubacker and Rao (1999) reported four potential biofuel crops for tropics. They reported that *Dodonea viscosa*, *Jatropha gossypifolia*, *Lantana camera* and *Prosopis juliflora* produced higher amount of biogas and were identified as biofuel crops. Jiang *et al.* (2000) observed effect of application of different plant residues to paddy field on $CH_4$ and $CO_2$ formation process. Alfalfa, clover, duckweed, water hyacinth, rice and wheat were added to the paddy soil on the basis of same carbon content. The amount of methane produced by the plant residues varied in the order of clover> alfalfa> water hyacinth> duck weed> rice> wheat. Singhal and Rai (2003) found that slurry of the two plants namely water hyacinth and channel grass used for phytoremediation produced significantly more biogas than that produced by the plants grown in deionized water; the effect being more marked with plants used for phytoremediation of 20 per cent pulp and paper mill effluent. Biogas production from channel grass was relatively greater and quicker (maximum in 6-9 days) than that from water hyacinth (in 9-12 days).

## Conclusion

Environmental contamination may be viewed as an ecosystem malaise, while phytoremediation can be regarded as a kind of environmental medicine. As such, phytoremediation can be used as preventive medicine as well as a post–contamination treatment option. Like most diseases, there is a need for case-specific evaluations and treatments. Phytoremediation of different contaminants must be based on a specific knowledge of the target chemicals, the requisite plant and the relevant environmental conditions. In this way, the aquatic weeds which are thought to cause nuisance and are typical to manage may be profitably utilized as environmental cleaners in waste water management in a controlled process and hence justify the conservation of aquatic biodiversity India.

## References

Abbasi, S. A. and Ramasami, E. (1999). Aquatic macrophytes in wastewater treatment systems: suitability, mechanism of action, design considerations, economics and environmental impact *In*: Biotechnological Methods of Pollution Control. Hyderguda, Hyderabad, India, University Press Ltd. pp. 8-51.

Abbasi, S. A.; Nipaney, P. C. and Schaumberg, G. D. (1990). Bioenergy potential of eight common aquatic weeds. *Biol. Wastes*. 34: 359-366.

Abubacker, M. N. and Rao, G. R. (1999). Four potential biofuel crops for tropics. *J. Environ. Stud. Pol.* 2 (2): 117-124.

Akcin, G.; Saltabas, O. and Afsar, H. Removal of lead by water hyacinth (*Eichhornia crassipes*). *J. Environ. Sci. Health* A29 (10): 2177-2184

Anderson, T. A. and Coats, J. R. (1995). An overview of microbial degradation in the rhizosphere and its implications for bioremediation. *In:* Skipper, H. D. and Turco, R. F. *eds.* Bioremediation: Science and Applications, SSSA Special Publication, 43, Madison, WI, Soil Science Society of America. pp. 135-143.

Anderson, T. A. (1996). Rhizosphere technology for phytoremediation. *In*: International Phytoremediation Conference, Arlington, VA, May 8-10, 1996. International Business Communications, Southborough, MA.

Baruah, B. K.; Baruah, D. and Das, M. (1996). Study on the effect of paper mill effluent on the water quality of receiving wetland. *Poll. Res.* 15 (4): 389-393

Black, Harvey (1995). "Absorbing Possibilities: Phytoremediation". *Environmental Health Perspectives.*

Bishop, J. (1997). Phytoremediation: A new technology gets ready to bloom. *Environ. Solu.* 10 (4): 29.

Boyajian, G. E. and Carreira, L. H. 1997. Phytoremediation: A clean transition from laboratory to marketplace. *Nat. Biotechnol.* 15 (2): 127-128.

Brown, K. S. (1995). The green clean: The emerging field of phytoremediation takes root. *Bioscience.* 45: 579-582.

Casabianca, C. M. L-de; Coma, C. and De-C. C., M. L. (1991). Treatment of paper industry effluents with *Eichhornia crassipes*: First result (Taratas factory, Landes). *Comptes-Rendus-de-l'Academic-des-Sciences.* 312 (11): 579-585.

Chandra, P.; Sinha, S. and Rai, U. N. (1997). Bioremediation of chromium from water and soil by vascular aquatic plants. *In*: Kruger, E. L.; Anderson, T. A. and Coats, J. R. *eds.* Phytoremediation of Soil and Water Contaminants. Washington, D. C., American Chemical Society. pp. 274-282.

Chawla, O. P. (1986). Advances in Biogas Technology. New Delhi, India, Publication and Information Division, Indian Council of Agricultural Research. pp. 75-95.

Chen, S.; Rahman, M.; Chatreck, R. H.; Jenny, B. F.; Malone, R. F. and Campbell, K. L. (1995). Constructed wetlands using black willow, duckweed and water hyacinth for upgrading dairy lagoon effluent. *In*: Versatility of Wetlands in the Agricultural Landscape. Florida, U. S. A., Sept. 17-20, 1995. pp. 273-282.

Chin, K. K. and Kee, K. C. (1989). Bioenergy recovery and conservation in Singapore. *In*: Wise, D. K. *ed*. International Biosystems. I, 1-15.

Chynoweth, D. P. and Reddy, K. R. (1990). Feasibility manual for aquatic plant wastewater treatement with energy recovery. TVA Fertilizer Publications. Z-260. 169p.

Comis, D. (1996). Green remediation: using plants to clean the soil. *J. Soil Water Conserv.* 51 (3): 184-187.

CPCB (2001). Biotechnologies for Treatment of Wastes. Information manual on pollution abatement and cleaner technologies series: Impacts/6/2000-01. Delhi, India, Central Pollution Control Board, Ministry of Environment and Forests. pp. 23-24.

Cunningham, S. D.; Anderson, T. A.; Schwab, A. P and Hsu, F. C. (1996). Phytoremediation of soils contaminated with organic pollutants. *Adv. Agron.* 56: 55-114.

Cunningham, S. D., Shann, J. R.; Crowley, D. E. and Anderson, T. A. 1997. Phytoremediation of contaminated water and soil. *In*: Kruger, E. L.; Anderson, T. A. and Coats, J. R. *eds*. Phytoremediation of Soil and Water Contaminants. Washington, D. C., American Chemical Society. pp. 2-17.

Cunningham, S. D. and Ow, D. W. (1996). Promises and prospects of phytoremediation. *Plant Physiol.* 110: 715-719.

Cunningham, W. P.; Copper, T. H.; Gorham, E. and Hepworth, M. T. (1999). Phytoremediation *In*: Environmental Encyclopedia. 2nd edition. Delhi, India, Jaico Publishing House. pp. 796-797.

Dash, B. (1999). Eco-Technology for sustainable development. *In*: Sharma, N. N. *ed*. Employment News. 24 (25): 1-2.

Debusk, T. A.; Reddy, K. R. and Clough, K. S. (1989). Effectiveness of mechanical aeration in floating aquatic macrophyte-based wastewater treatment systems. *J. Environ. Qual.* 18: 349-354.

Debusk, T. A.; Reddy, K. R. and Isaacson, R. (1991). Waste water treatment and biomass production by floating aquatic macrophytes. *In*: Methane from Community Wastes. FL, U. S. A. pp. 21-36.

Delgade, M.; Bigeriego, M. and Guardiola, E. (1992). Water hyacinth biomass production in Madrid. *Biomass Bioenergy*. 3 (1): 57-61.

Dushenkov, S.; Kapulnik, Y.; Blaylock, M., Sorochisky, B., Raskin, I. and Ensley, B. (1997). Phytoremediation: a novel approach to an old problem. In: Global Environmental Biotechnology Proceedings of the Third Biennial Meeting of the International Society for Environmental Biotechnology. Amsterdam, New York, July, 15-20, 1996. International Society for Environmental Biotechnology Meeting. pp. 563-572.

Ebbs, S. D.; Larat, M. M.; Brady, D. J.; Cornish, J.; Gordon, R. and Kochian, L. V. (1997). Phytoextraction of cadmium and zinc from a contaminated soil. *J. Environ. Qual.* 26: 1424-1430.

El-Shinnawi, M. M.; El-Din, A. M. N; El-Shimi, S. A. and Badawi, M. A. (1989). Biogas production from crop residues and aquatic weeds. *Res. Conserv. Recycl.* 3 (1): 33-45.

Eyini, M.; Jayakumar, M. and Pannirselvam, S. (1993). Distillery effluent induced changes in the epicuticular wax deposits of *Eichhornia crassipes. Ind. J. Ecol.* 20 (1): 1-4.

Granato, M. (1993). Cyanide degradation by water hyacinth, *Eichhornia crassipes* (Mart). Solms. *Biotechnol. Lett.* 15 (10): 1085-1090.

Gupta, A. and Sujatha, P. (1996). Treatment of tannery wastewater by water hyacinth application. *J. Ecot. Environ. Monit.* 6 (3): 209-212.

Gupta, M. and Chandra, P. (1994). Lead accumulation and toxicity in *Vallisneria spiralis* (L). and *Hydrilla verticillata* (l. f). Royle. *J. Env. Sci. Hlth. Part A–Env. Sci. Engg.* 29 (3): 503-516.

Gupta, M.; Tripathi, R. D.; Rai, U. N.; Chandra, P. and Gupta, M. (1998). Role of glutathione and phytochelatin in *Hydrilla verticillata* (l. f). Royle and *Vallisneria spiralis* L. under mercury stress. *Chemosphere.* 37 (4): 785-800.

Huang, J. W.; Chen, J.; Berti, W. R.; Cunningham, S. D. (1997). Phytoremediation of lead contaminated soils: role of synthetic chelates in lead phytoextraction. *Environ. Sci. Technol.,* 31: 800-805.

Jain, S. K.; Gujral, G. S. and Vasudevan, P. (1990). Production of biogas from aquatic biomass: a comparison with terrestrial biomass. *Res. Ind.* 35: 104-107.

Jebanesan, A. (1997). Biological treatment of dairy waste by *Eichhornia crassipes* solms. *Environ. Ecol.* 15 (3): 521-523.

JinHong, Q.; Zayed, A.; Yong-Liang, Z.; Mei, Y. and Terry, N. (1999). Phytoaccumulation of trace elements by wetland plants III: Uptake and accumulation of ten trace elements by twelve plant species. *J. Environ. Qual.* 28 (5): 1448-1455.

Kelley, C.; Curtis, A. J.; Uno, J. K. and Berman, C. L. (2000). Spectroscopic studies of the interaction of Eu (III) with the roots of water hyacinth. *Water Air Soil Poll.* 119: 171-176.

Madamwar, D.; Patel, A. and Patel, V. (1990a). Effect of temperature and retention time on methane recovery from water hyacinth-cattle dung. *J. Fermen. Bio-engg.* 70 (5): 340-342.

Madamwar, D.; Patel, V. and Patel, A. (1990b). Effect of agricultural and other wastes on anaerobic digestion of water hyacinth-cattle dung. *J. Fermen. Bioengg.* 70 (5): 343-344.

Mahabal, B. L. (1993). Principles and design of effluent treatment plants (ETP). *In*: Nagaraj, J. *ed.* Industrial Safety and Pollution Control Handbook. Secunderabad, India. A Joint Publication of National Safety Council and Associate (DATA) Publishers Pvt. Ltd. pp. 257-270.

Mallik, M. K.; Singh, U. K. and Ahmad, N. (1990). Batch digester studies on biogas production from *Cannabis sativa*, water hyacinth and crop wastes mixed with dung and poultry litter. *Biol. Wastes.* 31: 315-319.

Metcalf and Eddy, Inc. (1995). Waste Water Engineering-Treatment, disposal and reuse, Revised by Tchobanoglous, G. and Burton, F. L., Tata Mc-Graw Hill Publishing Company Ltd. New York, 1334p.

Misra, U.; Singh, S., Singh, A.; Pandey, G. N.; Mishra, U.; Singh, S. and Singh, A. (1992). A new temperature controlled digester for anaerobic digestion for biogas production. *Energy Conver. Manage.* 33 (1): 983-986.

Moenandir, J. and Murgito. (1994). Heavy metal absorption by aquatic weeds. Kemampuran penyerapanlogam berat oleh eceng gondok. *Agrivita*, 17 (2): 61-64.

Montien-art, B.; Kakazone, T.; Okamoto, C.; Odahara, T.; Kikuchi, M. and Kabata, K. (1998). Utilization of aquatic plants for turbidity treatment and feedstuff at Kumamoto zoo of Japan in 1997. *Thai J. Agri. Sci.* 31 (2): 252-254.

Moorhead, K. K. and Nordstedt, R. A. (1993). Batch anaerobic digestion of water hyacinth: Effects of particle size, plant nitrogen content, and inoculum volume. *Biores. Tech.* 44: 71-76.

Neralla, B.; Weaver, R. W.; Varvel, T. W. and Lesikar, B. J. (1999). Phytoremediation and on-site treatment of septic effluents in sub-surface flow constructed wetlands. *Environ. Tech.* 20 (11): 1139-1146.

Nyer, E. K. and Gatliff, E. G. (1996). Phytoremediation. *Ground Water Monit. Remediat.* 16 (1): 58-62.

Patel, V.; Datta, M.; Patel, V. and Madamwar, D. (1992). Two phase anaerobic fermentation of water hyacinth-cattle dung. *Fresenius Environ. Bull.* I Suppl.: S86-S92.

Patel, V. B.; Patel, A. R.; Patel, M. C. and Madamwar, D. B. (1993). Effect of metals on anaerobic digestion of water hyacinth-cattle dung. *Appl. Biochem. Biotechnol.* 43 (1): 45-50.

Rai, U. N.; Sinha, S.; Tripathi, R. D. and Chandra, P. (1995). Wastewater treatability potential of some aquatic macrophytes: Removal of heavy metals. *Ecol. Engg.* 5: 5-12.

Rana, B. C. and Kumar, J. I. N. (1988). Energy dispersal analysis by X-rays of certain aquatic macrophytes. *Ind. J. Weed Sci.* 20 (1): 46-49.

Raskin, I. And Ensley, B. D. (2000). Phytoremediation of toxic metals: using plants to clean up the environment, John Wiley and Sons, Inc., New York, 301 p.

Raskin, I.; Smith, R. D. and Salt, D. E. (1997). Phytoremediation of metals: using plants to remove pollution from the environment. *Curr. Opin. Biotechnol.* 8: 221-226.

Reddy, K. R. (1983). Fate of nitrogen and phosphorous in a wastewater retention reservoir containing aquatic macrophytes. *J. Environ. Qual.* 12 (1): 137-141.

Reddy, K. R. and D'Angelo, E. M. (1990). Biomass yield and nutrient removal by water hyacinth (*Eichhornia crassipes*) as influenced by harvesting frequency. *Biomass.* 21 (1): 27-42.

Reddy, K. R. and Debusk, W. F. (1985). Nutrient removal potential of selected aquatic macrophytes. *J. Environ. Qual.* 14 (4): 459-462.

Reddy, K. R. and Sutton, D. L. (1984). Water hyacinth for water quality improvement and biomass production. *J. Environ. Qual.* 13 (1): 1-8.

Rice, P. J.; Anderson, T. A. and Coats, J. R. (1997). Phytoremediation of herbicide-contaminated surface water with aquatic plants. *In*: Kruger, E. L.; Anderson, T. A. and Coats, J. R. *eds.* Phytoremediation of Soil and Water Contaminants. Washington, DC., American Chemical Society. pp. 133-151.

Saini, R. S. (1995). Evaluating the potential of water hyacinth to treat raw wastewater employing a short detention period of 48 hours. *Poll. Res.* 14 (1): 141-143.

Salt, D. E.; Blaylock, M.; Kumar, N. P. B. A.; Viatcheslav, D. and Ensley, B. D. (1995). Phytoremediation: a novel strategy for the removal of toxic metals from the environment using plants. *Bio-Technol.* 13: 468-474.

Salt, D. E.; Smith, R. D. and Raskin, I. (1998). Phytoremediation. *Annu. Rev. Plant Physiol. Plant Mol. Biol.* 49: 643-668.

Saltabas, O. and Akcin, G. (1994). Removal of chromium, copper and Nickel by water hyacinth (*E. crassipes*), *Toxicol. Environ. Chem.* 41 (3-4): 131-134.

Sarkar, S. K. (1997). Seasonal influence of the plant *Lemna major* in the treatment of eutrophic ponds. *Poll. Res.* 16 (4): 247-249.

Sasmal, C. (1992). Activation and inhibition of biogas production by aquatic weeds. *Orissa J. Agric. Res.* 5 (1-2): 75-80.

Satyakala, G. and Jamil, K. (1997). Studies on the effect of heavy metal pollution on *Pistia statiotes* L. (Water lettuce). *Ind. J. Environ. Hlth.* 39 (1): 1-7.

Selvapathy, P.; Jesline, J. J. and Prebha, S. (1998). Heavy metals removal from wastewater by water lettuce. *Ind. J. Env. Prot.* 18 (1): 1-6.

Sen, H. K. (1930). Water hyacinth as a source of power. *In*: Trans. 2nd World Power Conference. Berlin. 6. pp. 221-237.

Sen, H. K. and Chatterjee, H. N. (1931). Gasification of water hyacinth (*Eichhornia crassipes*). *J. Ind. Chem. Soc.* 8: 1-6.

Shoyakuov, R. S. and Aitmetova, K. I. (1999). Chemical composition of *Eichhornia crassipes* and *Pistia stratiotes*. *Chem. Natu. Compd.* 35 (2): 227-228.

Siciliano, S. D. and Germida, I. J. (1998). Mechanisms of phytoremediation: biochemical and ecological interactions between plants and bacteria. *Environ. Rev.* 6 (1): 65-79.

Singaram, P. (1994). Removal of chromium from tannery effluent by using water weeds. *Ind. J. Environ. Hlth.* 36 (3): 197-199.

Singhal, V. and Rai, J. P. N. (2003). Biogas production from water hyacinth and channel grass used for phytoremediation of industrial effluents. *Biores. Technol.* 86: 221-225.

Singhal, V. and Rai, J. P. N. (2003). Studies on phytoremediation potential of water hyacinth (*Eichhornia crassipes*) against pulp and paper mill and distillery effluents. *Acta Botanica Indica* (In Press)

Singhal, V.; Kumar, A. and Rai, J. P. N. (2003). Phytoremediation of pulp and paper mill and distillery effluents by channel grass (*Vallisneria spiralis*). *J. Sci. Ind. Res.* (In Press).

Sinha, S., Gupta, M. and Chandra, P. (1994). Bioaccumulation and toxicity of Cu and Cd in *Vallisneria spiralis* (L).. *Environ. Monit. Asses.* 33 (1): 75-84.

Srivastava, A. K. and Purnima (1998). Phytoremediation for heavy metals-a land plant based sustainable strategy for environmental decontamination. *Proc. Nat. Acad. Sci.* 68 (B) III and IV: 199-215.

Srivastava, P. K. and Pandey, G. C. (1999a). Effect of fertilizer effluent on total chlorophyll content and biomass of some aquatic macrophytes. *J. Ecot. Environ. Monit.* 11 (2): 123-127.

Srivastava, P. K. and Pandey, G. C. (1999b). Paper mill effluent induced toxicity in *Eichhornia crassipes* and *Spirodela polyrrhiza*. *J. Environ. Biol.* 20 (4): 317-320.

St-Cyr, L. and Campbell, P. G. C. (1994). Trace metals in submerged plants of the St. Lawrence river. *Can. J. Bot.* 72 (4): 429-439.

Terry, N. and Zayed, A. M. (1998). Phytoremediation of selenium *In:* Frankenbeyer, W. T. and Engberg, R. (ed) *Environmental Chemistry of Selenium*, Marcel Dekker Inc., pp. 633-655.

Thomas, T. H. and Eden, R. D. (1990). Water hyacinth-a major neglected resource. *In*: Sayigh, A. A. M. Energy and the Environment, Into the 90s. Proceedings of the Ist World Renewable Energy Congress. Reading, U. K., Sept. 23-28, 1990. pp. 2092-2096.

Trivedy, R. K. and Nakate S. S. (1999). Aquatic weeds based wastewater treatment plants in India. *J. Ind. Poll. Contl.* 15 (2): 275-279.

Vajpayee, P.; Rai, U. N.; Sinha, S.; Tripathi, R. D. and Chandra, P. (1995). Bioremediation of tannery effluent by aquatic macrophytes. *Bull. Environ. Contam. Toxicol.* 55 (4): 546-553.

Versters, A. (1994). Profitability of biogas from a Chinese type digester in Benin. *Tropiculture.* 12 (1): 27-28.

Vesilind, P. A.; Peirce, J. J. and Weiner, R. F. (1994). Environmental Engineering. III edition., Boston, U. S. A., Butter worth-Heinemann. p. 158.

Walsh, G. E.; Weber, D. E.; Nguyen, M. T. and Esry, L. K. (1991). Responses of wetland plants to effluents in water and sediment. *Env. Expt. Bot.* 31 (3): 351-358.

Wang, W. and Wang, W. (1991). Plants for toxicity assessment. *In*: Gorsuch, J. W.; Lower, W. R. and Lewis, M. A. *eds.* 2nd Symp. on use of plants for toxicity assessment. California, USA., April. 23-24, 1990. pp. 68-76.

Watanabe, M. E. (1997). Phytoremediation on the brink of commercialization. *Environ. Sci. Tech.* 31 (4): 182A-186A.

Welsh, R. H. and Denny, P. (1979). Translocation of lead and copper in two submerged aquatic angiosperm species. *J. Expt. Bot.* 30: 339-345.

Whiting, G. J.; Chanten, J. P. and Armstrong, J. (1996). Control of the diurnal pattern of methane emission from emergent aquatic macrophytes by gas transport mechanisms. *Aq. Bot.* 54 (2-3): 237-253.

Wolverton, B. C.; Mcdonald, R. C. and Gordon, J. (1975). Bioconversion of water hyacinth into methane gas and fertilizer. Part I. NASA Tech. Mem. TN-X-72725 12p.

Zayed, A.; Gowthaman, S. and Terry, N. (1998). Phytoaccumulation of trace elements by wetland plants I. Duckweed, *J. Environ. Qual.* 27: 715-721.

Zhu, Y. L.; Zayed, A. M., Qian, J. H., Souza, M. de and Terry, N. (1999). Phytoaccumulation of trace elements by wetland plants II. Water hyacinth. *J. Environ. Qual.* 28: 339-344.

**Aquatic Biodiversity in India: The Present Scenario, 2005** **184–201**
*Edited by*: D.R. Khanna, A.K. Chopra & G. Prasad
*Published by*: Daya Publishing House, New Delhi

# 11

# Biotic Communities in Aquatic System

**N. Rai, S.A. Ansari, S. Chauhan, Mukesh Ruhela* and R. Bhutiani****

*Division of Biological Standardization, IVRI, Izat Nagar, Bareilly*
** Dept. of Environmental Sciences, Guru Jambeshwar Univ., Hissar*
*** Department of Zoology and Environmental Sciences, Gurukul Kangri University, Hardwar.*

Water is one of the most unusual natural compounds found on our earth and it is also one of the most and main component for the survival of any living being since life on earth was began in the seas. Water is one of the main agent in pedogenesis (soil formation) and is the medium for several different ecosystems. On one hand it is required for irrigation, industrial and domestic consumption for human beings, and on the other hand it is the main component to maintain aquatic ecosystem. Oceans, rivers, glaciers, lakes, ponds, streams etc. are the main sources of water. About three fourth of the earth's surface (71 per cent) is covered with hydrosphere. It has been estimated that hydrosphere contains about $1.46 \times 10^9$ cubic kms of water, of this 93 per cent is in the oceans unsuitable for human use, 4.1 per cent is in the earth's crust, 2 per cent in the glaciers and polar ice, and 0.052 per cent is in fresh water lakes, river and atmospheric moisture.

## Water Resources in India

In India, rainfall is the main source of water, but its existing utilization capacity is very low. Because of deforestation, India can't hold on maximum water. The large portion of monsoon water disappears into the sea as the reason we have lost the capacity to restore and recharge the ground water. Because of degradation of forests, several old streams and springs are now without water and rest is becoming polluted due to improper management. Most of the villages in the Himalayas are facing acute water problem. At present India uses only a $10^{th}$ part of rainfall receives annually and there is an urgent need to learn to store and use this water without polluting it.

According to Central Ground Water Board and Natural Commission on Agriculture, the estimated total annual proportion was 400 millions hectare meter, out of which only 38 million hectare meter *i.e.* 9.5 per cent was in use in the year 1974. The rest 115 million hectare meter goes surface run off and 215 million hectare meter perculates into the soil and 70 million hectare meter evaporates immediately. Out of 215 million hectare meter water which perculates underground, 165 million hectare meter is used for soil moisture and remaining 50 million hectare meter enters the ground water table.

## Types of Water and Distribution of Microbes

The earth's moisture is in continuous circulation, a process known as the water cycle or hydrologic cycle (Fig. 11.1). It has been estimated that about 80,000 cubic miles of water from oceans and 15,000 cubic miles from lakes and land surfaces evaporated annually. The total evaporation is equalled by the total precipitation, of which about 24,000 cubic miles fall on land surfaces. Microorganisms of various kinds are present at different stages of this cycle.

Natural water is commonly grouped into four classes (a) Atmospheric water (b) Surface water (c) Stored water and (d) Ground water.

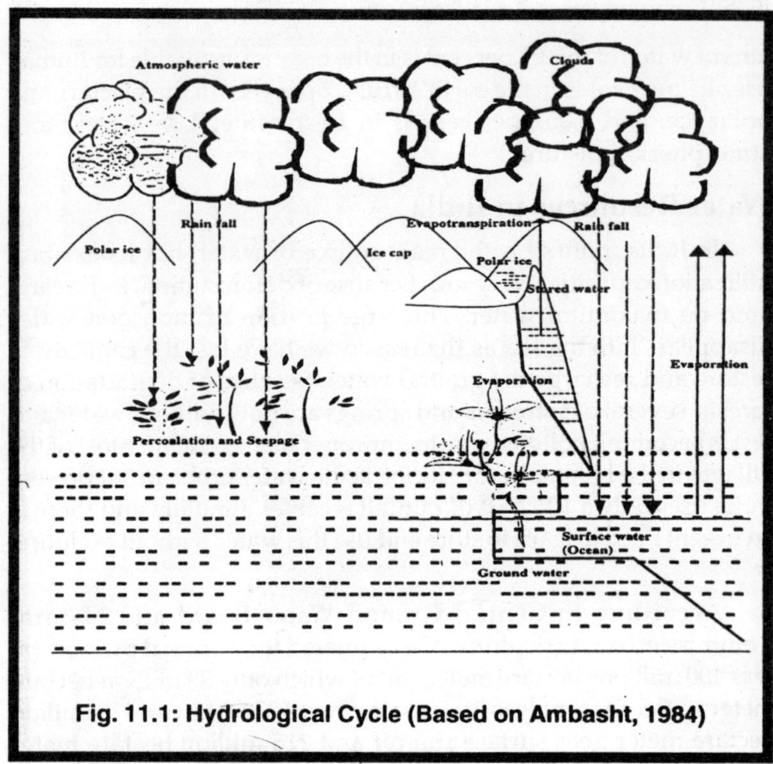


**Fig. 11.1: Hydrological Cycle (Based on Ambasht, 1984)**

## Atmospheric Water

The moisture contained in clouds are precipitated as snow, sleet, hail and rain constitutes atmospheric water. Microorganisms in the form of cells/dominant propagules and dust particles remain suspended in water and snow. Number of bacterias are present in the rain water, but after heavy rain or snow fall, these dust particles and bacteria are washed from the atmosphere.

## Surface Water

Bodies of water such as lakes, streams, rivers and oceans form surface water. As soon as raindrops or snow touches the earth, it becomes contaminated by the microorganisms present in soil. Total microbial number in water depends on microbial population of soil, types of soil, types of organic materials present in soil and also on types of microorganisms and their activities. Moreover, microorganisms of water are governed by climatic, chemical and biological conditions of water.

## Stored Water

The stagnant land water present in ponds, lakes etc. form stored water. During storage in general, the number of microorganism gets reduced, thus it establishes to some extent the purity and stability. However, in stored water, the microorganisms are affected by several factors such as sedimentation, UV light, temperature, osmotic effects, food supply and activity of other microorganisms as well.

## Ground Water

It is subterranean water that occurs where all pores in the soil or rock containing materials are saturated. The percentage, space and size of pores regulate the quantity of rain water that can be soaked and held in soil. As water moves down through soil pores, bacteria and other microorganisms with suspended particles are filtered. Therefore, the microorganisms are carried only to some distance thus making the deep wells with negligible number of microorganisms.

# The Structure of Aquatic Ecosystems

Global aquatic system fall into two broad classes definable by salinity or amount of material dissolved in water as–fresh water ecosystem and salt water (marine) ecosystem.

Fresh water ecosystem, the study of which is known as limnology are divided into two groups–lentic (standing or still water habitats) and lotic (running water habitats). On the other hand salt-water ecosystem include inland brackish water as well as marine and estuarine habitats. Estuarine represents a transitional zone between a river and the sea and it contains dissolved solid content intermediate between those of fresh and marine water.

### Distribution of Microorganisms Depending Upon Physico-chemical Nature of Fresh Water

The microbial population in natural water bodies is determined by the physical and chemical conditions which prevail in that habitat. Some of these conditions are described below:

## Pressure, Density and Buoyancy

In all the fresh water environments maximum pressure is much less than in the ocean, and organisms appear to adjust to them readily.

The density of water varies inversely with temperature and directly with the concentration of dissolved substances. Dissolved salts increases the density of water. However, when great evaporation occurs in a lake having no outlet, the lake may become hypersaline than the ocean. The few species capable of living in these salty lakes include some algae and protozoa, the brine shrimp *Artemia gracilis* and the immature stages of two brine flies, *Ephydra gracilis* and *E. hians*.

Buoyancy varies with the density of water. Viscosity, the measure of the internal friction of water, varies inversely with few parameters and also influences buoyancy. Most aquatic organisms keep stations by swimming movements or have special adaptations to decrease the specific gravity of the body. For this purpose, fresh water aquatic organisms have some swimming adaptations, clinging organs, absorption of large amount of water to form jelly-like tissues, storage of gas or air bubbles within the body, formation of light-weight fat deposits within the body, increase of surface area in proportion to body mass and so on.

## Temperature

Diurnal and seasonal variations of temperature are very much evident in fresh water environments than in marine environments. According to temperature relations lakes have been classified as Tropical lakes, in which surface temperature are always maintained above 4°C; Temperate lakes, in which surface temperature vary above and below 4°C; and Polar lakes, in which it never goes above 4°C. Decreasing temperature often cause a fall in metabolism, resulting in a lower rate of food consumption. So fluctuations in temperature of aquatic system regulate the breeding periods, initiate hibernation, gonadial activation etc. On the basis of their ability to tolerate thermal variations, most fresh water organisms are stenothermine with a narrow range temperature tolerance, but some are euthermic with a wide range of temperature tolerance. For example, the stenothermic oligochaetes includes stenominimothermal forms (*e.g.* Aeolosoma,

*Megascoles mauritii*), stenomaximothermal (*e.g. Dero limosa*) or steno-optimothermal forms (*e.g. Branchiodrilus semperi* and *B. menoni*).

## Light

It influences both fresh water and marine ecosystem greatly. The fresh water often have a lot of suspended materials. While affording protection to the light sensitive species, these substances more often obstruct the light, normally reaches the water. The degree of such obstruction of light influences the productivity of the fresh water ecosystems. A shallow lake receives light to its very bottom resulting in an abundant growth of vegetation both phytoplankton, and rooted vescular plant. Light also controls the orientation and changes in position of attached species and their nature of growth and it also caused into diurnal migration of planktonic species of fresh water.

## Oxygen

The oxygen, which is a most essential chemical component of life processes, remain dissolved in fresh and marine waters. Oxygen contents of a fresh water body are depleted in numerous ways.

Primarily oxygen is utilized in the respiration of organisms and decomposition of dead organisms in the aquatic environment. Photosynthesis remains restricted to the surface layer of water containing phytoplankton and exposed regions of rooted vascular plants.

## Carbon Dioxide

Aquatic vegetation and phytoplankton requires carbon dioxide for photosynthetic activity. The carbon dioxide of fresh water environments is produced as the end product of respiration and of decomposition. Carbon dioxide diffuses into water from atmosphere resulting in carbonic acid, which affects the pH of water. The high saturations levels of oxygen and carbon dioxide have been found to have toxic affects on aquatic biota.

## Salinity

Fresh water being efficient solvent contains many solutes in solution, but even then its salt contents remain under 0.2 per cent. Different dissolved salts reach the water by erosion, inflow and decay of aquatic forms. Dissolved substances have peculiar significance for floating aquatic vegetation and phytoplankton compounds of N, P, Si are also present in fresh water. Many other elements like Ca, mg, Mn, Fe, Na, K, S and Zn are found dissolved in water and influence the fauna variously.

Iron being a growth promoting element for plants exists as the compound of oxygen (Ferrous oxide). Calcium is essential element for plants. Deposition of calcium carbonate in water called marl is produced by the activity of plants. External coverings of arthopodes and the shell of molluscs and tubes of some worms need calcium carbonate. Bryozoans, sponges and cladocerans prefer an increased calcium content.

### Hydrogen Ion Concentration (pH)

In fresh water environments pH is a determining factor for the biota by becoming a limiting factor.The pH value of different fresh water bodies may fluctuate seasonally and annually. Aquatic microorganisms, in general, can be grown at pH 6.5 to 8.5.

## Biodiversity in Fresh Water Communities

Fresh water ecosystems are of two types-lentic ecosystem and lotic ecosystem.

### Lentic Ecosystems

It includes all standing water habitats such as lakes, ponds, marshes, swamps, bogs, meadows etc. Factors that contribute to the development of zones in aquatic systems are sunlight, temperature, aeration and dissolved nutrient content. These variations create numerous macro and micro environments for communities of organisms.

A lake is stratified vertically into three zones or strata (Fig. 11.2). The uppermost region called the photic zone, extends from the surface to the lowest limit of sunlight penetration. Directly beneath the photic zone lies the profundal zone, which extends from the edge of the photic zone to the lake sediment.The sediment itself, or benthic zone, is composed of organic debris and mud, and it lies directly on the bedrock that forms the lake basin.

A lake can be stratified horizontally. The horizontal zonation includes the shore-line or littoral zone. It contains upper warm and oxygen rich circulating water layer which is called epilimnion.

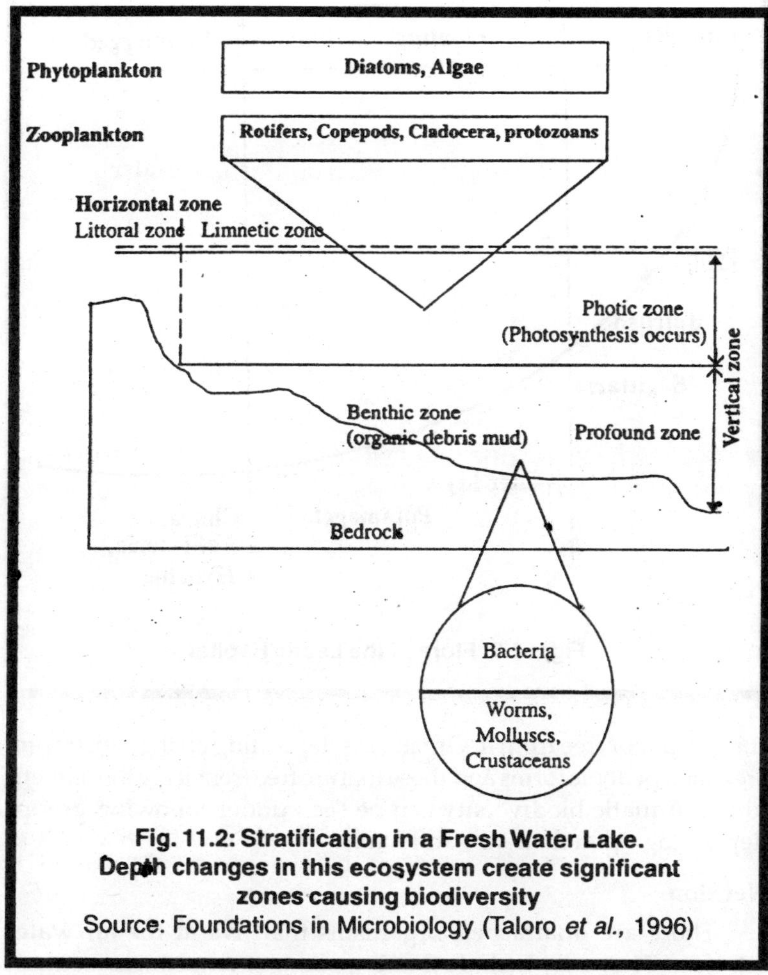


**Fig. 11.2: Stratification in a Fresh Water Lake. Depth changes in this ecosystem create significant zones causing biodiversity**

Source: Foundations in Microbiology (Taloro *et al.*, 1996)

Sublittoral zone extends from rooted vegetation to the non-circulating cold water with poor oxygen zone *i.e.* hypolimnion.

The open deeper water beyond the littoral zone is the limnetic zone.

## Flora and Fauna of Lentic Ecosystems

Different organisms of the lentic environment can be ecologically classified based on whether they are dependent on the

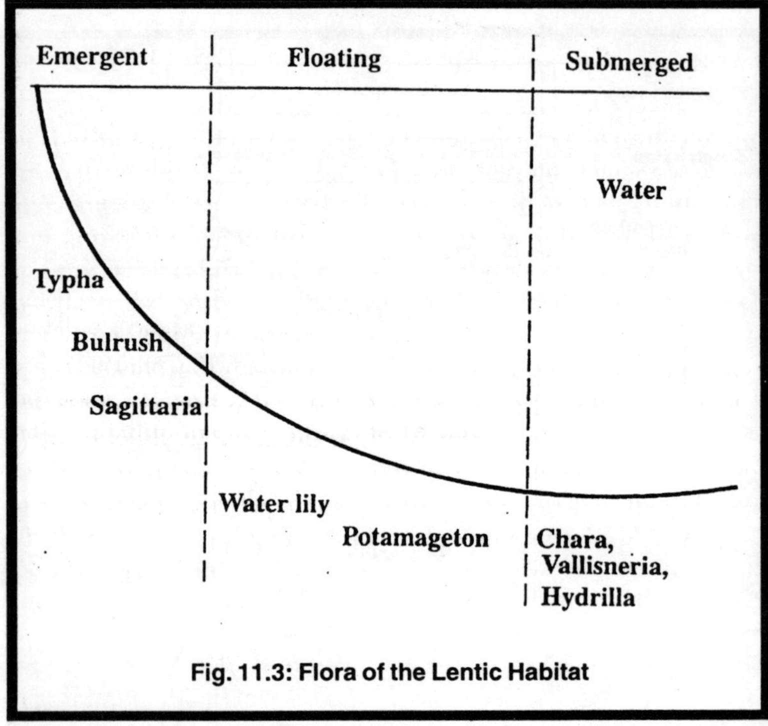


**Fig. 11.3: Flora of the Lentic Habitat**

substratum or free from it. Organisms depending on the substratum are called pedonic forms and those that are free from it are the limnetic forms. Aquatic biodiversity can be seen under following groups depending upon their size and habits.

### Neuston

There are unattached organisms that live at the air water interface. They may include floating plants such as duckweed. *e.g. Wolffia, Lemna, Spirodella, Azolla, Pistia, Eichhornia.*

### Plankton

These are forms which are found in all aquatic ecosystems. They are small plants and animals whose power of self locomotion are so limited that they can not overcome currents. Plankton may be composed primarily of algae (phytoplankton), or it may be predominantly protozoa and other minute animal life (Zooplankton).

### Nekton

Nektonic animals are swimmers. In order to overcome currents, these animals are relatively large and powerful.

### Benthos

These includes the organisms living at the bottom of water mass.

## Biota of Littoral Zone

The littoral zone of lake remains rich in pedonic flora especially upto the depth to which effective light penetration is possible facilitating the growth of rooted vegetation. Certain rooted emergent plant species of littoral zone are *ranunculus, monochoria, cyperus* and *rumex*. rooted plants with floating leaves such as the water lilies– *Nymphaea, Nelumbo, Aponogeton, Trapa,* and *Potamageton* etc. Submerged vegetation includes plants like *Elodea, Vallisneria, Hydrilla, Chara, Potamogeton* etc.

The phytoplankton of littoral zone of lake composed of diatoms (e.g. *Navicula, Cyclotella*), blue green algae (*e.g. Microcystis, Oscillatoria*), green algae (*e.g. Cosmarium,Staurastrum*) and holophytic flagellates.

The littoral zone also contains animals, which remain distributed in recognizable communities. These include protozoans like *Vorticella, Stentor, Laccotrophes,* climbing dragonfly, flat worms, *Hydra*, snails (lymnaea) and others. The zooplanktons consists of water fleas such as Daphnia, rotifers.

## Biota of Limnetic Zone

It is the region of rapid variation, with the water level, temperature and oxygen content varying from time to time.

Various protozoans which are capable of encystment during adverse ecological conditions, tardigrades like *Macrobiotus,* rotifers like *Rotaria, Philodina,* snails and frogs occupy the limnetic zone. The linnetic zone has autotrophs in abundance.

A large number of microorganisms both saprophytes and pathogens are found in water which fall under the groups bacteria, fungi, algae, protozoa and nematodes.

Ingole *et al.* (1998), in their study showed the presence of rich and diverse population of micro-invertebrates in the fresh water lake Priyadarshines at Schirmacher, Oasis.

The majority of bacteria found in water belongs to groups: fluorescent bacteria (*e.g. Pseudomonas*), chromogenic rods (*e.g. Xanthomonas*), coliform-group (*E. coli, Aerobacter*), proteus group, non-gas forming, non-chromogenic and non-spore forming rods, spore formers of the genus Bacillus, and pigmented and non-pigmented cocci (Tables 11.1 and 11.2).

**Table 11.1: Prokaryotic Genera Found in Aquatic Habitats**

| *Group* | *Genera* |
|---|---|
| Photoautotroph | Chromatium |
| | Chlorobium |
| Photoheterotrophs | Rhodospirillum |
| | Rhodoseudomonas |
| | Rhodomicrobium |
| Chemoautotroph | Beggiatoa |
| Chemoheterotroph | Caulobacter |
| | Gemmobacter |

**Table 11.2: Some of the Aquatic Fungi Found in Different Water Bodies**

| *Aquatic Fungi* | *Sources* |
|---|---|
| *Achlya Americana* | Tap water |
| *A. androcompocita* | Tank water |
| *A debaryana* | Water |
| *Allomyces Neo-moniliformis* | Lake water |
| *A. laevis* | Lake water |
| *B. rostata* | Water |
| *Chytridium brevipes* | Water |
| *Cladochytrium setigerum* | Water |
| *Dictyuchus pisci* | Pond water |
| *Isoachlya anisosporea* | Pond water |
| *Olpidiopsis luxurians* | Lake water |
| *Pythium undulatum* | Pond water |
| *P. echinulatum* | Water |
| *Saprolegnia spp* | Tap water |
| *Sapromyces indicus* | Stream water |

## Flora and Fauna of Lotic Ecosystems

The moving water ecosystem can be divided into rapidly flowing water and slowly flowing water.

### Rapidly Flowing Water

Some animals of rapidly flowing water streams live among the mosses and flowering plants like *Hydrolyum lichenoides*. Animals such as Cephalopteryx, Helodes, Gammarus etc. live among leaves and stems of aquatic plants.

Some are rock inhabitant forms like *Ferrissia*, *Baetis* larvae, caddishfly worms, mayfly nymphs Iron and Pesephenus etc.

Some are inhabitants occurring beneath the rocks *e.g.* Small loaches (*Noemacheilus*), limpet-like fishes (*e.g. Glyptosternum, Balitora*) etc.

### Slowly Flowing Water

The slow streams have a higher temperature, so protozoans occurs in large number in this ecosystem. The detritus feeding benthos of slow water ecosystem include those which either live on the bottom, such as isopods, molluscs (*Sphaenius, Pisidium*) and mayfly or which burrow into the sediment, such as tubeworms, Sialis (alderfly) nematodes, snails and rotifers.

Plant life is also abundant in a slow water ecosystem. It includes rooted vascular plants such as pond weeds and grasses, firmly attached aquatic mosses and large multicellular filamentous algae. Motile algae such as diatoms and flagellates are also found.

## Biodiversity in Marine Communities

The sea biota is not abundant but contains well marked diversity. Every major group of algae and almost every major group of animals can be found somewhere in the oceans. The diversity in marine biotic communities can be studied as follows:

### Biotic Communities of Oceanic Region

The oceanic or pelagic zone is less rich in species and numbers than the coastal areas, but it has its characteristics species. Many of these are transparent or bluish and since the sediment free water of the open sea is transparent, these animals are nearly invisible. Animals have locomotory organ due to continuity of the sea.

The pelagic plankton (epiplankton) are exceedingly diverse. Phytoplankton includes diatoms and dinoflagellates. Certain holoplanktons of pelagic zooplanktons are *Globigerina, Noctilucea, Protocystis, Valella, Diplyces, Euphausia, Calanus, Sagitta* etc. Zone planktonic forms are the Forminifera and Radiolaria, arrow worms, jelly fishes, swimming snails are most abundant. Zooplanktons of inshore waters of Mandapam (South India) has many animal species-flagellate protozoans *Noctiluca miliaris*, larval forms of coelenterates like *Planula, Ephogra*, Ctenophers like Pleurobrachia, heteropod and pteropod molluscs like *Careinaria, Cliona, Creseis*, species of copepods like *Acartia, Calanopia, Eucalanus* and *Hicrosetella* and so on.

The largest animals in the pelagic region are the nektons. These include cephalopods (squid) and marine vertebrates (bony fishes, sharks, sea turtles, whales). Certain fishes such as tuna, shark, sardine, herring, bonito live near the surface.

The abyssal benthic zone or deep sea of oceanic region is pitch dark and universal absence of light in this environment excludes the possibility of any growth of vegetation or other photosynthetic organisms.The bottom of the sea is a soft ooze, made of the organic remains and shells of Foraminifera, Radiolaria and other animals plants. Sea cucumbers, brittle star, sea lilies, sea urchins, crustaceans, sea anemones, clams are all found on the bottom.

## Biotic Communities of Continental Shelf

The communities of continental shelf are both richer and more diverse than those of open ocean. Several kinds of sea weeds are harvested from India.

The littoral, intertidal or eulittoral zone is the region of sea shore which exists between the high and low tide lines. The intertidal zone is the most variable zone in the entire sea. Any organism that is to survive in the intertidal must be either resistant to periodic dessication or able to burrow to water level. Many sea weeds have tough pliable bodies, able to blend with the waves without breaking while the animals either are encased in hard calcareous shells (*e.g.* Molluscs, Bryozoa, starfish, crabs) or are covered by a strong leathery skin that can be bend without breaking (*e.g.* sea anemone and octopus).

The rocky shore presents solid substratum for the attachment of many sessile animals which often remain abundant here *e.g.* Patella, Haliotis, Oysters, Barnacles, Hydrozoans.

The okha coast of Gujarat in India is a typical rocky sea coast which has well defined fauna and flora.

Dominant weeds of subtidal zone of okha coast are *Sargassum, Dictyopteris, Padina Ulva, Polysysiphonia* etc.

The sandy shore may be even more harsh than the rocky shore. It is subjected to all the extremes of the temperature, salinity, turbidity, wave action etc. Sandy beach fauna of Vishakhapatnam coast though lacks or contains small number of tubicolous polychaets and crustacea due to unstable substratum, but include following animal species–spinoid *Prionospio krusadiensis,* Aricia and cumbriconeries. The crustacean fauna of this zone includes *Emerita asiatica, Albunea symnista,* hermit crab *Clibanarius aretheustus,* eel *Ophichtys orientalis* and medusaie like Acquoria and Chiropsalmus.

### Biotic Communities of Coral Reefs

Coral reefs are the specialized ecosystem of ocean which are among the most productive of all ecosystems anywhere, with a diversity equalled only by tropical rain forests. The amount of oxygen in reef environment is very high. Coral reefs normally exist in water with mean temperature 20.5°C, but they can withstand very high temperatures. Thus coral reef ecosystems harbour a abundant and well diverse biota. Examples are *Porites, Montipora, Favia, Acropora, Symphylia, Platygyra,* hydroid coelenterates like *Syncoryne, Pennaria, Tabularia, Plumularia, Alcyonium, Teresto,* beautifully pigmented fishes like holocentrus *Lutianus sp., Chaetodon, Linophora, Thallasoma* and *Zanclus* etc.

## Biodiversity of Microorganisms in Marine Habitat

A large number of microorganisms are found in marine habitat (Table 11.3) of which extremophiles shows much diversity. Microorganisms capable of copying with low temperatures are diversed in the natural environments and are regarded as the most successful colonizers of our planet (Russel, 1990).

**Table 11.3: Distribution of Microorganisms in Marine Habitat**

| *Microorganism* | *Examples* |
|---|---|
| Bacteria | *Pseudomonas, Vibrio, Flavobacterium, Spirillum, Achromobacter, Hyphomicrobium, Cytophaga, Microcyclus, Actinomycetes, Bacillus, Desulfovibrio, Nitrosococcus, Nitrosomonas, Nitrospina, Nitrococcus, Nitrobacter, Cyanobacterium, Trichodesmium, erythreum* (red sea algae). |
| Fungi | *Candida, Torulopsis, Cryptococcus, Trichosporon, Saccharomyces, Rhodotorula, Rhodosporidium.* |
| Viruses | Viruses are known to occur in marine water. Marine plants, animals and microorganisms are subjected to infection by viruses. |
| Algae | *Chlorophycophyta, Euglenophycophyta, Phaeophycophyta, Chrysophycophyta, Cryptophycophyta, Rhodophycophyta.* |

## Psychrophiles

The term "Psychrophile" is used for those microorganisms which are able to grow and multiply at 0°C. The definition for psychrophilic has under gone a slight change based on the cardinal growth temperatures, according to which a physchrophiles exhibits an optimum growth temperature of 15°C or below, a maximum growth temperature of 20°C or below and a minimum growth temperature of 0°C or below.

Psychrophilic and Psychrotrophic microorganisms are of particular importance in global ecology since the majority of terrestrial and aquatic ecosystems of our planet is permanently or seasonally submitted to cold temperatures. The world's oceans occupy 71 per cent of the earth surface and 90 per cent of their volume is below 5°C (Hebrand and Potier, 1999). These microorganisms often represent the dominant flora and therefore, be regarded as the most successful colonizers of our planet. These microorganisms manage to survive by synthesizing specific proteins termed as cold shock proteins. Liquid water is generally thought necessary for life and some liquid water exists in snow at temperatures below freezing *i.e.* even at–10°C in Siberian permafrost 0.5 to 3 per cent of water is unfrozen and microorganisms have been isolated and cultured form there (Carpenter *et al.*, 2000).

Thus psychrophiles can be found in soil, in water (fresh and saline, still and flowing) and associated with plants and cold blooded animals.

The use of refrigeration for foods provides a great diversity for Psychrophiles to colonize and causing food spoilage and food poisoning (Russell and Gould, 1991).

## Barophiles

The occurrence of bacteria in the deep sea has been known for more than a century. The first barophilic bacterium Spirillum sp. was isolated in Japan. It was reported that a bacterium isolated from deep sea environment *Colwellia hadaliensis* which is a gamma proteobacteria, is an obligate barophile. Barotolerant or baroduric organisms are capable of growth over the ranges of 1 to 600 atmosphere whereas barophiles grow faster at elevated hydrostatic pressures.

## Acidophiles and Alkalophiles

The relative abundance of $H^+$ and OH– ions in a medium has a strong influence on the growth of nearly all organisms. Sulphur dependent archeans predominate in acidic hot springs. Thermoplasmas having optimum temperature 55°C to 59°C has optimum pH 2.

The other extreme of the pH scale is the pH of 10 and above. It was started after the discovery of alkalitolerant nitrifying bacteria and an alkalophilic *Streptococcus faecalis*. Alkalophilic bacteria are a diverse group ranging from eubacteria such as *Bacillus* sp to archaebacteria such as *Natronobacterium* sp.

Alkalophiles have pH optima for growth between 10 to 12 and grow very slowly at neutral pH (Madigan *et al.*, 1997).

## Halophiles

Halophiles are able to live in salty conditions through a fascinating adaptation, a cell suspended in a very salty solution will lose water and become dehydrated unless its cytoplasm contains higher concentration of salt than its environment.

Halophiles are generally those organisms which can grow optimally in 0.5 M NaCl (or other salts in addition to a minimum amount of NaCl). Extreme halophiles grow best in 2.5M to 5.2M (saturated) salt media. Halotolerant organism have growth optima in less than 0.5M NaCl though they can grow at higher salinity.

The most well known extremely saline environments are salterns (salt evaporating ponds) which often exhibit colourful red cultures of Halobacteria. *Halobacterium salinarum* concentrates KCL in its interior.

Some companies in India are marketing natural beta carotene capsules. These are derived from a Halobacterium *Dunaliella salina*.

## Thermophiles

Though the thermophiles have been variously described, the most accepted nomenclature describes an organism to be thermophile if its maximum growth temperature is > 60°C and the optimum growth temperature is > 50°C. The organism which have the maximum growth temperature of >90°C and optimum growth temperature of >65°C are called extreme thermophiles or caldocactive organisms and those organisms capable of growth above 100°C are called hyperthermophiles. As hyperthermophily was a redic of Archaeal environmental conditions and the anoxic state was the second important trait than in today's condition prevalent on the earth, the survival of the most primitive form of life is possible only under similar extreme conditions. Deep sea hydrothermalvents fit well this environmental description. (Prasad and Sankar, 1997).

## References

Carpenter, E. J., Lin, S. and Capone, D. G. (2000). Bacterial activity in south pole snow. *Appl. Environ. Microbiol.*, 66: 4514-4517.

Dubey, R. C. and Maheshwari, D. K. (1999). A textbook of Microbiology. S. Chand and Co. Ltd., New Delhi.

Hebrand, M. and Potier, Patnick (1999). Cold shock response and low temperature adaptation in psychrophilic bacteria. *J. Mol. Microbiol. Biotechnol.*, 2: 211-219.

Ingole, B. and Dhargalkar, V. (1998). Ecobiological assessment of a fresh water lake at Schirmacher Oasis, East Antartica, with reference to human activities. *Curr. Sci.*, 74: 529-533.

Madiyan, M. T. and Marrs, B. L. (1997). Extremophiles. Scientific American. 66-71.

Pelczar, M. J., Chan, E. C. S. and Krieg, N. R. (1997). Microbiology-9$^{th}$ reprint. Tata McGraw-Hill Publishing Co. Ltd.

Prasad, R. and Sarkar, S. (1997). Evolutionary extremophilic archaeal domain of life. *Curr. Sci.*, 73: 842-854.

Russell, N. J. (1990). Cold adaptation of microorganisms. *Philos Trans. Roy. Soc. Lond.* 326: 595-611.

Russell, N. J. and Gould, G. W. (1991). Factors affecting growth and survival. In Russell N. J., Gould, G. W. (eds). : Food Preservatives Glasgow: Blackie. 13-21.

Tortora, G. J., Funke, B. R. and Case, C. L. (1998). Microbiology–An Introduction, 6$^{th}$ edn. Benjamin/Cummings Publishing Company, California.

Verma, P. S. and Agarwal, V. K. (1990). Cell Biology, Genetics, Evolution and ecology-17$^{th}$ edn. S. Chand and Co. Ltd., New Delhi.

**Aquatic Biodiversity in India: The Present Scenario, 2005** **202–242**

*Edited by*: D.R. Khanna, A.K. Chopra & G. Prasad

*Published by*: Daya Publishing House, New Delhi

# 12

# Aquatic Diversity in Jharkhand with Reference to Macro Zoobenthos

**Arvind Kumar and Chandan Bohra**

*Pollution Research Unit, Department of Zoology, B.S.K. College, Barharwa – 816 101*

The organisms living on the bottom of the water bodies are termed as benthos. The benthos, an integral part of the food-web and production of water body, is biocoenoses of the solid-liquid inteface which has become an important aspect of the limnology. The macrobenthic community of an ecosystem, be it lentic or lotic, like other communities has a series of attributes that do not reside in its individual species components and have meaning only with reference to the community level of integration such as species diversity, growth in form and structure, dominance, relative abundance and trophic structure. One of these attributes or many of these or all, depending upon situation, may be changed with the changing ecology of the water body concerned. In fact, the present view of the nature of community lies closer to Gleason's individualistic view than to Clement's super organismic interpretation. Species are distributed individualistically according to their own genetic characteristics population of most of the species tend to change gradually along the environmental gradients. Most species are not in obligatory associations with other species which suggests that association are formed with many combinations of species and vary continuously in space and time. Hence, a study on macro-benthic community composition and dynamics of different population of the community becomes a reliable source to provide the picture of environmental status and influence of changing limnology of the water body concerned.

The benthos of a reservoir is not an harmonious unit, clearly defined by morphometric or functional considerations. To some, the fauna of a reservoir includes both a littoral and a benthic fauna, as though the bottom dwelling organisms along the shore were distinct from the spatially identified sublittoral and profundal regions. The area from the water's edge to the limit of rooted vegetation to the limit of light penetration is recognized as sublittoral zone. The littoral zone shows the greatest diversity of bottom conditions, being sandy, muddy or rocky, each type supporting characteristic biota. Benthic fauna is generally composed of sedentary or relatively slow moving animals. Benthic animals in general of littoral zone show different zonations with two major components: (a) Epifauna: which refers to the organisms that live on the surface either as attached or motile forms and (b) In fauna: these are the animals which dig into the substrate and live in tubes or burrows. The shore fauna is comprised of three basic assemblage: (a) that depends upon wave action on a stony shore, particularly for respiratory needs; (b) that dependent upon the shelter of macrophytic vegetation from which it derives its food supply, the aufwuchs or overgrowth on the weed surface; and (c) the true fauna of shallow water sediments.

Part of the infauna may exist in the shallowest part of the reservoir in and upon the sediment immediately beneath the weed associated littoral assemblage, but it is clearly less likely to find suitable accumulations of decomposing organic material on a wave washed shore. Hence, the benthos here is considered as the true infauna, and especially that inhabiting substrates ranging from sand through mud to silt.

Hutchinson (1967) has suggested the following terms for various benthic communities based on the niches they occupy. These are:

1. Rhizobenthos–rooted in substratum but well extended into the aqueous phase.
2. Haptobenthos (periphyton) attached to an immersed solid surface.
3. Herpobenthos–growing or moving through mud.
4. Psammon–growing or moving through sand.
5. Endobenthos–boring in solid substrate.

Thus, the organisms associated with solid/liquid interface were ordinarily termed as benthos and naturally two sub-headings, "Phytobenthos" and "Zoobenthos" were used. Although zoobenthos form essential food items of some cultivated fishes but their study has not received much attention in India. The benthic macroinvertebrate community, particularly standing stock of benthos has appealed to many biologists as a valuable index of productivity of water body. The fishes, apex of aquatic productivity, are known to depend on benthos directly or indirectly of most part of their food. The gut analysis of fish by different fishery biologists have revealed that benthos play a key role at times (Jhingran, 1977).

The body of knowledge relating to the quantitative and qualitative responses of benthic communities to change in habitat has grown to the point where an examination of benthic community structure has become a valuable tool for regulatory agencies, water resource managers, and aquatic ecologists in assessing and monitoring water quality and detecting pollution sources. The composition and abundance of benthic animals are commonly used to demonstrate the effects of pollution on the biological integrity of surface waters and changes in the biotic activities of man. The benthic macroinvertebrates fauna may offer good opportunities to study, the structure and dynamics of natural communities as this fauna is discretely located, and is relatively easily sampled and manipulated and is relatively quick to react to change.

The benthic community structure present an integral measure of both autotrophic and heterotrophic processes occurring in the aquatic ecosystem. These macroinvertebrate communities are usually dominated by two groups of organisms such as oligochaete worms and chironomid larvae (Bazzanti and Loret, 1982). Other groups such as molluscs and at times crustaceans and various insect larvae also occur in rather limited numbers. Many of the benthic forms are detritivores and depend to a large extent on organic detritus for food. As the detritus reach the bottom, it enters a web of energy transfer that sustains the benthic community. Thus, the benthic macroinvertebrates play a key role in the mineral recycling and, in turn serve as food for fish. Many benthic larvae and oligochaetes are the major food sources for small and big fishes and are therefore, form the links in the food web in aquatic system. Thus, these macroinvertebrates are of both scientific and practical interest.

The relative abundance of benthic macroinvertebrates have been related to the pollution source, and their values have been judged as indicators of eutrophication level and water quality. Benthological variables are particularly useful in measuring the water quality and such biological monitoring can provide resolution in space and time Tittizer and Kothe (1978) and Price (1978). The benthic macroinvertebrate communities in a lake or reservoir fulfill all of these requirements (Wiederholm, 1980). The species composition especially of the benthic community in a given aquatic ecosystem often reflects the environmental conditions, which may have prevailed during its course of development. Because of the adverseness of the environmental conditions, the sensitive species might have got eliminated. Such changes in the species composition persist for sometime, and thereby help in monitoring the imprints of the proceeding adverse environmental conditions. Thus, the property of indicating the conditions makes several groups of aquatic macroinvertebrates particularly the benthic organisms more informative than the physico-chemical indicators (Hellawell, 1986).

Benthic macroinvertebrates are useful bioindicators in pollution studies. The use of bioindicators belonging mostly to profundal benthic communities has so far given good result in studies on the water quality of lakes and reservoirs. However studies on the benthic fauna of littoral zone seem to be qualitatively and quantitatively mearge (Wetzel and Likens, 1991). Benthic communities have been the best indicators of water quality and organic pollution because of their constant presence and relatively long sedentary habitats, comparatively large size and varying tolerance to stress (Curry, 1962;

Hynes, 1962; Wass, 1967; Mylinski and Ginsburg, 1977; Reddy and Rao, 1991 and Petridis, 1993).

Apart from the normal ecological condition, in polluted systems, natural changes induced by pollution are well manifested by certain organisms. The benthic biota provides a method of biomonitoring of pollution and act as a bioindicator. Biological examinations have been stressed to provide a more accurate picture than the physico-chemical examinations (Hirsch, 1958,Hynes, 1960; 1964 a and b, 1965; Hakkari, 1972; Ivanova, 1976; Sarkar and Krishnamurthy, 1977; Singh, 1988; Sinha *et al.*, 1989; Singh and Roy, 1991; Roy and Dutta Munshi, 1993; Kumar, 1994 c; Bose and Lakra, 1994; Jana and Manna, 1995). In aquatic systems, the macrobenthic invertebrate community is more often investigated due to absence of mobility and sensitivity towards physico-chemical stress. Several workers have pointed out that benthic organisms provide a valuable indicator of past and present conditions of the water quality and prone to be the most useful in assessment of pollution (Hynes, 1965; Hussainy and Abdulappa, 1967; Milbrink, 1973 a and b; Cook and Jahnson, 1974; Rama Rao *et al.* 1978 a and b; Timms, 1982; Manna, 1986; Sinha *et al.* 1989; Sinha *et al.* 1990, Kumar, 1996 a, Sinha *et al.* 1997 and Reddy and Rao, 1998) because of their life-cycle length, central position in food chain and ease of collection, sorting and preservation (Cairns *et al.*, 1972; Brinkhurst, 1970 and 1972). Thus, the pollution ecology of the macrobenthic community becomes a very important biological tool for environmental impact assessment and management.

Considerable works has been carried out by several workers in other countries. Richardson (1928) studied the distribution, adundance, valuation and index value of bottom fauna in Iillinois river. Ludwig (1932) reported the bottom invertebrates of the Hocking river. (Murray,1938) studied the ecological aspect of macroinvertebrates of some northern Indian streams. (Forhne,1939) stressed the importance of macrozoobenthos while studying the biology of certain sub-aquatic flies reared from emergent water plants. Lyman (1943) gave a detailed spectrum of a preimpoundment bottom fauna of the Watts Bar Reservoir area in Tennessee. Shoup (1943) studied the distribution of freshwater gastropods in relation to total alkalinity of streams. Lyman and Dendy (1943) reported the community structure of bottom fauna of Cherokee reservoir. Pennak and Gerpen (1947) gave a clearcut picture of macrozoobenthos production in relation to physical nature of the substrate in a northern Colorado trout steam. Berg (1949 and 1950) advocated the limnological importance of aquatic insects in relation to hydrophytic plants and described the biology of certain chironomidae from potamageton. Cridland (1958) showed the impact of certain ecological factors affecting the numbers of snails in a permanent stream. Clarke and Berg (1959) studied in detail the fresh water mussels of central New York. Tebo and Hassler (1961) established a relation between aquatic insects and seasons in western-north Carolina trout streams. Oakland (1963) reviewed the quantity of bottom fauna in Norwegian lakes and rivers. Hynes (1960, 1964a and b and 1965) gave an exhaustive account of benthic invertebrate fauna of Welsh mountain stream and stressed on the significance of macroinvertebrates in the study of mild river pollution.

The geographical distribution of riverine macroinvertebrates in the Vaal river of Southern Africa has been recorded by Harrison (1963 and 1965). Bovbjerg (1964) has studied the dispersal of aquatic animals relative to density. Maitland (1964) has made a qualitative studies on the invertebrate fauna of sand and stony substrates in the river Endrick of Scotland. Morgan and Egglishaw (1965) have surveyed the bottom fauna of streams in the Scottish Highlands. Sowa (1965) elaborated the ecological characteristics of the bottom fauna of the Wielka Puszcza stream. (Waters,1965) has made an interpretation of invertebrate driff in stream. Brinkhurst and Kennedy (1965) studied the biology of the Tubificidae (oligochaeta) in a polluted stream. Eriksen (1966) recorded the interrelationship between benthic invertebrates and some substrate current-oxygen. Reed and Bean (1966) have reported benthic animals and food eaten by brook trout in Archuleta creek of Colorado. Coleman and Hynes (1970) had reported the critical distribution of the macroinvertebrate fauna in the bed of a stream. Mason *et al.* (1971) reported a good collections of macroinvertebrates and described its role in biomonitoring the water quality of Ohio river. Pratt and Coler (1976) developed a procedure for the routine biological evaluation of urban runoff in small rivers by using macroinvertebrates. Coon *et al.* (1977) studied the relative abundance and growth of mussels (Eulamellibranchia) in pools and Mississippi river. Mylinski and Ginsburg (1977) advocated the macroinvertebrates as bioindicator of pollution. Barton (1980) reported benthic macroinvertebrate communities in a large northern river, the Athabasca in Mackay. Whiting and Clifford (1983) established a relation between invertebrate and urban runoff in a small northern stream of Canada. Pascar (1987) and Perry and David (1987) observed the aquatic oligochaeta in some tributaries of the Rio da la plata in Argentina and longitudinal distribution of riverine benthos respectively. Armitage *et al.* (1987) studied the use of prediction to assess macroinvertebrate response to river regulation and Reynoldson (1987) observed an interaction between sediment contaminants and benthic organisms in lotic ecosystem.

Apart of these workers, some scholars have also done worthy work in the field of macroinvertebrate ecology (Shreevastava, 1956a and b, 1959; Schutte and Frank, 1964; Rajan, 1965, Krishnamoorthy, 1966; Kajak and Ryback, 1966; Micheal, 1968; Fretter, 1968; Grantham, 1969; McLachlan, 1970; John and Dickson, 1971; Cairns *et al.*, 1972; Harman, 1972; Jonasson, 1972; Fernando, 1973; Brinkhurst and Cook, 1974; Carter, 1976; Mackey 1976; Bowman, 1976; Lang, 1978; Pip, 1978; Vasisht and Bhandal, 1979; Moore, 1980; Lund and Peters, 1981; Rai *et al.* 1981; Crozet, 1982; Biggs, 1982; Dall *et al,*. 1984; Rao and Jain, 1985; Sarkar, 1989; Reddy and Rao, 1989 and 1991).

In India a few published records of work done on the study of benthic fauna of freshwater aquatic bodies are available (Ray *et al.* 1966; Bose, 1968; Vaidya, 1979; Pahwa, 1979; Bilgrami and Datta Munshi, 1979; Jhingran, 1983; Janakiram and Radhakrishnan, 1984; Mahadevan and Krishnaswamy, 1984; Patil *et al.* 1984; Reddy and Rao, 1987; Anwar and Siddiqui, 1988; Singh *et al.* 1988; Ahmad and Singh 1988; Singh, 1988; Sinha and Das, 1993; Kumar, 1996a and 1996f; Kumar and Singh, 1997 and Shah and Pandit, 2001).

Review of literature on the subject revealed a lack of adequate information about macrozoobenthos in Indian water bodies. Till date, no attempt has been made on the community structure and seasonality of macroinvertebrates in any water body of Santal Pargana, expect the work of Kumar (1996a), Kumar and Singh (1997) and Kumar *et al.* (2001) who have made simple attempt to monitor the water quality of the river. Therefore, the present investigation is an attempt to examine the composition and seasonal variations of benthic macroinvertebrates and to evaluate the impact of domestic sewage on the water quality of the pond water of Barharwa.

## Materials and Methods

Zoobenthos were collected with an indigenously fabricated sampler of 12.3 cm. Two haulings were made at each pond twice a month for a period of two years (January, 1999–December, 2000) from different zones. Dredged materials were mixed together in plastic bucket and washed with pond water in a metallic sieve no. 40 (ISS sieve hauling 0.545mm meshes). Sieved materials were placed in enamel tray and benthic macroinvertebrates were sorted out by hand picking method and preserved in jar with 10 per cent formalin.

In laboratory, the collected macroinvertebrates were identified upto species level with the help of Brinkhurst (1970), Needham and Needham (1972) and Tonapi (1980). The quantitative value of benthic organisms was obtained by Welch's formula (1948):

$$N = n/a.h$$

*Where,*

N = Number of macroinvertebrate.

n = Number of macroinvertebrate per sampled area.

a = Area of grab sampler in square meter, and

h = Number of hauls.

Species diversity (H) was also computed from the mathmetical expressions as suggested by (Shannon and Weaver,1964):

H =

*Where,*

N = The total number of individuals in S species.

ni = The number of individuals in $i^{th}$ species.

## Results

The zoobenthos community sampled for assessment of water quality of the ponds belongs to the following four major orders:

1. Oligochaeta
2. Insecta
3. Gastropoda and
4. Pelecypoda

The common and dominant species of macrozoobenthos of each order are as follows:

### Oligochaeta

1. *Tubifex tubifex*
2. *Limnodrilus sp.*

### Insecta

1. *Chironomus plumosa*
2. *Cryptochironomus* sp.
3. *Ephemeropteran* larvae
4. *Mesogomphus lineatus*
5. *Potamarcha obscura*
6. *Zyxomma petiolatum.*

### Gastropoda

1. *Brotia costula*
2. *Melonia tuberculata*
3. *Lymnaea accuminata*
4. *Vivipara bengalensis*
5. *Gyrulus* sp.

### Pelecypoda

1. *Corbicula striatella*
2. *Lamellidens consobrinus*
4. *Indonia coerulea*
5. *Parreysia sp.*
6. *Novaculina sp.*

The seasonal fluctuations of macrozoobenthos and their species diversity (H) in both the ponds are tabulated in Tables 12.1–12.6 and depicted in Figs. 12.1–12.8.

The maximum densities of macrozoobenthic invertebrates were $1484/m^2$ and $5825/m^2$ in May and June in 1999 at Nathu and Munshi pond respectively while in 2000 the maximum density was recorded $1445/m^2$ and $6639/m^2$ in May at Nathu and Munshi pond respectively. The percentage compostition of macrozoobenthos was found 75.80 per cent, 7.30 per cent and 16.89 per cent at Nathu pond and 34.86 per cent, 57.49 per cent and 7.65 per cent at Munshi pond

**Table 12.1: Fluctuation in Macrozoobenthos (Individuals/m$^2$) of Nathu Pond in 1999**

| Sl.No. | Macrozoobenthos | Jan. | Feb. | Mar. | Apr. | May. | Jun. | Jul. | Aug. | Sep. | Oct. | Nov. | Dec. | Total |
|---|---|---|---|---|---|---|---|---|---|---|---|---|---|---|
| **Insecta** | | | | | | | | | | | | | | |
| 1. | *Cryptochironomus* sp. | 0 | 0 | 41 | 83 | 124 | 0 | 0 | 0 | 0 | 0 | 41 | 83 | 372 |
| 2. | Ephemeropteran larvae | 0 | 0 | 165 | 248 | 330 | 84 | 0 | 0 | 0 | 0 | 0 | 0 | 827 |
| 3. | *Mesogomphus lineatus* | 0 | 0 | 85 | 125 | 207 | 289 | 247 | 331 | 413 | 578 | 249 | 123 | 2647 |
| 4. | *Potamarcha obscura* | 0 | 84 | 41 | 164 | 123 | 249 | 331 | 412 | 0 | 85 | 165 | 84 | 1738 |
| 5. | *Zyxomma petiolatum* | 42 | 164 | 247 | 124 | 83 | 84 | 0 | 41 | 124 | 205 | 83 | 41 | 1238 |
| | **Total** | **42** | **248** | **579** | **744** | **.867** | **706** | **578** | **784** | **537** | **86a** | **538** | **331** | **6822** |
| | **Percentage** | **25.77** | **66.85** | **70.18** | **62.16** | **58.42** | **58.74** | **100.00** | **100.00** | **100.00** | **100.00** | **93.08** | **80.34** | **75.81** |
| **Gastropoda** | | | | | | | | | | | | | | |
| 1. | *Mefonia tubercufata* | 0 | 0 | 0 | 41 | 40 | 0 | 0 | 0 | 0 | 0 | 0 | 0 | 81 |
| 2. | *Brotia costufa* | 0 | 0 | 83 | 41 | 160 | 41 | 0 | 0 | 0 | 0 | 0 | 0 | 325 |
| 3. | *Vivipara bengafensis* | 0 | 0 | 41 | 84 | 85 | 41 | 0 | 0 | 0 | 0 | 0 | 0 | 251 |
| | **Total** | **0** | **0** | **124** | **166** | **285** | **82** | **0** | **0** | **0** | **0** | **0** | **0** | **657** |
| | **Percentage** | **0** | **0** | **15.03** | **13.87** | **19.2** | **6.822** | **0** | **0** | **0** | **0** | **0** | **0** | **7.3008** |

*Contd...*

**Table 12.1–Contd...**

| Sl.No. | Macrozoobenthos | Jan. | Feb. | Mar. | Apr. | May. | Jun. | Jul. | Aug. | Sep. | Oct. | Nov. | Dec. | Total |
|---|---|---|---|---|---|---|---|---|---|---|---|---|---|---|
| **Pelecypoda** | | | | | | | | | | | | | | |
| 1. | *Lamellidens consobrinus* | 40 | 0 | 41 | 41 | 83 | 125 | 0 | 0 | 0 | 0 | 0 | 0 | 330 |
| 2. | *Novacufina* sp. | 41 | 83 | 0 | 40 | 83 | 165 | 0 | 0 | 0 | .0 | 0 | 0 | 412 |
| 3. | *Indonia coerufea* | 0 | 40 | 41 | 82 | 125 | 124 | 0 | 0 | 0 | 0 | 0 | 41 | 453 |
| 4. | *Parreysia* sp. | 40 | 0 | 40 | 124 | 41 | 0 | 0 | 0 | 0 | 0 | 40 | 40 | 325 |
| | **Total** | **121** | **123** | **122** | **287** | **332** | **414** | **0** | **0** | **0** | **0** | **40** | **81** | **1520** |
| | **Percentage** | **74.23** | **33.15** | **14.79** | **23.98** | **22.37** | **34.44** | **0.00** | **0.00** | **0.00** | **0.00** | **6.92** | **19.66** | **16.89** |
| | **Grand Total** | **163** | **371** | **825** | **1197** | **1484** | **1202** | **578** | **784** | **537** | **868** | **578** | **412** | **8999** |

**Table 12.2: Fluctuation in Macrozoobenthos (Individuals/m$^2$) of Nathu Pond in 2000**

| Sl.No. | Macrozoobenthos | Jan. | Feb. | Mar. | Apr. | May. | Jun. | Jul. | Aug. | Sep. | Oct. | Nov. | Dec. | Total |
|---|---|---|---|---|---|---|---|---|---|---|---|---|---|---|
| **Insecta** | | | | | | | | | | | | | | |
| 1. | *Cryptochironomus* sp. | 0 | 0 | 0 | 40 | 83 | 0 | 0 | 0 | 0 | 0 | 0 | 0 | 123 |
| 2. | Ephemeropteran larvae | 0 | 0 | 82 | 248 | 332 | 289 | 0 | 0 | 0 | 0 | 0 | 0 | 951 |
| 3. | *Mesogomphus lineatus* | 0 | 0 | 41 | 83 | 165 | 249 | 330 | 497 | 660 | 415 | 83 | 0 | 2523 |
| 4. | *Potamarcha obscura* | 0 | 40 | .42 | 124 | 164 | 207 | 289 | 363 | .82 | 0 | 40 | 125 | 1476 |
| 5. | *Zyxomma petiolatum* | 0 | 125 | 332 | 206 | 124 | 84 | 0 | 0 | 125 | 248 | 41 | 82 | 1367 |
| | **Total** | **0** | **165** | **497** | **701** | **868** | **829** | **619** | **860** | **867** | **663** | **164** | **207** | **6440** |
| | **Percentage** | **0.00** | **50.00** | **63.31** | **62.76** | **60.07** | **66.80** | **100.00** | **100.00** | **100.00** | **100.00** | **100.00** | **100.00** | **75.75** |
| **Gastropoda** | | | | | | | | | | | | | | |
| 1. | *Melonia tuberculata* | 0 | 0 | 0 | 0 | 40 | 41 | 0 | 0 | 0 | 0 | 0 | 0 | 81 |
| 2. | *Brotia costula* | 0 | 0 | 40 | 83 | 125 | 165 | 0 | 0 | 0 | 0 | 0 | 0 | 413 |
| 3. | *Vivipara bengalensis* | 0 | 0 | 0 | 84 | 123 | 166 | 0 | 0 | 0 | 0 | 0 | 0 | 373 |
| | **Total** | **0** | **0** | **40** | **167** | **288** | **372** | **0** | **0** | **0** | **0** | **0** | **0** | **867** |
| | **Percentage** | **0** | **0** | **5.096** | **14.95** | **19.93** | **29.98** | **0** | **0** | **0** | **0** | **0** | **0** | **10.198** |

*Contd...*

**Table 12.2–Contd...**

| *Sl.No.* | *Macrozoobenthos* | *Jan.* | *Feb.* | *Mar.* | *Apr.* | *May.* | *Jun.* | *Jul.* | *Aug.* | *Sep.* | *Oct.* | *Nov.* | *Dec.* | *Total* |
|---|---|---|---|---|---|---|---|---|---|---|---|---|---|---|
| **Pelecypoda** | | | | | | | | | | | | | | |
| 1. | *Lamellidens consobrinus* | 40 | 41 | 83 | 0 | 84 | 0 | 0 | 0 | 0 | 0 | 0 | 0 | 248 |
| 2. | *Novaculina* sp. | 41 | 0 | 40 | 83 | 41 | 0 | 0 | 0 | 0 | 0 | 0 | 0 | 205 |
| 3. | *Indonia coerulea* | 40 | 83 | 41 | 41 | 124 | 0 | 0 | 0 | 0 | 0 | 0 | 0 | 329 |
| 4. | *Parreysia* sp. | 83 | 41 | 84 | 125 | 40 | 40 | 0 | 0 | 0 | 0 | 0 | 0 | 413 |
| | **Total** | **204** | **165** | **248** | **249** | **289** | **40** | **0** | **0** | **0** | **0** | **0** | **0** | **1195** |
| | **Percentage** | **100.00** | **50.00** | **31.59** | **22.29** | **20.00** | **3.22** | **0.00** | **0.00** | **0.00** | **0.00** | **0.00** | **0.00** | **14.06** |
| | **Grand Total** | **204** | **330** | **785** | **1117** | **1445** | **1241** | **619** | **860** | **867** | **663** | **164** | **207** | **8502** |

**Table 12.3: Fluctuation in Macrozoobenthos (Individuals/m$^2$) of Munshi Pond in 1999**

| Sl.No. | Macrozoobenthos | Jan. | Feb. | Mar. | Apr. | May. | Jun. | Jul. | Aug. | Sep. | Oct. | Nov. | Dec. | Total |
|---|---|---|---|---|---|---|---|---|---|---|---|---|---|---|
| **Oligochaeta** | | | | | | | | | | | | | | |
| 1. | *Limnodrilus* sp. | 58 | 173 | 231 | 347 | 578 | 230 | 59 | 0 | 0 | 0 | 58 | 115 | 1849 |
| 2. | *Tubifex tubifex* | 115 | 230 | 462 | 924 | 1385 | 1848 | 173 | 0 | 0 | 58 | 174 | 175 | 5544 |
| | **Total** | **173** | **403** | **693** | **1271** | **1963** | **2078** | **232** | **0** | **0** | **58** | **232** | **290** | **7393** |
| | **Percentage** | **23.01** | **30.30** | **29.25** | **38.59** | **38.13** | **35.67** | **33.33** | **0.00** | **0.00** | **100.00** | **40.00** | **35.76** | **34.86** |
| **Insecta** | | | | | | | | | | | | | | |
| 1. | *Potamarcha obscura* | 0 | 0 | 59 | 58 | 175 | 114 | 230 | 345 | 0 | 0 | 115 | 57 | 1153 |
| 2. | *Chironomus plumosa* | 346 | 463 | 809 | 1040 | 1795 | 2710 | 175 | 0 | 0 | 0 | 175 | 290 | 7803 |
| 3. | *Cryptochironomus* sp. | 175 | 405 | 575 | 520 | 810 | 578. | 59 | 0 | 0 | 0 | 0 | 115 | 3237 |
| | **Total** | **521** | **868** | **1443** | **1618** | **2780** | **3402** | **464** | **345** | **0** | **0** | **290** | **462** | **12193** |
| | **Percentage** | **69.2819** | **65.26** | **60.91** | **49.12** | **54** | **58.4** | **60.667** | **100** | **0** | **0** | **50** | **56.967** | **57.492** |
| **Gastropoda** | | | | | | | | | | | | | | |
| 1. | *Vivipara bengalensis* | 0 | 0 | 0 | 59 | 58 | 115 | 0 | 0 | 0 | 0 | 0 | 0 | 232 |
| 2. | *Gyrulus* sp. | 0 | 0 | 58 | 115 | 58 | 116 | 0 | 0 | 0 | 0 | 0 | 0 | 347 |
| 3. | *Lymnaea accuminata* | 58 | 59 | 175 | 231 | 289 | 114 | 0 | 0 | 0 | 0 | 58 | 59 | 1043 |
| | **Total** | **58** | **59** | **233** | **405** | **405** | **345** | **0** | **0** | **0** | **0** | **58** | **59** | **1622** |
| | **Percentage** | **7.71** | **4.44** | **9.8** | **412.30** | **7.87** | **5.92** | **0.00** | **0.00** | **0.00** | **0.00** | **10.00** | **7.27** | **7.65** |
| | **Grand Total** | **752** | **1330** | **2369** | **3294** | **5148** | **5825** | **696** | **345** | **0** | **58** | **580** | **811** | **24208** |

**Table 12.4: Fluctuation in Macrozoobenthos (Individuals/m$^2$) of Munshi Pond in 2000**

| *Sl.No.* | *Macrozoobenthos* | *Jan.* | *Feb.* | *Mar.* | *Apr.* | *May.* | *Jun.* | *Jul.* | *Aug.* | *Sep.* | *Oct.* | *Nov.* | *Dec.* | *Total* |
|---|---|---|---|---|---|---|---|---|---|---|---|---|---|---|
| **Oligochaeta** | | | | | | | | | | | | | | |
| 1. | *Limnodrilus* sp. | 115 | 231 | 290 | 580 | 693 | 173 | 0 | 0 | 0 | 58 | 114 | 170 | 2424 |
| 2. | *Tubifex tubifex* | 230 | 345 | 580 | 1040 | 1617 | 1963 | 230 | 0 | 0 | 57 | 171 | 231 | 6464 |
| | **Total** | **345** | **576** | **870** | **1620** | **2310** | **2136** | **230** | **0** | **0** | **115** | **285** | **401** | **8888** |
| | **Percentage** | **33.30** | **38.45** | **30.11** | **39.01** | **34.79** | **57.84** | **30.83** | **0.00** | **0.00** | **25.00** | **38.26** | **38.74** | **37.96** |
| **Insecta** | | | | | | | | | | | | | | |
| 1. | *Potamarcha obscura* | 0 | 0 | 115 | 170 | 114 | 230 | 346 | 462 | 58 | 0 | 0 | 0 | 1495 |
| 2. | *Chironomus plumosa* | 405 | 404 | 924 | 1155 | 3060 | 981 | 170 | 0 | 0 | 0 | 115 | 230 | 7444 |
| 3. | *Cryptochironomus* sp. | 230 | 461 | 750 | 693 | 924 | 346 | 0 | 0 | 0 | 345 | 230 | 346 | 4325 |
| | **Total** | **635** | **865** | **1789** | **2018** | **4098** | **1557** | **516** | **462** | **58** | **345** | **345** | **576** | **13264** |
| | **Percentage** | **61.2934** | **57.74** | **61.92** | **48.59** | **61.73** | **42.16** | **69.169** | **100** | **100** | **75** | **46.309** | **55.652** | **56.65** |
| **Gastropoda** | | | | | | | | | | | | | | |
| 1. | *Vivipara bengalensis* | 0 | 0 | 58 | 170 | 58 | 0 | 0 | 0 | 0 | 0 | 0 | 0 | 286 |
| 2. | *Gyrulus* sp. | 0 | 0 | 57 | 115 | 58 | 0 | 0 | 0 | 0 | 0 | 0 | 0 | 230 |
| 3. | *Lymnaea accuminata* | 56 | 57 | 115 | 230 | 115 | 0 | 0 | 0 | 0 | 0 | 115 | 58 | 746 |
| | **Total** | **56** | **57** | **230** | **515** | **231** | **0** | **0** | **0** | **0** | **0** | **115** | **58** | **1262** |
| | **Percentage** | **5.41** | **3.81** | **7.96** | **12.40** | **3.48** | **0.00** | **0.00** | **0.00** | **0.00** | **0.00** | **15.44** | **5.60** | **5.39** |
| | **Grand Total** | **1036** | **1498** | **2889** | **4153** | **6639** | **3693** | **746** | **462** | **58** | **460** | **745** | **1035** | **23414** |

in 1999; while their percentage composition in 2000 were 75.75 per cent, 10.20 per cent and 14.06 per cent in Nathu pond and 37.96 per cent, 56.65 per cent and 5.39 per cent in Munshi pond.

**Table 12.5: Seasonal Fluctuations in Species Diversity Index (H) of Benthic Macroinvertebrates in the Ponds During 1999**

| *Month* | *Nathu Pond* | *Munshi Pond* |
|---|---|---|
| January | 2.55 | 0.183 |
| February | 2.71 | 0.264 |
| March | 2.75 | 0.141 |
| April | 2.66 | 0.136 |
| May | 2.72 | 0.851 |
| June | 2.88 | 0.416 |
| July | 2.67 | 0.543 |
| August | 2.17 | 0.317 |
| September | 2.51 | 0.335 |
| October | 2.87 | 0.758 |
| November | 2.36 | 0.813 |
| December | 2.92 | 0.189 |

**Table 12.6: Seasonal Fluctuations in Species Diversity Index (H) of Benthic Macroinvertebrates in the Ponds During 2000**

| *Month* | *Nathu Pond* | *Munshi Pond* |
|---|---|---|
| January | 2.47 | 0.134 |
| February | 2.61 | 0.855 |
| March | 2.73 | 0.424 |
| April | 2.61 | 0.841 |
| May | 2.65 | 0.422 |
| June | 2.81 | 0.331 |
| July | 0.62 | 0.785 |
| August | 2.17 | 0.807 |
| September | 2.61 | 0.428 |
| October | 283 | 0.185 |
| November | 2.47 | 0.284 |
| December | 2.77 | 0.257 |

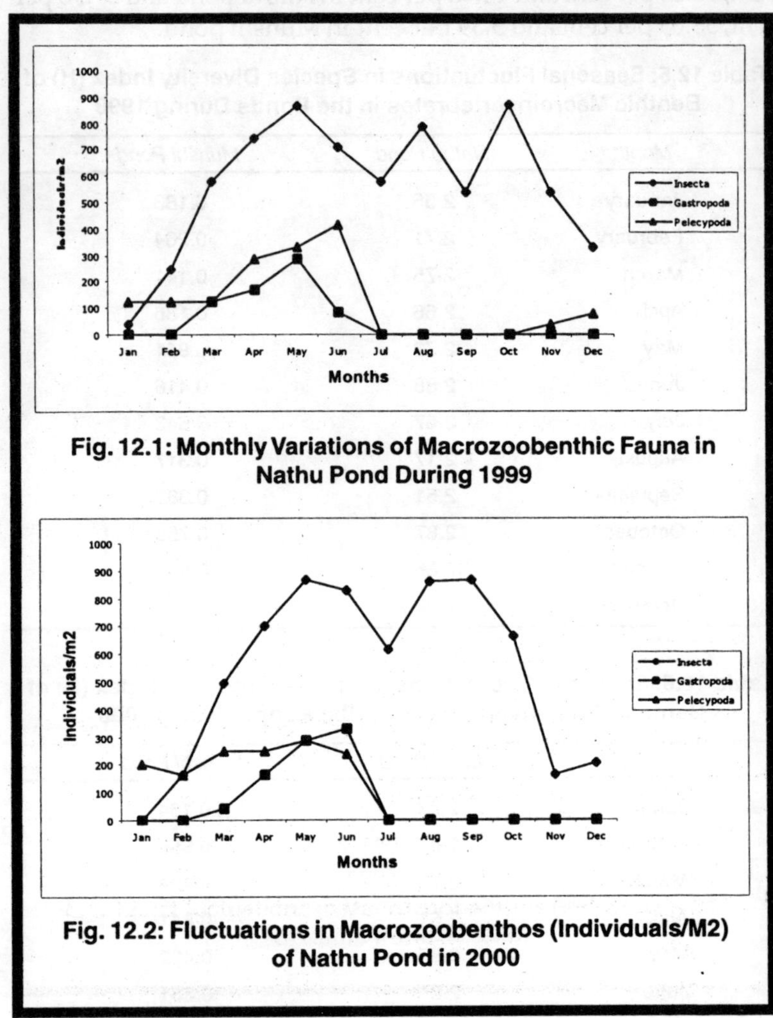


**Fig. 12.1: Monthly Variations of Macrozoobenthic Fauna in Nathu Pond During 1999**



**Fig. 12.2: Fluctuations in Macrozoobenthos (Individuals/M2) of Nathu Pond in 2000**

## Oligochaeta

The oligachaeta constituted one of the major portion of the macrozoobenthic invertebrate density. It was represented by only two species *i.e. Tubifex tubifex and Limnodrilus sp.* But these oligochaetes were remarkably absent from the Nathu pond.

The number of oligochaeta was higher in the summer months and lower during the winter. Generally the monsoon months, July–

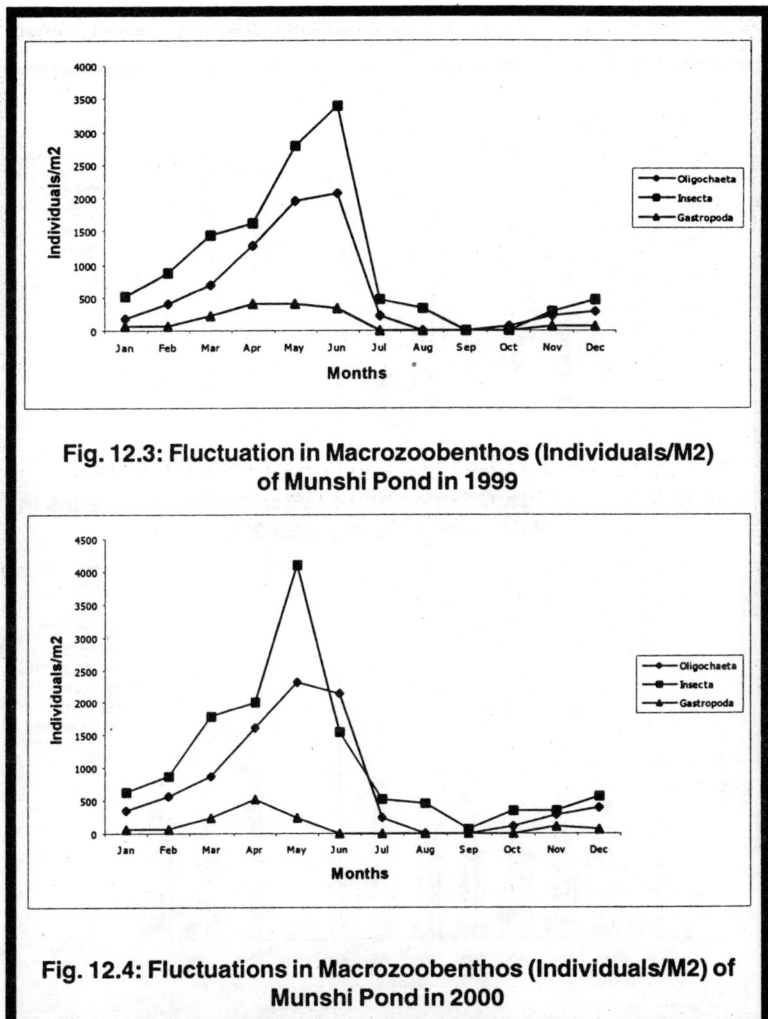


**Fig. 12.3: Fluctuation in Macrozoobenthos (Individuals/M2) of Munshi Pond in 1999**

**Fig. 12.4: Fluctuations in Macrozoobenthos (Individuals/M2) of Munshi Pond in 2000**

October formed the lean period for oligochaete production. Regarding the monthly variations of oligochaeta at Munshi pond, the maximum number was 2078/m$^2$ in June during the first year of observation while in the next year, the maximum number was 2310/ m$^2$ in May. The average percentage of oligochaetes in total macrobenthic invertebrate population was 34.86 per cent in 1999 and 37.96 per cent in 2000.

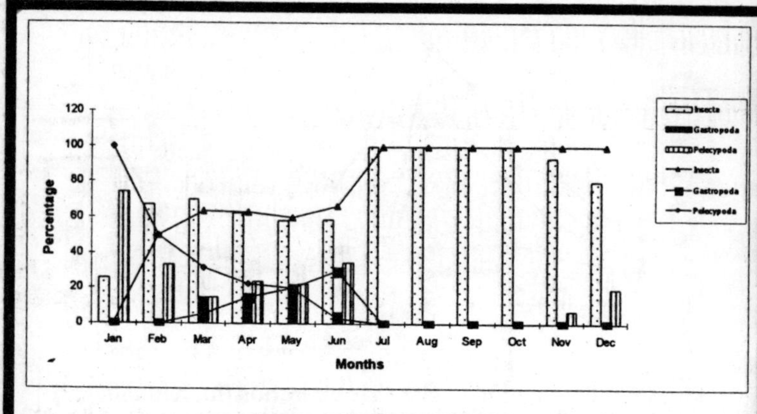


**Fig. 12.5: Percentage, Composition of Macrozoobenthic Fauna in Nathu Pond During 1999-2000**

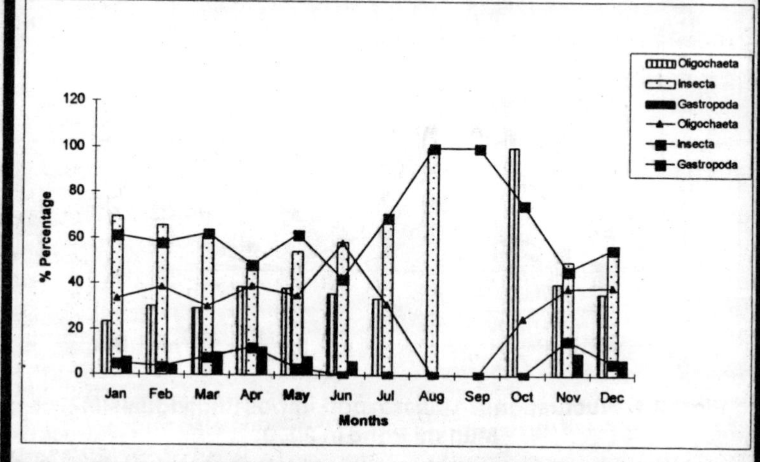


**Fig. 12.6: Percentage Composition of Macrozoobenthic Fauna in Munshi Pond During 1999-2000**

The analysis of oligochaeta indicated that the maximum count of *Tubifex* sp was $1848/m^2$ and $1963/m^2$ in the month of June of both the years. They were found absent in the months of August and September of both the years.

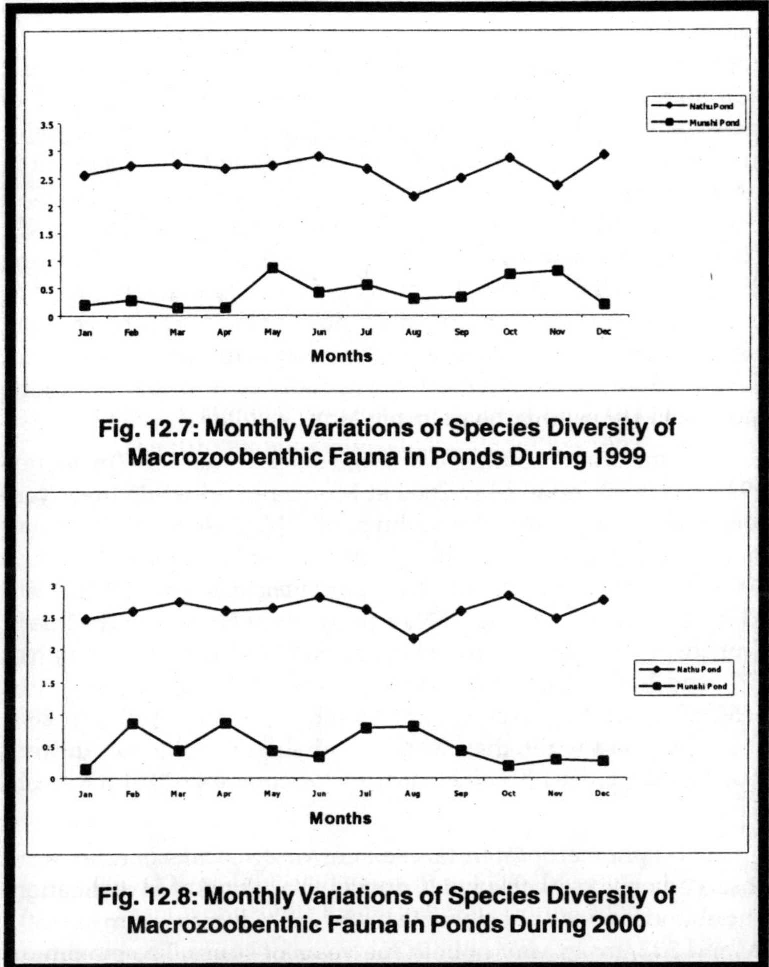


**Fig. 12.7: Monthly Variations of Species Diversity of Macrozoobenthic Fauna in Ponds During 1999**



**Fig. 12.8: Monthly Variations of Species Diversity of Macrozoobenthic Fauna in Ponds During 2000**

*Tubifex* sp. was responsible for determining the pattern of fluctuation of total oligochaetes as such they also showed annual maxima.

*Limnodrilus* sp. was recorded maximum (578/m$^2$) in the month of May and minimum (58/m$^2$) in the month of January and November of 1999. Their maximum count was recorded 693/m$^2$ in May and minimum in the month of October (58/m$^2$) of 2000. But they were recorded nil from August to October of 1999 and from July to September of 2000.

## Insecta

The insecta constituted major portion of the macrozoobenthic community. They were observed in both the ponds. Among the zoobenthos, the insecta are quallitatively and quantitatively dominated the scene at both the ponds during the period of investigation. The average percentage of insecta in total macrobenthic invertebrate community was 75.80 per cent and 54.79 per cent in 1999 while in 2000, the percentage composition of insects was 75.75 per cent and 56.65 per cent at Nathu and Munshi pond respectively. They exceeded all other macrobenthic groups by numbers during the majority of the months of study period at both the ponds. The maximum population of insects was 868/$m^2$ in October of 1999 and also 868/$m^2$ in May of 2000 at Nathu pond; while at Munshi pond, the maximum number recorded was 3402/$m^2$ and 4098/$m^2$ in the month of May of both the years of study.

The maximum count of *Chirnomus plumosa* was 2710/$m^2$ in June 1999 and 3060/$m^2$ in May 2000 at Munshi pond while they were remarkably absent from the Nathu pond. The presence of *Chirnomus plumosa* was nil in the months from August to October 1999 and 2000. The maximum count of *Cryptochironomus* was 124/$m^2$ and 83/$m^2$ in May of 1999 and 2000 respectively at Nathu pond while in Munshi pond it was recorded 810/$m^2$ and 924/$m^2$ in May month of 1999 and 2000 respectively. They were recorded only in the months of March-April-May-November-December 1999 and April-May 2000 at Nathu pond while they were found absent in the month from August to October 1999 and from July to September 2000 at Munshi pond.

The Ephemeropteran larvae and *Mesogomphus lineatus* were observed only in Nathu pond during the period of investigation. The abundance of ephemeropteran larvae was maximum *i.e.* 330/$m^2$ and 332/$m^2$ in May of both the years of study. The maximum count of *Mesogomphus lineatus* was 578/$m^2$ in October 1999 and 660/$m^2$ in September 2000. *Potamarcha obscura* was found in both the ponds during the study period. The maximum count was 412/$m^2$ and 363/$m^2$ in the month of August in 1999 and 2000 respectively in Nathu pond while it was recorded 365/$m^2$ and 462/$m^2$ also in August of both the years of investigation at Munshi pond. *Zyxomma petiolatum* was found nil at Munshi pond throughout the period of investigation. Its maximum count was 247/m$^2$ and 332/m$^2$ in March of 1999 and 2000 respectively at Nathu pond.

## Gastropoda

The Gastrpoda were generally recorded from March to June of the study years while they were either absent or present in very negligible number during rest of the month. The population of gastropoda was maximum in Munshi pond in comparison of Nathu pond. The average percentage composition of gastropoda in total macrobenthic invertebrate population was 7.30 per cent and 7.65 per cent in the first year of study and 10.20 per cent and 5.39 per cent in the second year of investigation in Nathu and Munshi pond respectively. The maximum density of gastropoda recorded 285/m$^2$ in May 1999 and 372/m$^2$ in June 2000 in Nathu pond while it was noted 405/m$^2$ in April-May 1999 and 515/m$^2$ in April,2000 in Munshi pond. The gastropoda was represented in samples by *Brotia costula, Vivipara bengalensis, Melonia tuberculata, Lymnaea accuminata and Gyrulus sp.*

The abundance of *Brotia costula* was recorded maximum *i.e.* 160/m$^2$ in May, 1999 and 165/m$^2$ in June, 2000 in Nathu pond. They were found absent in Munshi pond throughout the study period. The maximum count of *Vivipara bengalensis* was 85/m$^2$ in May,1999 and 166/m$^2$ in June,2000 in Nathu pond while it was recorded 115/m$^2$ in June,1999 and 170/m$^2$ in April, 2000 in Munshi pond.

They were remarkably recorded in both the pond. The maximum count of *Melonia tuberculata* was 41/m$^2$ in April, 1999 and also 41/m$^2$ in June 2000 in Nathu pond. They were absent in Munshi pond. *Lymnaea accuminata* was recorded only in Munshi pond. Its maximum count was recorded 289m$^2$ in May 1999 and 230/m$^2$ in April, 2000 The maximum coun tof *Gyrulus* sp. was recorded 116/m$^2$ in June, 1999 and 115/m$^2$ in May, 2000 in Munshi pond. It was found absent in Nathu pond.

## Pelecypoda

It was found only in Nathu pond. The average percentage composition of pelecypods in total macrobenthic invertebrate population were 16.89 per cent and 14.06 per cent in 1999 and 2000 respectively. The maximum density of pelecypod recorded was 414/m$^2$ in June, 1999 and 289/m$^2$ in May, 2000.

The population density of *Parreysia* sp was maximum in April in both the years. Its maximum count was recorded 124/m$^2$ and 125/m$^2$ in 1999 and 2000 respectively.

They were only observed in January and March to May during 1999 while during 2000, they were recorded from January to June. The maximum count of *Indonia coerulea* was 125/m$^2$ in May,1999 and 124/m$^2$ in May, 2000. The maximum count of *Lamellidens consobrinus* was 125/m$^2$ in June, 1999 and 84/m$^2$ in May, 2000. The maximum count of *Novaculina sp.* was recorded 165/m$^2$ in June,1999 and 83/m$^2$ in May, 2000.

The values of species diversity (H) of macrozoobenthos ranges between 2.17 and 2.92 and 0.136 to and 0.851 in 1999 whereas in 2000, its value varied from 2.17 to 2.83 and 0.134 and 0.855 in Nathu and Munshi pond respectively. The result showed that the values of species diversity were always lesser than 1.0 in Munshi pond and more than 2.0 in Nathu pond (Tables 12.5 and 12.6).

## Discussion

Macrobenthic fauna is an integral and inseparably associated biotic component of an aquatic ecosystem may it be a lotic or lentic Bias *et al.* (1992). The benthos are extremely important component of water body from various limnological view point. They serve as the primary source of food for fish and higher aquatic organisms. Many species are extremely sensitive to pollution and respond to it very quickly. The use of macrobenthic fauna as water quality indicator has been advocated by many authors Bias *et al.* (1992). In the productive capacity of water bodies, the importance of bottom fauna as a link in the "energy-flow" from primary production to fish yield has been stressed by many authors, not only in temperate Grimas (1965) Hynes (1965) and Brinkhurst (1970 and 1972) but also in tropics namely Srivastava (1959 a and b) in a number of lakes of Northern India, Govind (1969) in Tungabhadra reservoir: Krishnamurthy (1971) in Pulicat lakes, Mandal and Moitra (1975) and Raman *et al.* (*1975*) in fish ponds of Eastern India, Gupta (1976) in Loni reservoir; Abraham (1978) in Bhawanisagar reservoir of South India and Haniffa (1977) and Haniffa and Pandian (1978) in South Indian fish ponds. On the other hand, the importance of biological examination of benthic macrofauna could provide a more accurate picture of the trophic status of the biotope than the routine assessment of prevailing physico-chemical parameters (Hirsch 1958; Hynes 1964 a and b, 1971; Sarkar and Krishnamoorthy, 1977; Sinha 1988; Sinha *et al.* 1989,1990,1991; Barbhuyan and Khan, 1992). Several workers have also pointed out that macrobenthic fauna provides a valuable tools as indicator of past and present water quality from pollution and in general stress view point (Hussainy and Abdulappa, 1967; Ram Rao *et al.* 1978 a and b; Sinha *et al.* 1989 and 1990). The benthic fauna becomes very important for pollution studies because they have long life cycle, they respond to changes in physico-chemical condition, they have central position in food-chain and they can be easily sorted out, preserved and handled for experimental work (Cairns and Dickson, 1971 and Brinkhurst 1970 and 1972).

Much of the evolution of impact assessment has occurred with reference to aquatic system rather than terrestrial systems (NRC, 1981). This sequence in development is understandable because human settlement has tended to follow waterways. Before a highway system developed, rivers and lakes were convenient means of moving about the country. Population centres that developed along these waterways found it convenient to discharge their wastes into the water where the wastes were then diluted and for carried away downstream. Sewage was transformed by natural mechanism within the stream and was often completely assimilated before it searched the next downstream uses. When population centres were small and scattered and the wastes were simple organics, this was sustainable use of the waterway. But with increased population and waste loads, problems become obvious.

The analysis of adverse effect of sewage on the structure of communities of aquatic organisms was among the first environmental assessment method to be formalized. The well known saprobian system of Kolkwitz and Marsson (1908 and 1909) worked on the principle of association. Organisms were censused in degraded and healthy system. Those organisms able to tolerate sewage discharge were categorized as saprobic, those that could only survive in relatively pristine conditions were labelled oligotrophic, and those somewhere in between were known mesotrophic.

Ample evidence exists that the macrobenthic community acts as a bioindicator and has become a very important biological tool for the assessment of water quality (Hynes 1965; Hussainy and Abdulappa 1967; Sarkar and Krishnamoorthy 1977; Rama Rao *et al.* 1978; Sinha *et al.* 1990; Sinha and Das 1993; Kumar 1994c and 1997c).

During the present investigation, the oligochaeta was only present in Mushi pond where DO content is comparatively low. Thus, their ability to withstand in considerable oxygen depletion environment is an essential adaptation to their niche within the community. Studies by Brinkhurst (1972) also supports the present findings. He has further suggested that the competition is avoided by selective digestion of the bacteria within the sediment, which leads to a degree of collaboration as the faeces of one species of worm become the preferred food of another species. This is probably the ecological basis of the very close clumping of individuals which makes quantitative ecological studies difficult.

The abundance of *Tubifex tubifex* in only polluted scenario might be due to quantity and quality of organic matter reaching the sediment which plays a more dominant role in determining which oligochaete species will be found in any given locality then do all of the commonly measured physical and chemical parameters of the water. Specific organic inputs might be excepted to affect the spectrum of bacterial species present, which in turn determines the worm species and their relative abundance.

In the present study, the insects have dominated the macrobenthic fauna both qualitatively and quantitatively. The ephemeropteran larvae are found in Nathu pond only. They are perhaps the prime grazer in the aquatic food-web. Most are algal feeders, some are scavengers, feeding both on algae and vegetational detritus. Odonates were also found in great variety of habitats. These are predaceous, feeding on protozoa, cladocera, other aquatic insects etc. (Kumar, 1996a). These are also not very sensitive group.

The present study shows that the macrobenthic community is clearly dominated by *Chironmus* sp. population in Munshi pond. The availability and distribution of chironomids on intra-lake level have been attributed to be relative to many factor Bowman (1976) Carter (1976). Ramchandran and Paterson (1978) demonstrated that the chironomids could be segregated with reference to food, space and time. However, one could regard the last two categories as further subdivisions of the food resource thereby enabling more species to co-exist, such as spatial and temporal division will be more pronounced and advantageous in a patchy environment Southwood (1977) and less marked in a uniform habitat.

Kajak and Wards (1968) while studying the relative abundance of algal cells in detritus and gut contents of *Chironomus*, postulated that *Chironomus* sp. has selective feeding habit and prefer algae. He further calculated that *Chironomus* larvae could live purely on bacteria ingested with the detritus. It is likely that selective feeding on any resource that is not very abundant is incompatible with a stationary, tubiculous life in the benthos. According to Johannsson and Beaver (1983), among algae, diatoms are the most important group of chironomid diet. But chironomids require much more energy than that provided by algae at all times so the altered food sources are bacteria, some protozoans and free organic molecules. Bacteria have found to be sufficient food for growth and emergence of chironomids.

Taylor (1961) while studying chironomids indicated that owing to trophic factor, *Chironomus* may be randomly distributed. Edgar and Moadows (1969) also reported random distribution in *Chironomus* larvae. McLachlan (1976) in a series of laboratory experiments using a shallow layer of sediment, showed that the distribution of some species of chironomids larvae was density depended. But during the present investigation, chironomids similar to tubificids have been found to pollution forms. The high population of chironomids is always associated with high bacterial activites and hence polluted conditions. Chironomid larvae have also been used as pollution indicators by a number of workers Gaufin (1957) Curry (1962). Thus, the high presence of chironomids in the benthic population is due to an impact of altered nature of substrate due to organic pollution.

Chironomids and oligochaetes are to some extent alike as they get their chief source of energy, *i.e.* bacteria by ingesting large volume of sediments. Recent studies on oligochaete abundance in Indian condition have shown a positive correlation between population densities of tubificids and bacteria. Thus, in a habitat where chironomids are abundant, a high population of tubificids may also be expected owing to similarities in the factors influencing their distribution. The similar result during the present study might be due to the same reason.

During the present investigation, the number of benthic organisms at Nathu pond is less in comparison to Munshi pond. It is also observed that the pond bed in Nathu pond is sandy as a result of which less nutrients are available there. The tubificid oligochaetes, which is the most tolerant species of pollution, were found totally absent in Nathu pond. In contrast to it, the intolerant species such as ephemeropteran larvae and *Novaculina* sp., a sensitive bivalve were present. This observation is the manifestation of either less or no pollution in Nathu pond.

At Munshi pond, the total type of species (08) was lower than that of Nathu pond. Only two species of oligochaetes (*Tubifex* sp. and *Limnodrilus* sp.) and one species of insecta (*Chironomus plumosa*) contributed predominantly towards the benthos at Munshi pond.

In general, as reported by Brinkhurst and Cook (1974), the heavy silting, high organic content of silt and associated with deoxygenation below organic effluents tend to increase the share of tubificid habitat on bed of river and lake. In polluted condition, a decline in oligochaete (tubificid) species with high increase in total number in contrast to normal habitat has been reported by many authors. Thus, relative abundance of tubificids may be related to simultaneous decrease in other benthic species. A similar spectrum has been observed so far the benthic community of polluted water bodies is concerned. Hence, the conclusion drawn by Brinkhurst and Cook (1974) that the abundance of tubificids relative to other organisms as well the relative abundance of various species of worms is of particular importance in pollution studies, stands well in the present study. Further, the two tubificids (*Tubifex tubifex* and *Limnodrilus* sp.) have been reported to be indicator of pollution by their mere presence in considerable number by many workers in temperate and tropical conditions.

Not only this, the diversity index (H) has also dropped to less than I at the Munshi pond, as compared to Nathu pond showing value higher than I. Mason (1981) opines that when an environment becomes stressed, the species sensitive to that particular stress will be eliminated, thus, reducing the richness of community and certain species may be favoured so that they become abundant compared with the other members of the community. In the present study, the dominance of tubificid may be attributed to the organic pollution caused by the sewage discharge, favouring them and eliminating other intolerant species.

The water current remains sluggish at Munshi pond for the greater part of the year and at this site, bottom is mainly muddy mixed with fine silt and detritus carries with sewage discharge. Such conditions are favourable for greater organic production of insect fauna Needham and Llyod (1916). The production of insect fauna, specially *Chironomus and Cryptochironomus* at this site confirms this observation. It is observed that those organism associated with the silted regions of the rivers are the most tolerant to organic pollution (Mason, 1981). The presence of *Tubifex* sp., *Limnodrilus* sp., *Lymnaea accuminata* and *Chironomus plumosa* at Munshi pond corroborates with the work of (Mason,1981). The absence of oligochaetes and appearance of intolerant species, such as ephemeropteran larvae, reflects the absence of organic load and the pond has recorded from the stress, which it had at Munshi pond due to various factors.

Wilhm and Dorris (1968) proposed a close relationship between degree of pollution and species diversity index of benthic macroinvertebrates from the variety of clean water and polluted stream that a value of diversity index greater than 3 indicates clean water, values in the range of 1-3 are characteristics of moderately polluted conditions and value of less than I is characterized as heavily polluted conditions. Mason (1981) is of opinion that diversity indices can be used to measure the stress in the environment. It is considered that unpolluted environments are characterised by a large number of species with no single species making up the majority of the community and a maximum diversity is obtained when a large number of species occur in relatively low number in a community.

Diversity is found to be low during the period of high abundance and also due to rapid increase and numerical dominance of one or two species in the community (Sagar and Hasler, 1969). Pollutants cause generally adjustment and alterations in species abundance and community species composition in aquatic systems. A polluted system is simplified and those species that survive encounter less competition and therefore, may increase in number (Dennis and Patil, 1977).

In the present investigation, at Munshi pond, the diversity index was always less than I but at Nathu pond, it was always more than 2. This study reveals that Munshi pond is heavily polluted, while Nathu pond is mildly polluted. The heavy pollution at Munshi pond may be attributed to sewage discharge. Thus, it can be concluded that the diversity index can be used as a measure of water pollution in the pond water and when studied in combination with physico-chemical factors, it provides a more realistic assessment of the water quality.

## References

Abraham, M. (1978): Studies on the bottom macrofauna of Bhawanisagar reservoir (Tamilnadu). J. Inland Fish Soc. India. 11: 41-48.

Ahmad, S. H. and Singh, A. K. (1988): Pollution effect on ecobiology of benthic macroinvertebrates in the river Ganga at Patna (Bihar), India. First International Seminar on "Bioassay techniques and their application" held at University of Lancaster, England from July 11-14 (Abstract).

Ahmad, M. S. and Siddiqui, E. N. (1990): Blue-green algae of Darbhanga. Biojournal. 2: 133-136.

Armitage, P. P., Gunn, R. J. M., Furse, M. T., Writeght, J. F. and Moss, D. (1987): The use of prediction to assess macroinvertebrate response to river regulation. Hydrobiologia. 48: 25-32.

Barbhuyan, S. I. and Khan, A. A (1992): Studies on structure and function of benthic ecosystem in an eutrophic body of water, temporal and special distribution of benthos J. Freshwater Biol. 4: 239-247.

Barton, D. R. (1980): Benthic macroinvertebrate communities of the Athabasca river near Mackay, Alberta. Hydrobiologia. 74: 151-160.

Bazzanti, M. and Loret, E. (1982): Macrobenthic community structure in a polluted Lake, Lac Nemi, Central Italy. Boll. Zool. 49: 79-91.

Berg, C. O. (1949): Limnological relations of insects to plants of the genus *Potomageton*. Tran. Am. Micr. Soc. 68: 279-291.

Bias, V. S., Yatheesh, S. and Agarwal, N. C. (1992): Benthic macroinvertebrates in relation to water and sediment chemistry. J. Freshwater Biol. 4: 183-191.

Biggs, B. J. E. (1982): Macroinvertebrates associated with various aquatic macrophytes in backwater and lakes of the upper Clutha valley, New Zealand J. of Mar. Freshwat. Res. 16: 81-88.

Bilgrami, K. S. and Dutta Munshi, J. S. (1979): Limnological survey and impact of human activities on the river Ganges (Barauni to Farakka), Technical Report (MAB project). pp. 91.

Bose, S. K. (1968): Ecological studies of some food fishes of Ranchi with reference to Hydrology, plankton and pedon (Bottom fauna), Ph. D. thesis, Ranchi Univ. Ranchi.

Bose, S. K. and Lakra, M. P. (1994): Studies on macrozoobenthos of two fresh water ponds of Ranchi. J. Freshwater Biol. 6: 135-142.

Bovbjerg, R. V. (1964): Dispersal of aquatic animals relative to density. Verh. Int. Ver. Theo Angew. Limnol. 15: 879-884.

Bowman, C. M. T. (1976): Factors affecting the distribution and abundance of Chironomids in three shorpshre meres, with special reference to larval tracheal system. Ph. D thesis, Univ. Keele.

Brinkhurst, R. O. (1970): Distribution and abundance of tubificid (oligochaeta) species in Toranto Harbour, Lake Ontario, J. Fish. Res. Bd. Can. 27: 1961-1969.

Brinkhurst, R. O. (1972): The role of sludge worms in eutrophication. U. S. Environmental Protection Agency, Washington D. C. Ecol. Res. Ser. EPA-R3–72-004, August 1972.

Brinkhurst, R. O. and Cook, D. G. (1974): Aquatic Earthworm (Annelida: Oligochaeta) In: pollution ecology of freshwater invertebrate (Ed. Hart, J. and Fuller, T). Academic press, New York. pp. 143-155.

Brinkhurst, R. O. and Kennedy, C. R. (1965): Studies on the biology of the Tubificide (Annelida, Oligochaeta) in a polluted stream. J. Anim. Ecol. 34: 429-443.

Cairns, J. and Dickson, K. L. (1971): A simple method for the biological assessment of the effect of waste discharge on the aquatic bottom dwelling Organisms. J. Wat. Pollut. Control Fed. 43: 755-772.

Cairns, J. Jr., Dickson, K. L. and Crossman, J. S. (1972): The response of aquatic communities to spills of hazardous materials. Proc. Nat. Conf. Hazardous Mater. Spills. 179-197.

Carter, C. E. (1976): A population study of the chironomidae (Diptera) of Lough Neagh. Oikos. 27: 346-358.

Clarke, A. H. and Berg, C. O. (1959): The fresh water mussels of Central New York. Cornell Univ. Agr. Sta. Ithaca. Exp.

Coleman, M. J. and Hynes, H. B. N. (1970): The Critical distribution of the invertebrate fauna in the bed of a stream. Limnol. Oceanogr. 15: 31-40.

Cook, D. G. and Jhanson, M. G. (1974): Benthic macroinvertebrates of the Lawrence Great Lakes. J. Fishes Res. Board. Can. 31: 763-782.

Coon, T. B., Eckblod, J. W. and Trygstad, P. M. (1977): Relative abundance and growth of mussels (Mollusca: Eulamellibranchia) in pools 8, 9 and 10 of the Mississippi river Fresh wat. Biol. 7: 279-285.

Cridland, C. C. (1958): Ecological factors affecting the number of snails in a permanent stream. J. Trop. Med. Hyg. 11-20.

Crozet, B. (1982): Contribution altetucle des communalites littorales demacro invertebrates benthiques du Laman (Petit-Loc) en relation quecleur Environment. Ph. D. Thesis, University of Geneva. pp. 215.

Curry, L. L. (1962): A survey of environmental requirements for the midge (Diptera-Tendipendidae). In: Biological problems in water pollution. Transaction of 3rd seminar (Ed. Trazell, U and Robert, P). A Traft Sanitary Engineering Centre, Cincinnati.

Dall, P. C., Lindegaad, C and Jonsson, G. (1984): Invertebrate communities and their environment in the exposed littoral zone of lake, Esron (Denmark). Arch. Hydrobiol. Suppl. 69: 477-524.

Dennis, B. and Patil, G. P. (1977): The use of community diversity indices for monitoring trends in water pollution impacts. Tropical Ecology. 19: 36-56.

Edgar, W. D. and Moadows, S. (1969): Case construction, movement, spatial distribution and substrate selection in the larvae of *Chironomus riparius*. J. Exp. Bio;. 50: 247-253.

Eriksen, C. H. (1966): Benthic invertebrates and some substrate current oxygen inter relationship. Spec. Publs. Pymatuning Lab. Fld. Biol. 4: 98-115.

Fernando, C. H. (1973): Seasonality and dynamics of aquatic insects colonizing small habitats. Verh. Int. Theo. Angew. Ver. Limnol. 18: 1564-1575.

Forhne, W. C. (1939): Biology of certain sub-aquatic flies reared from emergent water plants. Mich. Acad. Sci. Arts. 24: 139-147.

Fretter, V. (1968): Studies in the structure, physiology and ecology of mollusca. Academic Press, New York. pp. 410.

Gaufin, A. R. (1957): The use and value of aquatic insects as indicators of organic enrichment. Biological problems in water pollution. U. S. Public Health. Serv. Washington, D. C. pp. 139-149.

Govind, B. V. (1969): Preliminary studies on the Tungabhadra Reservoir. Indian J. Fish. 10: 148-158.

Grantham, B. J. (1969): The fresh water pelecypod fauna of Mississippi. Univ. of Southern Mississippi, Hattiesburg. pp. 1-243.

Grimas, V. (1965): Effects of impoundment on the bottom fauna of high mountain lakes. Acta Universitatis Uppsaliensis 51: 5-24.

Gupta, S. D. (1976): Macrobenthic fauna of Loni Reservoir. J. Inland Fish. Soc. India. VIII: 49-59.

Hakkari, L. (1972): Zooplankton species as indicator of environment. Aqua. Fenn. 46-54.

Haniffa, M. A. (1977): Secondary productivity and energy flow in a tropical pond. Hydrobiologia. 59: 49-65.

Haniffa, M. A. and Pandian, T. J. (1977): Morphometry, primary productivity and energy flow in a tropical pond. Hydrobiologia. 59: 23-48.

Harman, W. N. (1972): Benthic substrates: The effect on fresh water mollusca. J. Ecology. 53: 271-277.

Harrison, A. D. (1963): Hydrological studies on the Vaal river in the Verrenigung area. Part. II. the chemistry, bacteriology and invertebrate of the bottom muds. Hydrobiologia. 21: 66-89.

Harrison, A. D. (1965): Geographical distribution of riverine invertebrate in Southern Africa. Arch. Hydrobiol. 61: 387-394.

Hellawell, J. M. (1986): Biological indicators of freshwater pollution and environmental management. Elsevier. London.

Hirsch, A (1958): Biological evaluation of organic pollution in New Zealand streams. N. Z. J. Sci. 1: 500-553.

Hussainy, S. V. and Abdulappa, M. K. (1967): Expression of biological data in water pollution research. Environ. Hlth. 9: 210-219.

Hynes, H. B. N. (1960): The biology of polluted water. Liverpool Univ. Press Liverpool. pp. 431.

Hynes, H. B. N. (1962): The significance of macroinvertebrates in the study of mild river pollution In: Biological problems in water pollution. Third seminar USPHAS, Washington, D. C. pp. 235.

Hynes, H. B. N. (1964a): Interpretation of biological data with reference to water quality. U. S. Public Health Service Publ. 999-AP-15.

Hynes, H. B. N. (1964b): The use of biology in the study of water pollution. Chem. Ind. London. 435-436.

Hynes, H. B. N. (1965): The significance of macroinvertebrates in the study of mild river pollution. In: "Biological problems in water pollution" 3rd seminar, U. S. Deptt. Health, Education and Welfare, Cincinnati.

Hynes, H. B. N. (1971): Biology of polluted waters. Univ. of Toranto Press. Toranto pp. 202. Ivanova, M. B. (1976): The effect of pollution on planktonic crustacea and the possibility of using them for determining the pollution in a river. In: Metody Biol. Analiza presnykh vod, Leningrad. pp. 68-80.

Jana, B. B. and Manna, A. K. (1995): Seasonal changes of benthic invertebrates in two tropical fish pond. J. Freshwater. Biol. 7: 129-136.

Jankiram, K. and Radhakrishnan, K. (1984): The distribution of freshwater mollusca in Guntur district (India), with a description of Scaphula nagar *Junai* sp. (Srcidae), Hydrobiologia. 199: 49-55.

Jhingran, V. G. (1977): Fish and fisheries in India. Hindustan Publ. Corp. of India, New Delhi.

Jhingran, V. G. (1983): Fish and fisheries of India. 2nd Ed. Hindustan Publishing Corporation, Delhi (India). pp. 665.

Johason, O. E. and Beaver, J. L. (1983): Role of algae in the diet of *Chironomus plumosus* from the Bay of Quinte lake Ontario. Hydrobiol. 107: 237-247.

Johason, P. M. (1972): Ecology and Production of the profundal benthos in relation to Phytoplankton in lake Esrom. Uikos. 14 (Suppl). pp. 148.

John, C. and Dickson, K. I. (1971): A simple method for the biological assessment of the effects of waste J. WPCF. 143: 755-772.

Kajak, Z. and Rybak, J. I. (1966): Production of some trophic dependence in benthos primary production and zooplankton production of several Masurain lakes. Verh. Int. Ver. Limnol. 16: 441-451.

Kajak, Z. and Wards, J. (1968): Feeding of benthic non predatory chironomids in lakes. Annales Zoologici Fennici. 5: 57-64.

Kolkwitz, R. and Marsson, M. (1908): Okologie Der Pflanzkichen Saprobien. Ber. Deutsch Bot. Ges. 26: 505-519.

Kolkwitz, R. and Marsson, M. (1909): Okologie Der Pflanzkichen Saprobien. Int. Rev. Ges. Hydrobiol. 2: 126-152.

Krishnamurthy, K. N. (1971): Primary studies on the bottom biota of Pulicat lake J. Mar. Biol. Assoc. India. 13: 264-269.

Krishnamurthy, K. P. (1966): Preliminary studies on the bottom macrofauna of the Tungabhadra reservoir. Proc. Ind. Acad. Sci. B. 65: 96-103.

Kumar, A. (1994c): Role of species diversity of aquatic insects in the assessment of pollution in wetlands of Sanhal Pargana (Bihar). J. Environ. Pollut. 1: 117-120.

Kumar, A. (1996a): Impact of organic pollution on macro zoobenthos of the river Mayurakshi of Bihar. Poll. Res. 15: 85-88.

Kumar, A. (1996f): A comparative study on stomach-content and forage ratio of zygopteran and anisopteran larvae in a fish pond of Santal Pargana (Bihar), India. Proc. Nat. Acad. Sci. 66: 360-367.

Kumar, A. (1997c): Biomonitoring of pollution by aquatic insect community and microorganisms in freshwater ecosystem. In: Ecotechnology for pollution control and Environment Management (Eds. R. K. Trivedy and A. Kumar) Enviromedia. pp. 255-263.

Kumar, A. and Bohra, C. (2001a): Assessment of nutrient budget of benthos in lentic freshwater ecosystem. In: Current Topics

Environmental Sciences (Eds. G. C. Pandey and G. Tripathi), ABD Publishers, Jaipur. pp 137-158.

Kumar, A. and Singh, A. K. (1997): Role of macrozoobenthos in monitoring the water quality of the river Mayurakshi in South Bihar. J. Mendel 15: 145-151.

Lang, C. (1978): Factorial correspondence analysis of oligochaete communities according to eutrophication level. Hydrobiologia. 57: 241-247.

Ludwig, W. B. (1932): The bottom invertebrates of the Hocking river Bull. Ohio. Biol. Surv. 26: 223-249.

Lund, J. C. and Peters, E. J. (1981): Production rates of aquatic insects in a turbid reservoir. Trans. Nabr. Acad. Sci. 9: 23-34.

Lyman, F. E. (1943): A pre-impoundment bottom fauna study of the watts Bar reservoir area (Tennessee). Trans. Am. Fish. Soc. 72: 52-62.

Mackey, A. P. (1976): Quantitative studies of chironomidae (Diptera) of the river Thames and Kennet. I. The Acorus Zone. Arch. Hydrobiol. 78: 240-267.

Mahadevan, A. and Krishnaswamy, S. (1984): Chironomid larval population size: an index of pollution in river Vaigai. Poll. Res. 3: 35-38.

Maitland, P. S. (1964): Quantitative studies on the invertebrate fauna of sand and stony substrates in the river Endrick, Scotland. Proc. R. Sco. Endinb. B. 68: 277-301.

Mandal, B. K. and Moitra, S. K. (1975): Studies on the bottom fauna of a fresh water fish pond at Burdwan. J. Inland Fish. Soc. India. 7: 43-48.

Manna, A. K. (1986): Seasonal changes of some biochemical components of plankton and bottom biota in relation to fish growth in the fish farming ponds. Ph. D. thesis University of Kalyani, Kalyani, West Bengal. pp. 220.

Mason, C. F. (1981): Biology of fresh water pollution. Longman, London. pp. 209.

Mason, W. T., Lewis, P. A. and Anderson, J. B. (1971): Macroinvertebrate collections and water quality monitoring in the Ohio river basin 1963-1967. Co-operative report, Office Tech.

Programs. Ohio Basin region and analytical quality control laboratory, WOO, USEPA, NERO Cincinnati.

McLachlan, A. J. (1970): Some effects of annual fluctuation in water level on the larval chironomid communities of lake Kariba. J. Anm. Ecol. 39: 79-90.

McLachlan, A. J. (1976): Factors restricting the range of *Glytotendipes paripes* (Diptera Chironomidae) in a Bog lake. J. Anm. Ecol. 45: 105-113.

Michael, R. G. (1968): Studies on the bottom fauna in a tropical freshwater fish pond. Hydrobiologia. 31: 203-230.

Milbrink, G. (1973a): Communities of oligochaete as indicators of the water quality in lake Hjalmare. Zoon. 1: 77-88.

Milbrink, G. (1973b): On the use of indication communities of tubificidae and some lumbericulidae in the assessment of water pollution in Swedesh lakes. Zoon. 1: 125-139.

Moore, J. W. (1980): Factors influencing the composition, structure and density of a population of benthic invertebrate. Arch. Hydrobiol. 88: 207-216.

Morgan, N. C. and Egglishaw, H. J. (1965): A survey of the bottom fauna of streams in the Scottish highlands Part I. Composition of the fauna Hydrobiologia. 25: 181-211.

Murray, M. J. (1938): An ecological study of invertebrate fauna of some Northern Indiana streams. Invest. Indiana lake streams. 1: 101-110.

Mylinsky, E. and Ginsburg, W. (1977): Macroinvertebrates as indicators of pollution. J. AWWA. 69: 538-548.

Needham, J. G. and Llyod, J. T. (1916): The life in inland waters, New York pp. 411.

Needham, J. G. and Needham, P. R. (1972): A guide to the study of fresh water biology. Holden Day Inc. Sanfransisco. pp. 160.

N. R. C. (1981): Testing for the effect of chemicals on ecosystems. National Research Council. National Academy Press, Washington, D. C.

Oakland, J. (1963): A review of the quantity of bottom fauna in Norwegion lakes and rivers. (English Summary). Fauna. Oslo. Suppl. 16: 1-67.

Pahwa, D. V. (1979): Studies of the distribution of benthic macrofauna in the stretch of river Ganga. Indian J. Anim. Sci. 49: 212-219.

Pascar, D. C. B. (1987): Aquatic oligochaeta in some tributaries of the Rio-de-la-plata, Buenos, Oires, Argentina. Hydrobiologia. 144: 125-130.

Patil, S. G., Harshey, D. K. and Singh, D. F. (1984): Benthic organism as indicators of pollution in lentic and lotic environment. Geobios. 11: 77-80.

Pennak, R. W. and Gerpen, E. D. V. (1947) Bottom fauna production and physical nature of the substrate in Northern Colorado trout stream. Ecology. 28: 42-48.

Perry, J. A. and David, J. S. (1987): The longitudinal distribution of riverine benthos. A river discontinum. Hydrobiologia. 148: 257-268.

Petridis, D. (1993): Macroinvertebrate distribution along organic pollution gradient in lake Lysimachia (West Greece). Arich. Flir. Hydrobiologia. 128: 367-389.

Pip, E. (1978): A survey of ecology and composition of submerged aquatic snail-plant communities. Can. J. Zool 56: 2263-2279.

Pratt, J. M. and Coler, R. A. (1976): A procedure for the routine biological evaluation of Urban run-off in small river. Water Res. 10: 1019-1025.

Price, D. R. H. (1978): Fish as indicators of water quality. Paper presented at the International symposium on biological indicators of water quality in New Castle.

Rai, D. N., Roy, S. P. and Sharma, U. P. (1981): Freshwater gastropod (Mollusca) community structure in relation to the macrophytes of littoral zone of a fish pond at Bhagalpur, Bihar. Indian J. Ecol. 81: 88-95.

Rajan, S. (1965): Environmental studies on Chilka lake benthic animal community. Indian J. Fish. 122: 492-499.

Raman, K. K., Ramakrishna, V., Radhakrishnan, S. and Rao, G. R. M. (1975): Studies on the Hydrology and benthic ecology of Pulicat lake. Bull. Deptt. Mar. Sci. Univ. Cochin. 7: 855-884.

Rama Rao, S. V. Singh, V. P. and Mall, L. P. (1978a): Pollution studies of river Khan (Indore), India. I. Biological assessment of pollution. Water research 12: 555-559.

Rama Rao, S. V. Singh, V. P. and Mall, L. P. (1978b): Biological methods for monitoring water pollution levels: Studies at Ujjain In.: Glimpses of Ecology. (Eds. J. S. Singh and B. Gopal). Internat. Sci. Pub. Jaipur pp. 341-348.

Ramachandran, V. and Paterson, C. G. (1978): A partial analysis of ecological segregation in the chironomid community of a Bog lake. Hydrobiologia 52: 129-135.

Rao. K. S. and Jain, S. (1985): Comparative quantitative studies on macrozoobenthic organisms in some central Indian freshwater bodies with relation to their utility in water quality monitoring. J. Hydrobiol. 12: 73-82.

Ray, P., Singh, S. B. and Sehgal, K. L. (1966): A study of some aspects of ecology of the river Ganga and Yamuna at Allahabad (U. P). in 1958-59. Proc Nat. Acad. Sci. India. Sec. B: 235-272.

Reddy, V. M. and Rao, B. M. (1987): Structure of benthic macroinvertebrates populations particularly the tubificidae and chironomid larvae in a sewage polluted Urban Canal. Pollu. Res. 6: 65-68.

Reddy, M. V. and Rao, B. M. (1989): Community structure of benthic macroinvertebrates of fishpond and sewage irrigated tanks in an Urban Ecosystem. Env. Ecol. 73: 713-716.

Reddy, M. V. and Rao, B. M. (1991): Benthic macroinvertebrates as indicators of organic pollution of aquatic ecosystems in a semi-arid tropical Urban system. In: Bioindicators and Environmental Management. Academic Press Ltd., Dublin. pp 65-77.

Reed, E. B. and Bean, G. (1966): Benthic animals and food eaten by brook trout in Archuleta Creek. Colorado. Hydrobiologia 27: 227-237.

Reynoldson, T. B. (1987): Interaction between sediment contaminants and benthic organisms. Hydrobiologia 149: 53-66.

Richardson, R. E. (1928): The bottom fauna of the middle Illinois river, 1913-1925, Its distribution, abundance, valuation and index value in the study of stream pollution. Bull. III. St. Nat. Hist. Surv. 17: 387-475.

Roy, S. P. and Dutta Munshi, J. S. (1993): Aquatic insect community structure as indicator of quality assessment of water of freshwater system of Bhagalpur (Bihar). In: Advances in Limnology (Ed. Singh, H. R). Narendra Publishing House, New Delhi. pp. 69-78.

Sagar, P. E. and Hasler, A. D. (1969): Species diversity in lacustrine phytoplankton. I. the components of the index from Shannon's formula. Amer. Nat. 103: 51-59.

Sarkar, S. K. (1989): Seasonal abundance of benthic macrofauna in a freshwater pond. Env. Ecol. 71: 113-116.

Sarkar, R and Krishnamoorthy, K. P. (1977): Biological methods for monitoring water pollution level at Nagpur. Indian. J. Environ. Hlth. 19: 132-139.

Schutte, C. H. J. and Frank, G. H. (1964): Observation on the distribution of freshwater Mollusca and chemistry of natural water in the South-Eastern Transvaal and adjacent Northern Swaziland. Bull. W. H. O 30: 398-400.

Shah, K. A. and Pandit, A. K. (2001): Macro invertebrates associated with macrophytes in various freshwater bodies of Kashmir Journal of Res. and Dev. I: 44-53.

Shanon, C. E. and Weaver, W. (1964): The mathematical theory of Communication. Univ. Illinois Press, Urbana III. pp. 125.

Shrivastava, V. (1956a): Benthic organism of a freshwater fish tank. Curr. Sci. 25: 159-169.

Shrivastava, V. (1956b): Studies on the freshwater bottom fauna of North India. Quantitative fluctuations and qualitative composition of benthic fauna in a lake in Lucknow Proc. Nat. Acad. Sci. B. 25: 406-416.

Shrivastava, V. (1959): Studies of freshwater bottom fauna-II. Quantitative composition and variation of the available food supply of fishes. Proc. Nat. Acad. Sci. India 29: 207-216.

Singh, D. K. (1988): Limnological survey of Subernarekha in and around Ranchi. Ph. D. Thesis (Patna University). pp. 210.

Singh, N. K. (1980): Systematic survey of algal flora of the river Ganges-Impact of some ecological factors. Ph. D. Thesis, Bhagalpur University, Bhagalpur.

Sinha, M. P. (1988): Studies on polluted water ecosystem. Ph. D. Thesis (unpublished) of Ranchi University.

Sinha, M. P., Mahato, B., Pandey, P. N. and Mehrotra, P. N. (1990): Impact of Coal mining and coal washing practices on macrobenthic community. In: Impact of mining on environment (Eds. R. K. Trivedi and M. P. Sinha). Ashish Pub. House, New Delhi. pp 203-226.

Sinha, M. P, Pandey, P. N. and Mehrotra, P. N. (1989): Some aspects of biological studies of organically polluted Urban stream in Ranchi. II. Macrobenthic fauna. The Indian Zoologist. 13: 79-83.

Sinha, M. P., Sinha, R. and Mahato, P. N. (1991): Composition and dynamic of freshwater macrobenthic community. I Oligochaete. Oikoassay. 8: 21-24.

Sinha, R., Ojha, N. K., Das, S. K. and Sinha, M. P. (1997): Community composition, population dynamics and species diversity of macrobenthic fauna of a tropical freshwater lenthic body. In.: Recent Advances in Ecobiological Research Vol. I (Ed. M. P. Sinha). APH Publishing Corporation, New Delhi pp. 123-174.

Sinha, R. K. and Das, N. K. (1993): Organic waste and its effect on macro-zoobenthos in Ganga at Patna (Bihar), India J. Freshwater Biol. 5: 33-40.

Southwood, T. R. E. (1977): Habitat, the templet for ecological strategies J. Am. Ecol. 46: 337-365.

Sowa, R. (1965): Ecological characteristics of the bottom fauna of the Wielka puszcza stream. Acta. Hydrobiol. 7: 61-92.

Taylor, L. R. (1961): Aggregation variance and the mean. Nature 189: 732-735.

Tebo, L. B. and Hassler, W. W (1961): Seasonal abundance of aquatic insect in western North Carolina trout streams. J. Elisha Mitchell Scient. Soc. 77: 249-259.

Timms, B. V. (1982): A study of the benthic communities of 20 lakes in the South Island, New Zealand. Freshwater Biol. 12: 123-138.

Tittizer, T. and Kothe, P. (1978): Possibilities and limitations of biological methods of water analysis. Paper presented at the International Symposium on biological indicators of water quality. New Castle.

Tonapi, G. T. (1980): Freshwater animal of India: An ecological approach. Oxford and IBH Publishing Company, New Delhi pp. 341.

Vaidya, D. P. (1979): Substratum as a factor in the distribution of two prosobranch fresh water. Hydrobiologia 65: 17-18.

Vasisht, H. S. and Bhandal, R. S. (1979): Seasonal variation of benthic fauna in some North Indian lakes and ponds. India J. Ecol 62: 33-37.

Wass, M. L. (1967): Indicators of pollution. In: Pollution indicators and marine ecology (Eds. T. A. Olson and F. J. Burgees). John Wiley and Sons. New York. pp. 271.

Waters, T. F. (1965): Interpretation of invertebrate drift in stream. Ecology. 46: 327-334.

Welch, P. C. (1948): Limnological Methods. Blakiston, Philadelphia pp. 228.

Wetzel, R. G. and Likens, G. E. (1991): Limnological Analysis. 2nd ed. Springer-Verlag, New York pp. 391.

Whitting, I. R. and Clifford, H. G. (1983): Invertebrates and Urban runoff in a small Northern stream. Edmonton (Alberta), Canada. Hydrobiologia 102: 73-80.

Wiederholm, T. (1980): Use of benthos in lake monitoring J. Wat. Poll. Control 52: 537-547.

Wilhm, R. L. and Dorris, T. C. (1968): The biological parameters for water quality criteria. Bio. Science 18: 477-492.

**Aquatic Biodiversity in India: The Present Scenario, 2005** **243–284**
*Edited by*: D.R. Khanna, A.K. Chopra & G. Prasad
*Published by*: Daya Publishing House, New Delhi

# 13

# Study of Cultural Eutrophication in Relation to Plant Diversity of Wetland: Ratheshwar in Central Gujarat

**Nirmal Kumar, J.I, Rita N. Kumar and Ira Bhatt**
*Nathu Bhai V. Patel College of Pure and Applied Sciences, Vallabh Vidyanagar – 388 120, Gujarat, India*

Wetlands serve a good source of water for various human activities but they are also receiving various contaminants in different ways.

Eutrophication is a natural process and all lakes and ponds with passage of time from a stage of oligotrophic to eutrophic through mesotrophic phases. The appearance of 'nuisance' algal blooms, discolouration of water, fish kills and the emergence of marshland or swamps are the common time lag phenomenon, besides water bodies experiencing acceleration of the natural aging process. Cultural eutrophication is the result of adding massive amounts of a wide range of contaminating substances including solid wastes, organic and in-organic nutrients and heated wastes etc.

According to Vollenweider (1968) eutrophication is seen as increasing enrichment of surface waters with plant nutrients, with all its applied consequences. Increasing nutrient supply results in accelerated productivity, affecting all compartments of limnetic system (Wetzel,1975). Wuhrmann (1976) defined eutrophication as development of an aquatic ecosystem into a state in which aerobic microbial decomposition of organic matter consumes more oxygen than is naturally introduced to the system, resulting in steady decrease of oxygen.

Phosphorus is generally considered as the decisive limiting factor of primary production in oligotrophic and mesotrophic freshwater ecosystems (Stumm and Stumm-Zollinger, 1972 Vollenweider, 1979). Nitrogen has been reported as a limiting factor, mainly in eutrophic still waters, with high amounts of available phosphorus (Goreing, 1972; Claesson and Ryding, 1977; Forsberg, 1977). Carbon may become limiting, especially in waters of low alkalinity, with excessive amounts of nitrogen and phosphorus (King, 1970; Likens, 1972). Micronutrients limitation of primary production has also been demonstrated (Goldman, 1972). Light may be considered as an ultimate factor (Wetzel and Hough, 1973).

In a eutrophic water, the total potential concentration of various nutrients are high as compared to oligotrophic waters. Increased nutrients lead to growth of algae and the Cyanophycean blooms (Morgan, 1970; Thomas, 1973). Major quantitative and qualitative changes in macrophytes, algal and bacterial populations are involved in the process of eutrophication.

The freshwater ecosystems are therefore, stressed by the addition of increasing number of foreign substances. These substances lead to far-reaching qualitative and quantitative alterations in the inhabitants and their bio-coenosis. The unwanted nuisance growth of flora is proliferated with respect to the eutrophic nutrient status. This abundant growth of weed is controlled by chemicals, which is an effective and fast method.

The wetlands which occupy a vital position in the rural areas serving the source of drinking, bathing, washing, irrigation and for various other purposes are prone to pollution threat. These wetlands receive the discharges of washing, bathing, faecal wastes from human and cattle. These results in the degradation of water quality, excessive growth of algae and aquatic weeds and hinders the beneficial utilization of these wetlands. These ponds are getting contaminated by human activities, run off water from the agricultural fields and also receive discharges from the surrounding localities. This leads to add nutrients and affects undesirable biotic and abiotic changes in water quality. The use of such water causes many water–born diseases in rural population. In an attempt to study the quality of water and its biotic components of certain Wetlands, the present study was undertaken.

Scanty literature is available on the trophic and pollution status of various wetlands. Among them Shaji (1989) and Jose (1990) have evaluated the algae as pollution indicators in running waters, where as Nirmal Kumar (1990, 1991 and 1992) and Nirmal Kumar and Rana (1989 and 1994) and Sreenivas (1991) have extensively studied the seasonality of Physico–Chemical properties of water and sediments, plankton, macrophyte diversity, density and productivity of static waters Nirmal Kumar and Rana (1994); Rana and Nirmal Kumar (1993) and Rana *et al.* (1995) Nirmal Kumar *et al.* (2002) have studied the various anthropogenic pressures in various wetlands. The present study was undertaken to assess the problems of eutrophication in relation to plant diversity of wetland Ratheshwar, Anand (Gujarat). Following were the main objectives of the present study:

1. To study the physico–chemical characters of water on monthly basis for one year.
2. Characterization of sediment on monthly basis
3. To study the periodic changes of phytoplankton population, their productivity, qualitative and quantitative changes to correlate the chemistry of water.
4. Aquatic macrophytes and their seasonal variation.

## Study Area

The Anand district lies between 72° 12'and 73° 17' east longitude and between 22° 7'and 23° 18' Latitude. The district has rather an oblong shape and more or less lies between the river Mahi (South–East) and Sabarmati (North–West). Ratheshwar Wetland is located in village Pariej, at about 45 km. South-west to Anand in Matar taluka of Anand district of Gujarat. The water body occupies about 33 hectares and depth of about 4-5 meters. The main water source of this wetland is channel of Nava talav. This water is used for drinking, washing, household and irrigation purposes by the surrounding villagers. Three stations were marked from where the surface water samples were collected during the investigation period.

## Materials and Methods

For the determination of hydrobiological characters of this wetland on monthly basis, the surface water samples were collected for only one year (June 1988 to May 1989). Three sites were selected where the surface water samples were collected throughout the investigation period. The study sites were selected so as to collect the representative samples from the different sites of the wetland. Composite surface water samples were collected by using one litre polyethylene and glass bottles. All samples were collected at monthly intervals during morning hours in first week of every month. Temperature and pH were recorded on the spot whereas to estimate the dissolved oxygen. Samples were taken in BOD bottles and immediately fixed using manganous sulphate and alkali–iodide azide solution for further analysis. For other parameters, the samples were brought to the laboratory and analyzed for chemical properties as early as possible, using the methods as suggested in APHA (1976; Trivedy and Goel (1984) and Adoni (1985).

The sediment samples were collected from and Ratheswar wetland on monthly basis from June 1988 to May 1989 for one year. The samples were collected from three respective sites from where the water samples were collected. Sediment samples were collected in polyethylene bags and transported to laboratory. The samples were mixed in equal proportions to make a single sample. A homogenous sample was prepared by grinding in mortar and pestle and was passed through 2 mm nylon sieve. Soil suspension was prepared 1: 5 (soil: water) by shaking 20 g of dried sediment in 100 ml of aerated distilled water for about one hour. Suspension was filtered through Whatman No. 50 filter paper using Buchner funnel and vacuum pump. Filtrate was used as soil sample. The chemical properties like total nitrogen, phosphorus sulphate, potassium, sodium, calcium, magnesium, chloride and organic matters were analyzed following the standard methods as suggested in (APHA,1976 and Trivedy and Goel,1984).

The monthly phytoplankton samples were simultaneously collected along with the water samples. A known quantity of water sample was concentrated by filtration through Whatman filter paper and centrifuged to make a total volume of 50 ml. The marked bottle was fixed by adding 1 ml of 4.0 per cent formaldehyde solution. The *Camera lucida* diagrams were drawn in a light microscope under different magnifications. The systematic identification of phytoplankton were carried out by using various monographs, books and published literature (Fritch, 1935; Smith, 1950; Desikachary, 1959; Prescott, 1962; Philipose, 1967; Iyengar and Desikachary, 1981; Patel, 1989).

The quantitative estimation of phytoplankton and other algae were made with the help of haemocytometer. Unicellular forms, filaments and colonies were counted as single unit, and an average of ten readings were recorded. Values per litre of original samples were computed by following formula:

$$Phytoplankton, Unit\ l^{-1} = \frac{\text{No. of phytoplankton in central chamber } \times 10^7}{\text{Concentration factor}}$$

**Species diversity index**: Species diversity index of phytoplankton was computed following (Shannon and Weavers,1949) as given in (Trivedy and Goel,1984). This index is expressed by the following formula.

$$\overline{H} = -\sum \left(\frac{ni}{N}\right) \log e \left(\frac{ni}{N}\right)$$

*Where,*

$\overline{H}$ = Index of species diversity;

Ni = Number of individuals in the species;

N = Total number of individuals.

## Chlorophyll

Biomass of phytoplankton is determined by analyzing the chlorophyll content. The chlorophyll content of phytoplankton was measured as described by (Trivedy and Goel,1984).

## The Macrophytes

Aquatic macrophytes collected in polyethylene bags for one year were brought to the laboratory and washed under tap water.

The plants were treated with 10 per cent mercuric chloride (90 per cent absolute alcohol) for one minute to withstand the fungal and bacterial infestation. Herbarium sheets were made and identified with the help of preserved herbaria. The monthly variation of aquatic macrophytes were recorded. Statistical synopsis: A multi–factorial correlation of the data was made to study the inter-relationship of the various parameters of water sediment, plankton density and chlorophyll content. The mean and standard deviation were calculated for showing general statistical relations. Data were computer processed at the department of Computer Science, Sardar Patel University, Vallabh Vidyanagar.

## Results

The physico–chemical characters of the water of Wetland–Ratheshwar for the site one for the year 1988-89 have been shown in Fig. 13.1.The data for sites two and three and the average values for the three sites for the year 1988-89 have been tabulated in Tables 13.1–13.3.

Temperature is one of the vital factor which controls the biotic nature of tropical and sub-tropical aquatic ecosystems both in lentic and lotic waters. The temperature of wetland water varied from 18.5° to 29.5° c. The maximum temperature (29.5°c) was registered during June, whereas January recorded the lowest temperature (18.5°c). The pH of water samples ranged from 7.1 to 9.2 for the study period. The maximum pH (9.2) was observed in the month of June, while January recorded the lowest pH (7.1). The seasonal variation indicates the lowest pH (7.1). The seasonal variation indicates that summer season showed higher values, whereas winter has lower values of pH.

The total alkalinity of the wetland water varied from a minimum of 110 $mgl^{-1}$ to maximum of 335 $mg\ l^{-1}$. June 1988 recorded the highest values of total alkalinity, while lower value was observed in January 1989. The average value indicates that site one was more alkaline than site two and three. Total solids varied from a maximum of 895 $mgl^{-1}$ and a minimum of 190 $mg\ l^{-1}$. Higher values were found in summer, whereas lower values were found in winter. The total solids were more at sample site one as compared to other sampling sites.

The carbonates varied from negligible to a maximum of 136 $mgl^{-1}$. The higher values were found in June, while lower values

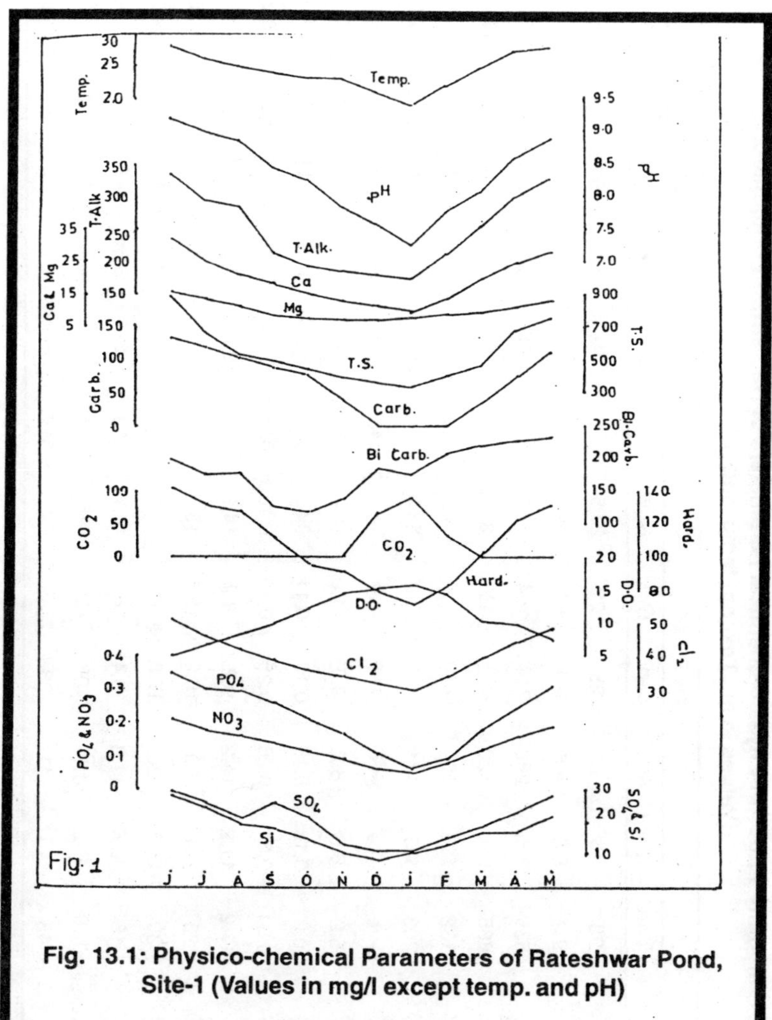


**Fig. 13.1: Physico-chemical Parameters of Rateshwar Pond, Site-1 (Values in mg/l except temp. and pH)**

were observed in winter. The average bicarbonate values were registered more at site one compared to other sites. It varied from 112 to 231 mgl$^{-1}$ for the year 1988-89. Higher bicarbonates were recorded in summer, while lower values were noticed in winter.

**Table 13.1: Monthly Values of Physico-chemical Parameters of Rateshwar Pond, Site-2 (Values in mg/l except temp. and pH)**

| Sl.No. | Parameters | 1988 | | | | | | | 1989 | | | | |
|---|---|---|---|---|---|---|---|---|---|---|---|---|---|
| | | *June* | *July* | *Aug.* | *Sep.* | *Oct.* | *Nov.* | *Dec.* | *Jan.* | *Feb.* | *March* | *April* | *May* |
| 1. | Temperature | 28 | 26 | 27 | 25 | 23 | 23.5 | 21 | 19 | 22 | 24.5 | 26.5 | 27 |
| 2. | pH | 9.2 | 9.0 | 8.6 | 8.4 | 8.15 | 7.e5 | 7.75 | 7.2 | 7.7 | 8.1 | 8.5 | 9 |
| 3. | Total alkalinity | 335 | 295 | 280 | 214 | 194 | 186 | 155 | 110 | 205 | 250 | 295 | 320 |
| 4. | Total solids | 885 | 640 | 520 | 480 | 430 | 395 | 350 | 320 | 380 | 440 | 645 | 765 |
| 5. | Carbonate | 132 | 118 | 104 | 88 | 72 | 42 | –ve | –ve | –ve | 31 | 64 | 99 |
| 6. | Bi-carbonate | 203 | 177 | 171 | 126 | 122 | 144 | 155 | 110 | 105 | 221 | 231 | 210 |
| 7. | Free $CO_2$ | –ve | –ve | –ve | –ve | –ve | –ve | 64.2 | 96 | 34.6 | –ve | –ve | –ve |
| 8. | Dissolved oxygen | 5.2 | 6.6 | 8.2 | 9.6 | 11.e | 14 | 15.2 | 16.4 | 13.2 | 10.6 | 9.2 | 7.6 |
| 9. | Nitrate | 0.2 | 0.18 | 0.15 | 0.13 | 0.11 | 0.09 | 0.07 | 0.06 | 0.08 | 0.11 | 0.14 | 0.17 |
| 10. | Phosphate | 0.34 | 0.31 | 0.28 | 0.25 | 0.20 | 0.16 | 0.10 | 0.05 | 0.09 | 0.19 | 0.28 | 0.32 |
| 11. | Silica | 17.2 | 11.2 | 14.2 | 9.2 | 8.2 | 14.2 | 13.2 | 9.2 | 7.6 | 11.2 | 13.2 | 15.2 |
| 12. | Sulphate | 28 | 24 | 20 | 18.2 | 15.6 | 13.2 | 11.2 | 10 | 13.8 | 17.6 | 21.6 | 26 |
| 13. | Calcium | 30 | 25.5 | 20 | 17.6 | 14.5 | 12 | 10.5 | 9.5 | 11.2 | 18 | 21 | 27 |
| 14. | Magnesium | 16 | 13.6 | 11.8 | 10.4 | 14.6 | 8.9 | 7.3 | 6 | 8.3 | 9.8 | 11.4 | 13.6 |
| 15. | Hardness | 140 | 132 | 124 | 108 | 92 | 86 | 78 | 70 | 82 | 98 | 120 | 132 |
| 16. | Chloride | 48 | 45 | 41 | 38 | 35 | 33 | 31.5 | 26.4 | 31 | 37 | 42 | 45 |

**Table 13.2: Monthly Values of Physico-chemical Parameters of Rateshwar Pond, Site-3 (Values in mg/l except temp. and pH)**

| Sl.No. | Parameters | 1988 | | | | | | | 1989 | | | | |
|---|---|---|---|---|---|---|---|---|---|---|---|---|---|
| | | June | July | Aug. | Sep. | Oct. | Nov. | Dec. | Jan. | Feb. | March | April | May |
| 1. | Temperature | 29.5 | 27.5 | 26 | 25 | 24.5 | 23 | 20 | 18.5 | 21 | 23 | 25 | 26 |
| 2. | pH | 8.9 | 8.7 | 8.4 | 8.1 | 7.6 | 7.4 | 7.25 | 7.1 | 7.4 | 7.9 | 8.3 | 8.6 |
| 3. | Total alkalinity | 302 | 251 | 209 | 1B9 | 164 | 151 | 139 | 122 | 112 | 148 | 189 | 249 |
| 4. | Total solids | 630 | 580 | 450 | 380 | 290 | 245 | 210 | 190 | 290 | 355 | 495 | 575 |
| 5. | Carbonate | 102 | 86 | 70 | 56 | 38 | –ve | –ve | –ve | –ve | 24 | 32 | 78 |
| 6. | Bi-carbonate | 200 | 165 | 139 | 133 | 126 | 151 | 139 | 122 | 112 | 124 | 157 | 171 |
| 7. | Free $CO_2$ | –ve | –ve | –ve | –ve | –ve | 42 | 68 | 85 | 38 | –ve | –ve | –ve |
| 8. | Dissolved oxygen | 6 | 8.4 | 10.2 | 13 | 15.2 | 16.8 | 17.2 | 18.6 | 17.2 | 11.8 | 9.8 | 7.8 |
| 9. | Nitrate | 0.11 | 0.09 | 0.07 | 0.05 | 0.03 | 0.03 | 0.02 | 0.01 | 0.02 | 0.05 | 0.08 | 0.1 |
| 10. | Phosphate | 0.10 | 0.08 | 0.09 | 0.07 | 0.06 | 0.04 | 0.03 | 0.02 | 0.04 | 0.05 | 0.07 | 0.01 |
| 11. | Silica | 11.2 | 7.6 | 4.2 | 6.2 | 7.2 | 9.2 | 5.4 | 6.4 | 8.2 | 4.4 | 5.6 | 9.2 |
| 12. | Sulphate | 21 | 18.5 | 15.2 | 12 | 10.5 | 6.8 | 7.6 | 6 | 9.4 | 13.5 | 16.8 | 18.1 |
| 13. | Calcium | 14.6 | 12.4 | 9.6 | 8.1 | 7.5 | 6.9 | 5.5 | 4.3 | 6.3 | 8.6 | 10.4 | 13.1 |
| 14. | Magnesium | 10.8 | 9.8 | 9.1 | 8.6 | 7.2 | 6.7 | 5.4 | 4.1 | 6.4 | 8.8 | 9.4 | 9.9 |
| 15. | Hardness | 95 | 91 | 86 | 72 | 61 | 53 | 49 | 43 | 52 | 59 | 78 | 88 |
| 16. | Chloride | 33 | 31 | 28 | 25 | 23 | 20 | 17 | 14 | 18 | 23 | 28.4 | 32 |

**Table 13.3: Average Values of Physico-chemical Parameters of Three Sites of Rateshwar Pond, with Standard Deviation (Values in mg/l except temp. and pH)**

| Sl.No. | Parameters | 1988–89 | | | | | |
|---|---|---|---|---|---|---|---|
| | | Site 1 | | Site 2 | | Site 3 | |
| | | Average | S.D. | Average | S.D. | Average | S.D. |
| 1. | Temperature | 24.04 | 2.48 | 24.33 | 2.63 | 24.08 | 3.02 |
| 2. | pH | 8.29 | 0.58 | 8.28 | 0.57 | 7.97 | 0.59 |
| 3. | Total alkalinity | 246.33 | 56.61 | 236.58 | 67.14 | 185.41 | 55.62 |
| 4. | Total solids | 529.16 | 166.85 | 520.83 | 169.18 | 390.83 | 146.42 |
| 5. | Carbonate | 65.58 | 46.81 | 62.58 | 45.56 | 40.50 | 35.68 |
| 6. | Bi–carbonate | 164.58 | 53.92 | 173.33 | 39.71 | 144.91 | 24.00 |
| 7. | Free $CO_2$ | 16.23 | 30.3 | 16.23 | 30.78 | 19.41 | 29.62 |
| 8. | Dissolved oxygen | 10.68 | 3.55 | 10.65 | 3.36 | 12.66 | 4.09 |
| 9. | Nitrate | 0.13 | 0.04 | 0.12 | 0.04 | 0.05 | 0.03 |
| 10. | Phosphate | 0.21 | 0.08 | 0.21 | 0.19 | 0.06 | 0.02 |
| 11. | Silica | 18.55 | 6.3 | 11.98 | 2.89 | 7.06 | 2.02 |
| 12. | Sulphate | 19.07 | 5.99 | 1B.26 | 5.57 | 13.19 | 4.66 |
| 13. | Calcium | 19.36 | 6.85 | 18.06 | 6.59 | 8.01 | 1.95 |
| 14. | Magnesium | 10.39 | 2.58 | 10.06 | 2.95 | 8.95 | 3.04 |
| 15. | Hardness | 107.91 | 23.14 | 105.16 | 22.9B | 69.75 | 17.26 |
| 16. | Chloride | 39.58 | 6.30 | 37.14 | 6.34 | 24.33 | 6.24 |

The free-carbon dioxide was found only during the winter (November to January). High free–$CO_2$ was noticed at site two compared to sites one and three. The dissolved oxygen (D.O) –one of the vital hydrobiological factor, ranged from 5.0 to 18.0 $mgl^{-1}$. The peak value of D.O. was noticed during January 1989, whereas lower value was recorded in June 1988. An average value of D.O was higher at site three compared to sites one and two. The interesting point is that seasonal variation of highest D.O. and free-carbon dioxide was recorded during winter season.

The nitrate, one of the important nutrient for the growth of phytoplankton, varied from 0.021 and 0.212 $mgl^{-1}$. The maximum amount of nitrate was observed at station two and the minimum nitrate was found at station three. The seasonal variation indicates that summer season favours the higher concentration of nitrate. The phosphate content ranged from 0.018 to 0.35 $mgl^{-1}$. The highest phosphate content was found in June 1988. The higher values of phosphate were noticed at site one compared to sites two and three.

The sulphate in the pond water ranged from 6 to 29.1 $mgl^{-1}$. The peak values were occurred in the month of June, while lower average value of sulphate were noticed in the month of January. The average value of sulphate was higher at site one followed by sites two and three. The silica content varied from 5.4 to 29.8 $mgl^{-1}$. Higher silica was recorded during June of summer season, while lowest silica was noticed during January 1989 of winter season. Stations one and two were found as highest silicate sites compared to site three.

The calcium content of wetland water ranged from 4.3 to 32.0 $mgl^{-1}$ for the 1988-89. The higher quantity of calcium was found in the summer season, while lower quantity was found in the winter season. The concentration of magnesium varied considerably in all sites. Minimum value of 4.1 $mgl^{-1}$ at station three, while the maximum value of 16 $mgl^{-1}$ at site one were observed. The higher and lower amounts of magnesium were found in summer and winter seasons, respectively.

Hardness of water of all the three study sites exhibited a minimum of 43 $mg1^{-1}$ and a maximum of 142 $mgl–^{1}$ at sites three and one, respectively. The seasonal variation denotes that in summer season higher hardness was observed as compared to winter season. The chloride varied from a maximum of 51 $mg1^{-1}$ to a minimum of 14.2 $mg1^{-1}$ during the year. The higher chloride was obtained during summer, while lower concentration was found during winter season.

The Table 13.4 shows the chemical composition of the Ratheshwar wetland sediments. It was observed that maximum amount of all the nutrients except organic matter was found in the month of January. However the amount of organic matter was maximum in June. The maximum total per cent composition of the nutrients showed the following pattern:

$$Ca > Mg > O.M. > T.N. > K > Na > Cl > P > SO_4$$

While cation pattern exhibited as follows:

$$Ca > Mg > K > Na$$

The sediment composition of Ratheshwar wetland is rich in cations like calcium and magnesium. However organic matter was also more in the sediment.

## Phytoplankton

The phytoplankton association not only reflects the environmental conditions, but can also be used as indicators of the aquatic ecosystems. In the present decade, several studies have been made using phytoplankton as indicators of surface waters. In the current investigation, phytoplankton composition, variation, density and diversity were studied to know the trophic status of water bodies.

The composition and monthly variation of phytoplankton in the Ratheshwar wetland of the year 1988-89 has been shown in Tables 13.5 and 13.6. In all 29 species belonging to 26 genera were recorded. The percentage of groups and their density have been plotted in Figs. 13.2 and 13.3. The total density of plankton ranged from 1.7 to $20.5 \times 19^7$ $ul^{-1}$. The maximum density of phytoplankton was recorded in June, while July recorded the lowest numbers.

## Cyanophyceae

This group from density point of view occupied third position among the planktonic groups. It was represented by five species belonging to four genera. It constitute only 5.8 to 21 per cent of total plankton population. Maximum quantity ($3.0 \times 10^7$ $\mu l^{-1}$) was recorded in June, while the lowest ($1.0 \times 10^6$ $\mu l^{-1}$) quantity was found in July. *Oscillatoria* and *Merismopedia* were recorded throughout the study period, while *Microcystis* disappeared during rains.

**Table 13.4: Sediment Composition of Ratheshwar Pond (per cent dry weight)**

| Sl.No. | Parameters | 1988 | | | | | | | 1989 | | | | |
|---|---|---|---|---|---|---|---|---|---|---|---|---|---|
| | | June | July | Aug. | Sep. | Oct. | Nov. | Dec. | Jan. | Feb. | March | April | May |
| 1. | Sulphate | 0.03 | 0.05 | 0.06 | 0.08 | 0.09 | 0.11 | 0.13 | 0.15 | 0.13 | 0.10 | 0.07 | 0.05 |
| 2. | Phosphate | 0.08 | 0.09 | 0.11 | 0.12 | 0.16 | 0.18 | 0.21 | 0.22 | 0.20 | 0.16 | 0.12 | 0.09 |
| 3. | Total nitrogen | 0.06 | 0.91 | 1.34 | 1.56 | 0.82 | 2.01 | 2.36 | 2.76 | 2.21 | 1.96 | 1.68 | 1.06 |
| 4. | Sodium | 0.21 | 0.29 | 0.36 | 0.48 | 0.54 | 0.61 | 0.69 | 0.76 | 0.70 | 0.58 | 0.49 | 0.32 |
| 5. | Potassium | 0.74 | 0.82 | 0.96 | 1.04 | 1.10 | 1.22 | 1.26 | 1.36 | 1.24 | 1.11 | 0.96 | 0.86 |
| 6. | Calcium | 8.09 | 8.30 | 8.62 | 8.91 | 8.99 | 9.18 | 9.34 | 9.56 | 9.24 | 8.98 | 8.71 | 8.59 |
| 7. | Magnesium | 6.91 | 6.97 | 7.11 | 7.32 | 7.55 | 7.71 | 7.82 | 7.92 | 7.78 | 7.64 | 7.32 | 7.04 |
| 8. | Organic matter | 5.69 | 5.26 | 5.14 | 4.78 | 4.40 | 4.25 | 4.11 | 4.08 | 4.65 | 4.91 | 5.28 | 5.34 |
| 9. | Chloride | 0.36 | 0.39 | 0.47 | 0.45 | 0.53 | 0.58 | 0.61 | 0.64 | 0.59 | 0.51 | 0.46 | 0.41 |

**Table 13.5: Monthly Variation of Phytoplankton in Ratheshwar Pond**

| Sl.No. | Phytoplankton | 1988 | | | | | | | 1989 | | | | |
|---|---|---|---|---|---|---|---|---|---|---|---|---|---|
| | | June | July | Aug. | Sep. | Oct. | Nov. | Dec. | Jan. | Feb. | March | April | May |
| **Cyanophyceae** | | | | | | | | | | | | | |
| 1. | *Microcystic aeruqinosa* | + | – | – | – | + | + | + | + | + | + | + | + |
| 2. | *M. marginata* | + | – | – | – | + | + | + | + | – | – | + | + |
| 3. | *Merismopedia minima* | + | + | + | + | + | + | + | + | + | + | + | + |
| 4. | *Arthrospira platensis* f. *granulata* | + | – | – | – | + | + | + | – | – | – | + | + |
| 5. | *Oscillatoria subtillisima* | + | – | + | + | + | + | + | + | + | + | + | + |
| **Chlorophyceae** | | | | | | | | | | | | | |
| 6. | *Pandorina morum* | + | + | + | – | + | + | + | + | + | + | + | + |
| 7. | *Schroederia setigera* | – | + | + | – | – | + | + | + | + | – | + | + |
| 8. | *Gollenkinia radiata* | + | – | – | – | – | + | + | + | – | + | + | + |
| 9. | *Micractinium pusillum* | + | – | – | – | + | + | + | + | – | + | + | + |
| 10. | *Pediastrum simplex* | + | + | – | – | – | + | + | + | + | + | – | – |
| 11. | *Tetradron tricoum* | + | + | – | – | – | + | + | + | + | – | + | + |
| 12. | *Dictyospaerium pulchellum* | + | + | + | – | – | + | + | + | – | + | + | + |
| 13. | *Actinastrum hantzschii* | + | – | – | – | + | + | + | + | – | – | + | + |
| 14. | *Tetrastrum triconum* | + | + | – | – | + | + | + | + | – | – | + | + |
| 15. | *Scenedesmus abundans* | + | – | – | – | + | + | + | – | + | + | + | + |

*Contd...*

**Table 13.5–Contd...**

| *Sl.No.* | *Phytoplankton* | *1988* | | | | | | | *1989* | | | | |
|---|---|---|---|---|---|---|---|---|---|---|---|---|---|
| | | *June* | *July* | *Aug.* | *Sep.* | *Oct.* | *Nov.* | *Dec.* | *Jan.* | *Feb.* | *March* | *April* | *May* |
| 16. | *S. dimorphous* | + | – | + | + | + | + | + | + | – | – | + | + |
| 17. | *Cosmarium subtumidum* | + | – | – | – | + | + | + | + | – | – | + | + |
| 18. | *Closterium* sp. | + | – | – | – | + | + | + | + | – | – | + | + |
| 19. | *Eusatrum* sp. | + | – | – | – | + | + | + | + | – | – | + | + |
| 20. | *Spirogyra* sp. | + | + | – | – | – | + | + | + | + | – | – | + |
| 21. | *Zygnema* sp. | + | – | – | – | – | + | + | + | + | – | – | + |
| 22. | *Oedogonium* sp. | + | – | – | – | + | + | + | + | – | – | + | + |
| 23. | *Chara* sp. | + | – | – | – | – | – | + | + | – | + | + | + |
| **Bacillariophyceae** | | | | | | | | | | | | | |
| 24. | *Fragillaria* | + | + | + | + | + | + | + | + | + | + | + | + |
| 25. | *Gomophonema* | + | + | – | + | + | + | + | – | – | + | + | + |
| 26. | *Nitzschia gracillis* | + | + | + | + | + | + | + | + | + | + | + | + |
| 27. | *Pinnularia* sp. | + | + | + | + | + | + | + | + | + | + | + | + |
| **Euglenophyceae** | | | | | | | | | | | | | |
| 28. | *Euglena acus* | + | + | – | – | + | + | + | + | – | – | + | + |
| 29. | *Phacus caudatus* | + | – | – | – | + | + | + | – | – | + | + | + |

**Table 13.6: Phytoplankton Population in Rateshwar Pond for the Year 1988-89 (x $10^7$ μ/l)**

| Sl.No. | Phytoplankton | 1988 | | | | | | | 1989 | | | | |
|---|---|---|---|---|---|---|---|---|---|---|---|---|---|
| | | *June* | *July* | *Aug.* | *Sep.* | *Oct.* | *Nov.* | *Dec.* | *Jan.* | *Feb.* | *March* | *April* | *May* |
| **Cyanophyceae** | | | | | | | | | | | | | |
| 1. | *Microcystis* | 1.2 | – | – | – | 0.4 | 0.3 | 0.8 | 0.7 | 0.2 | 0.4 | 0.6 | 0.9 |
| 2. | *Merismopedia* | 0.7 | 0.1 | 0.1 | 0.2 | 0.1 | 0.1 | 0.2 | 0.1 | 0.3 | 0.2 | 0.4 | 0.6 |
| 3. | *Oscillatoria* | 0.8 | – | 0.3 | 0.2 | 0.1 | 0.1 | 0.4 | 0.2 | 0.3 | 0.4 | 0.7 | 0.7 |
| 4. | *Arthrospira* | 0.3 | – | – | – | 0.2 | 0.1 | 0.3 | – | – | – | 0.2 | 0.1 |
| **Chlorophyceae** | | | | | | | | | | | | | |
| 5. | *Pandorina* | 0.9 | 0.1 | 0.2 | – | 0.3 | 0.2 | 0.4 | 0.3 | 0.1 | 0.4 | 0.7 | 0.8 |
| 6. | *Pediastrum* | 0.7 | 0.1 | 0.2 | – | – | 0.2 | 0.4 | 0.1 | 0.4 | 0.2 | 0.5 | 0.9 |
| 7. | *Dictyospaerium* | 0.5 | 0.1 | 0.2 | – | – | 0.1 | 0.2 | 0.1 | – | 0.4 | 0.1 | 0.3 |
| 8. | *Golenkinia* | 0.5 | – | 0.1 | 0.2 | – | 0.1 | 0.2 | 0.1 | – | 0.2 | 0.3 | 0.3 |
| 9. | *Micractinium* | 0.3 | – | – | – | 0.2 | 0.1 | 0.1 | 0.2 | – | 0.1 | 0.2 | 0.2 |
| 10. | *Schroederia* | 0.4 | – | – | – | – | 0.1 | 0.2 | 0.1 | – | 0.1 | 0.2 | 0.3 |
| 11. | *Scenedesmus* | 1.4 | 0.2 | 0.4 | 0.3 | 0.5 | 0.4 | 0.6 | 0.2 | 0.4 | 0.6 | 0.9 | 1.2 |
| 12. | *Actinastrum* | 0.1 | – | – | – | 0.3 | 0.2 | 0.5 | 0.1 | – | – | 0.1 | 0.2 |

*Contd...*

**Table 13.6–Contd...**

| Sl.No. | Phytoplankton | 1988 | | | | | | | 1989 | | | | |
|---|---|---|---|---|---|---|---|---|---|---|---|---|---|
| | | June | July | Aug. | Sep. | Oct. | Nov. | Dec. | Jan. | Feb. | March | April | May |
| 13. | *Tetrastrum* | 0.5 | 0.2 | – | – | 0.1 | 0.2 | 0.3 | 0.2 | – | – | 0.3 | 0.4 |
| 14. | *Tetraedron* | 0.5 | 0.1 | – | – | – | 0.1 | 0.3 | 0.2 | 0.2 | – | 0.3 | 0.8 |
| 15. | *Cosmarium* | 0.6 | – | – | – | 0.2 | 0.4 | 0.8 | 0.1 | – | – | 0.1 | 0.2 |
| 16. | *Closterium* | 0.4 | – | – | – | 0.1 | 0.2 | 0.4 | 0.1 | – | 0.1 | 0.2 | 0.3 |
| 17. | *Euastrum* | 0.6 | – | – | – | 0.2 | 0.4 | 0.0 | 0.3 | – | 0.2 | 0.3 | 0.5 |
| **Bacillariophyceae** | | | | | | | | | | | | | |
| 18. | *Nitzschia* | 1.7 | 0.1 | 0.5 | 0.9 | 1.1 | 1.4 | 1.2 | 0.6 | 0.3 | 0.7 | 1.1 | 1.3 |
| 19. | *Fragillaria* | 2.6 | 0.2 | 0.9 | 1.2 | 1.5 | 1.9 | 1.1 | 0.9 | 1.7 | 1.4 | 1.8 | 2.1 |
| 20. | *Gomphonema* | 2.9 | 0.2 | – | 0.5 | 1.1 | 1.6 | 1.2 | – | – | 0.7 | 1.4 | 2.1 |
| 21. | *Pinnularia* | 1.9 | 0.2 | 0.8 | 1.0 | 0.7 | 1.2 | 1.7 | 0.2 | 0.9 | 1.2 | 0.9 | 1.5 |
| **Euglenophyceae** | | | | | | | | | | | | | |
| 22. | *Euglena* | 0.6 | 0.1 | 0.2 | 0.2 | 0.3 | 0.4 | 0.3 | 0.2 | 0.2 | – | 0.3 | 0.5 |
| 23. | *Phacus* | 0.3 | – | 0.1 | 0.1 | 0.2 | 0.1 | 0.1 | 0.2 | 0.1 | 0.2 | 0.1 | 0.3 |
| | **Total Count** | **20.5** | **1.7** | **3.9** | **4.6** | **7.5** | **9.9** | **12.5** | **5.2** | **4.1** | **7.5** | **11.7** | **16.0** |

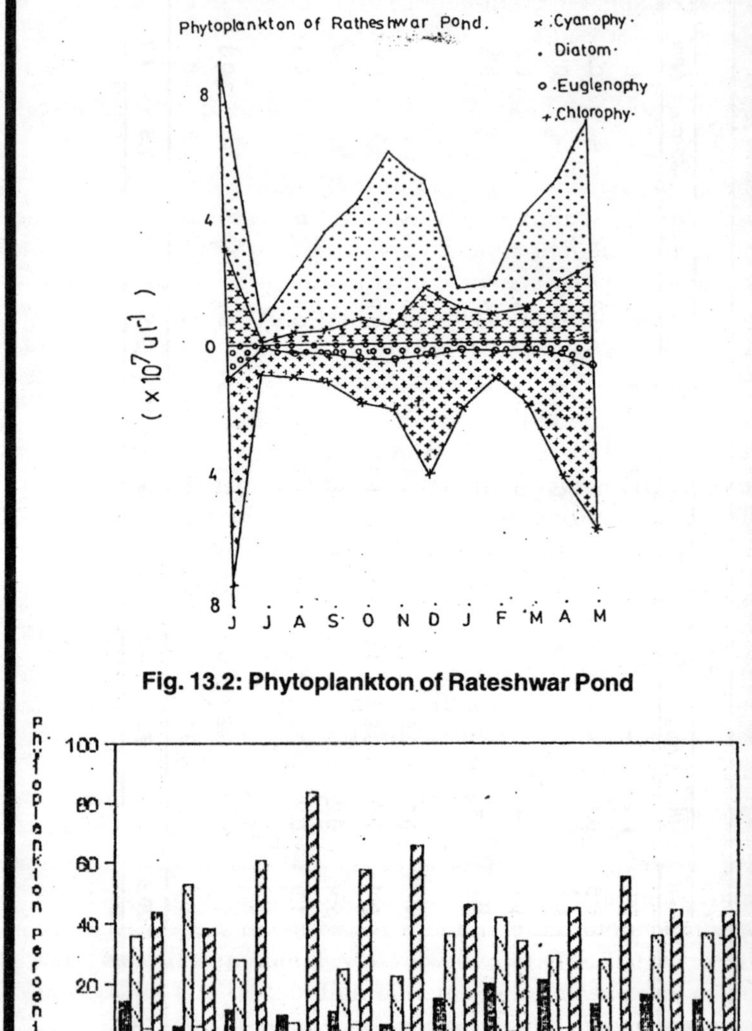


Fig. 13.2: Phytoplankton of Rateshwar Pond



Fig. 13.3: Rateshwar Pond

## Chlorophyceae

The members of chlorophyceae though present throughout the study period occupied second position and represent 6.9 to 52.9 per cent of the total population. This group represented by 18 species belonging to 16 genera in the year 1988-89. The largest number of Chlorophyceae were found in the months of May-June and lowest in July and September, respectively. The major bulk was contributed by *Scenedesmus* followed by *Pandorina* and *Pediastrum*. The filamentous algae such as *Spirogyra, Chara, Oedogonium* and *Zygnema* were also found during winter months.

## Bacillariophyceae

The diatoms constituted 34 to 83.7 per cent of the total phytoplankton population. The highest peak was observed in the month of June [9.0 × 10 $\mu l^{-1}$], however, population declined immediately thereafter in the month of July when a minimum of population [7.0 × $^{10}$ $\mu l^{-1}$ ] was recorded. It recorded four genera, *Fragillaria, Nitzschia, Gomphonema* and *Pinnularia* with one species each.

## Euglenophyceae

This group was represented by *Euglena acus* and *Phacus caudatus* and contributed nil to 6.5 per cent only to the total phytoplankton population. The maximum population was recorded in June, while rainy months has poor growth.

## Chlorophyll Content

The total biomass of phytoplankton can be measured by extracting the chlorophyll content of the plankton. To determine the phytoplankton biomass, chlorophyll content was extracted from water samples. Table 13.7 shows the chlorophyll-a, chlorophyll-b and total chlorophyll of Ratheshwar wetland for the year 1988-89. The total chlorophyll content of phytoplankton has been plotted in Fig. 13.4. This wetland recorded highest amount of chlorophylls in the month of June followed by December, while July recorded the lowest amount of these pigments.

## Diversity Index

The Shannon's diversity index of the wetland is shown in Table 13.6. Species diversity index ranged from 1.1 to 2.35 in

**Table 13.7: Monthly Variation of Chlorophyll Content of Ratheshwar Pond (mg/m³)**

| Site | Pigment | 1988 | | | | | | | 1989 | | | | |
|---|---|---|---|---|---|---|---|---|---|---|---|---|---|
| | | *June* | *July* | *Aug.* | *Sep.* | *Oct.* | *Nov.* | *Dec.* | *Jan.* | *Feb.* | *March* | *April* | *May* |
| I | Chl. a | 0.161 | 0.019 | 0.045 | 0.059 | 0.071 | 0.098 | 0.121 | 0.048 | 0.029 | 0.075 | 0.098 | 0.129 |
| | Chl. b | 0.089 | 0.012 | 0.025 | 0.034 | 0.050 | 0.058 | 0.064 | 0.026 | 0.018 | 0.046 | 0.064 | 0.074 |
| | Total chl. | 0.250 | 0.028 | 0.070 | 0.093 | 0.121 | 0.156 | 0.185 | 0.074 | 0.047 | 0.121 | 0.162 | 0.203 |
| II | Chl. a | 0.134 | 0.014 | 0.041 | 0.054 | 0.062 | 0.072 | 0.085 | 0.037 | 0.028 | 0.074 | 0.092 | 0.114 |
| | Chl. b | 0.083 | 0.005 | 0.023 | 0.032 | 0.043 | 0.051 | 0.060 | 0.019 | 0.016 | 0.034 | 0.056 | 0.072 |
| | Total chl. | 0.217 | 0.019 | 0.064 | 0.087 | 0.106 | 0.123 | 0.146 | 0.059 | 0.044 | 0.108 | 0.148 | 0.186 |

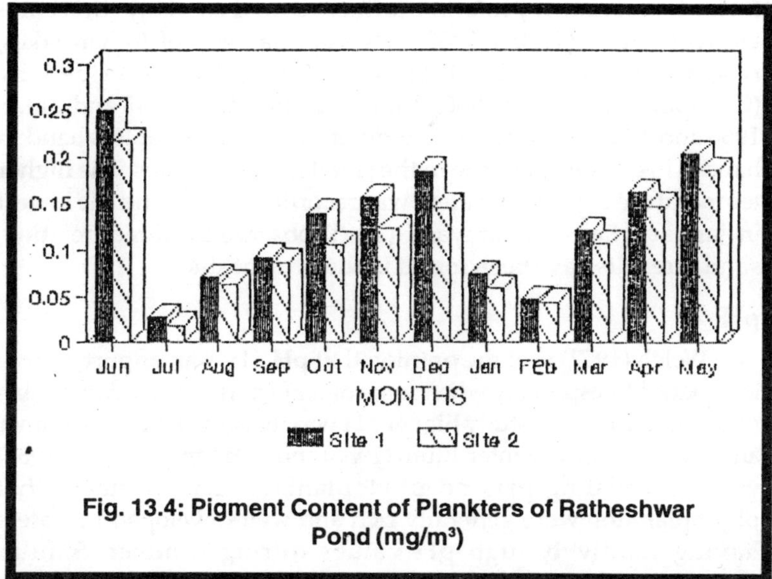


**Fig. 13.4: Pigment Content of Plankters of Ratheshwar Pond (mg/m³)**

Ratheshwar wetland. The maximum diversity index values were observed in July, whereas minimum diversity index values were found in June. The diversity index values showed monthly variation.

## Discussion

### Physico-chemical Characterization of Water

The quality of physico-chemical and biological characterizations are index to provide a complete and reliable picture of the conditions prevailing in the water bodies. The assessment of water quality lies on the delicate interface between physics, chemistry and biology. Eutrophicated or nutrient enriched water bodies are nothing but polluted water bodies in a more correct sense. The phenomenon invariably leads to change in water quality and phytoplankton, particularly their types. The most important parameters found for eutrophication have been the phosphorus and nitrogen (Schelaske, 1979). The results of physico–chemical and biological characterization of the wetland have been discussed here.

### Temperature

The temperature is generally regarded as one of the important factor in the aquatic ecology (Ruttner, 1953) and no other single factor has so much profound influences and so many direct and indirect effects (Welch, 1952). Thermal patterns of Indian lakes, reservoirs, tanks and Wetlands have been described by (Ganapati,1960; Singh,1960,. Munawar and Zafar,1967 and Zafar, 1966 and 1968). Water and air temperature go more or less hand in hand. This is specially true to the smaller water bodies. The higher temperature of surface waters was recorded in June and the lowest in January. Lakshminarayana (1965) observed a direct relation, similar results have been found in the present study.

## pH

Welch (1952) is of the opinion that pH plays an important role in aquatic life, especially when some other factors are in unfavourable range. In the present study higher pH was registered during summer and low pH during winter. Blum (1956) and Lakshminarayana (1965) while correlating pH and phytoplankton density stated that phytoplankton were generally rich and well-developed in waters having relatively high pH values during summer. Similar observations are also recorded in present study.

## Total Solids

The total solids in the present investigation showed appreciable seasonal variation. The high content of total solids during May and June in Ratheshwar ponds are probably affected by nutrients and phytoplankton density.

## Total Alkalinity

The total alkalinity may be due to carbonate and bi-carbonates only. Hardness of water depends on total alkalinity as also reported by (Barnett,1953 and Moyle,1940).The present study indicates that the wetland is of hard type. Higher values of total alkalinity were recorded during the summer season. The summer peak was attained in the month of May-June.

## Carbonate and Bi-carbonate

Zafar (1966) found higher quantities of carbonates during the summer, which is coinciding with the current investigation. The amount of bi-carbonate in the pond shows similar pattern, in that higher concentration was recorded during summer season. Rao (1972) also found high bi-carbonate content in summer. A direct relationship with carbonate, pH, temperature and other nutrients have been observed while D.O. and free $CO_2$ show inverse relationship.

## Free-Carbon Dioxide

The free-carbon dioxide was always recorded more during January and less during June in the study wetland (Rao *et al.*, 1982 and Unni, 1984). However, Welch (1952) found high free–$CO_2$ amount in summer and stated its occurrence due to accelerated rate of organic matter decomposition. Free-carbon dioxide appears to be an important component of the buffering system of water bodies. Free–$CO_2$ shows a positive correlation with dissolved oxygen and inverse relationship with all other nutrients.

## Dissolved Oxygen

The dissolved oxygen is one of the important parameter to know the health of water bodies. It has been found that polluted or eutrophic waters usually have low concentration of oxygen, indicating rapid utilization of oxygen by various biotic components. During the present investigation, Ratheshwar recorded highest D.O. The concentration of D.O. was always more during January–February and less during June. Our results support the observations of (Zutshi and Vas,1984and Hegde,1985).

## Nitrate

In the present investigation it was observed that nitrate content increased during summer, whereas declined during winter. Verma and Shukla (1970), Khan and Zutshi (1985) and Paramasivam and Sreenivasan (1981) also observed an almost similar trend of seasonal variation of nitrate. Statistical analysis showed that nitrate exhibited a significant positive correlation with most of the factors (parameters) and an inverse correlation with free-$CO_2$ and D.O.

## Phosphate

Of the elements present in living organisms ecologically phosphorus is quite important, because the ratio of phosphorus to other elements in organisms tends to be considerably greater than the ratio in the primary sources of the biological elements. A deficiency of phosphorous is therefore, more likely to limit the plant productivity of any region of the earth's plant than is a deficiency of any other materials except water (Hutchinson, 1957).

In the trophic status of freshwaters, phosphate is one of the vital constituent to monitor the plankton growth. Higher values of orthophosphates were found during summer and lower values during winter season. Similar trend of seasonal variation was observed by (Qasim *et al.*,1969 and Saran and Adoni,1984). The higher values during summer season could be ascribed to a higher rate of microbial decomposition of macrophytes and phytoplankton at high temperature (summer). The low amount of phosphates during winter season, might be due to the uptake of phosphates by developing flora and a slow rate of decomposition. Significant positive correlation of phosphate was observed with nitrate, carbonate, temperature and inverse correlation with free-$CO_2$ and D.O. during the study period.

Nitrates and phosphates are known to stimulate the autotrophic productivity. Vollenweider (1968) demonstrated that the amount of total phosphorus generally increase with lake productivity. Qasim *et al.* (1969) and Khan and Siddique (1971) have found a significant positive correlation between phosphate and productivity, which is in agreement with the present investigation. Schindler *et al.* (1971) concluded from the nutrient studies on small lakes of various types that phosphorus often act as a limiting nutrient for productivity.

## Sulphate

Beeton (1965) stated sulphate as an index of eutrophication and its dynamic in many lake eco-systems is subjected to transformation by microorganisms. Higher values were found during summer in the study wetland (Welch, 1952; Qadri *et al.*, 1981). Hutchinson (1957) stated that when oxygen deficiency prevails, total sulphates increase during summer due to anaerobic decomposition of organic matter and released carbon dioxide and hydrogen sulphate at the bottom and get oxidize in the entire mass of water. This might be the reason in the current study that sulphate found maximum during summer. It is also interesting that sulphates exhibit more or less positive correlation with other nutrients, except D.O. and free-$CO_2$. However, Munawar (1970) reported inverse relationship with calcium, pH and iron.

## Silica

In the present study, a direct relationship exists between silica and temperature. It might be that silica accumulates in water during summer partly because the diatoms, which are main consumers of this elements and are comparatively less in the summer, whereas in winter demands for silica increases manifold due to increase in the diatom population.

## Hardness

Hardness of water reflects the higher concentrations of many cations. The concentration of these cations are more in summer as compared to winter in the study wetland. Sreenivasan (1966) also found similar results in their studies. A positive correlation was observed between total hardness and alkalinity in the present study.

## Calcium

The seasonal variation was pronounced in the study wetland. Statistical analysis revealed a positive correlation between calcium and phosphate, which is in agreement with Hergenrader (1980). During the present investigation calcium did not show any fixed pattern. The results are supported by the study of (Munawar,1970 and Swarup and Singh,1979). Sreenivasan (1964, 1968) observed better growth of biota when calcium content exceed 25 ppm which coincide with the present study.

## Magnesium

The wetland waters did not exhibit any appreciable seasonal variation of magnesium in the current study. The higher values of magnesium suggest a close affinities between magnesium concentration and inorganic pollution. Zafar (1964, 1966); Munawar (1970); and Rao (1971) have also found a direct correlation between these two parameters.

## Chloride

Chloride concentration is an index, not only of eutrophication, but also of pollution, caused by cattle, sewage and other wastes (Singh, 1960; Sreenivasan 1965; Munawar, 1970; Misra and Yadav, 1978). The statistical analysis revealed that chloride content shows a positive correlation with temperature and other nutrients, except free-$CO_2$ and D.O. A higher content of chloride was observed in summer, while January recorded lower values. Similar observations have been made by many workers (Rana 1977; Khan and Zutshi, 1980; Dakshini and Gupta 1984).

## Sediment Composition

Sediment analysis revealed a higher concentration of sulphate, phosphorus, total nitrogen, potassium, sodium and chloride in Ratheshwar Wetlands. Further, most of the nutrients of the sediment, except organic matter, were higher during winter season (January) while they decrease in the summer. This indicates that the concentrations of most nutrients were poorer in water when sediment contain higher quantities of these nutrients. The poor growth of most plankton observed during winter (January) can be explained as most of the nutrients are locked up in the sediment during this period. It has been observed by many workers that nutrients like phosphorus, nitrogen and sulphate etc. are recycled through sediment and water of wetland during different seasons (Haslam, 1978; Anderson and Jacob, 1988; Ali *et al.*, 1988).

In contrast when a peak of most nutrients in water together with phytoplankton were observed during summer, a low nutrients level of sediment was observed. These observations support the view that sediment chemistry controls the aquatic flora (Berendse *et al.*, 1989). The recycling of nutrients can also be explained by temperature gradient. During winter when water temperature is low, the sediment decomposition is slow and release of nutrients in the surface water also decreases, resulting in low concentration of nutrients and poor growth of plankton. With the increase in water temperature the rate of decomposition is low and release of sediment nutrients increase and a growth peak is observed in summer. The increase in nutrients coincide well with the increase in phytoplankton and a peak is observed in summer.

## Phytoplankton

Many indices have been proposed and are being tested for their validity in using the biota for monitoring the environment. In the present study to evaluate the trophic status of water body, Shannon and Weaver diversity index was applied. The occurrence, composition, monthly variation and diversity indices of the phytoplankton have been discussed here.

## Cyanophyceae

It is encountered as the most important group both from diversity and density point of view. With respect to number of individuals

*Microcystis* dominated while *Raphidiopsis, Arthrospira* and *Oscillitoria* were more common and *Raphidiopsis* was common, *Merismopedia* and *Oscillatoria* dominated in Ratheswhar wetland. In Ratheshwar wetland, only one peak was observed, *i.e.* in June (summer). (Munawar,1970) also reported similar observations in different wetlands of Hyderabad.

Many workers have emphasized the importance of water temperature in the periodicity of blue–greens. (Ganapati,1960 and Seenayya,1972) have pointed out that Cyanophyceae develops in summer due to high temperature. Similarly in the present study the pond exhibited abundant growth of Cyanophyceae during summer (June).

Seenayya (1971) and Vijayvergia (1988) recorded high pH during the thick growth of Cyanophyceae. In the present study also the alkaline pH favours the growth of Cyanophyceae. (Ruttner,1953 and Prescott,1960) found that water with higher bicarbonate stimulate Cyanophycean growth, an observation which coincide with the present results.

Smith (1983) and many others have discussed the role of phosphate in the formation of Cyanophycean bloom. There was a marked decrease in the concentration of nitrate and phosphate at the time of disappearance of Cyanophyceae, probably higher concentrations of these substances are pre-requisites for the luxuriant growth of this group of algae. The importance of nitrate and phosphate in ecology of Cyanophyceae have been emphasized by (Lannineer *et al.*,1982 and Henry *et al.*,1984)

## Chlorophyceae

The dominance of this group during winter (November–December) coinciding with low temperature and high dissolved oxygen content has been reported by Venkateshwarlu (1969). Dhakar (1979) also observed that the green algae prefers waters with comparatively higher concentration of dissolved oxygen. In the present study, Chlorophycean dominance was observed generally during low temperature and high dissolved oxygen.

The temperature is considered to be an important factor in the periodicity of the Chlorococcales population. In the present study water temperature plays vital role in the periodicity of this group. Well-developed Chlorococcales population was found in November–

December, when the temperature was low in the wetland. This is an agreement with the results of (Munawar,1974). The desmids are known to prefer low nitrate and phosphate concentrations as observed by Singh (1960), Zafar and Munawar (1974). In the present study also Desmids were observed to grow in Ratheshwar wetland having low concentrations of nitrates. Rao (1955); Zafar (1968) and Munawar (1974) found that a temperature range of 22-30°C favours the growth of desmids and peak was observed in the winter. In the present study also desmids were abundantly observed during November–December, when the temperature ranges from 21-28° C. The low values of calcium and magnesium in Ratheshwar wetland harbours rich growth of desmids. Our observations also find support from the work of (Zafar, 1967; Munawar, 1974 and Singh and Swarup, 1979). Desmids are known to grow in waters having less pollutants and diversity of desmids indicates the oligotrophic nature of water. On the contrary in eutrophic and polluted waters, the diversity of desmids decreases (Venkateswarlu, 1981,1986). In the present study Ratheshwar wetland supports the maximum desmids flora which is comparatively less polluted as supported both by chemical and biological properties.

## Euglenophyceae

This group as a whole is facultatively heterotrophic and generally abundant in waters rich in organic matter (Hutchinson, 1975). In the present investigation, a few number of euglenoids were recorded. Species like *Euglena acus*, and *Phacus caudatus* grew profusely in pond. Munawar (1970, 1974) concluded from the studies that high values of carbon dioxide, phosphate, nitrate, chloride and low values of dissolved oxygen favour the growth of euglenoids. Ratheshwar wetland with low nutrients and high D.O. supports only a few species of Euglenoids.

## Bacillariophyceae

Although, a few species of diatoms in the present investigation have been recorded, they differ both in quality and quantity in the study wetland. Ratheshwar with less concentrations of nutrients showed highest density of diatoms. This confirms with Zafar (1964); Fjerdingstad (1965); Seenayya (1972) and Munawar (1974), who observed that diatoms do not prefer highly polluted waters. The data further suggests that even the higher concentrations of phosphorus and nitrates were not helpful in sustaining a rich growth of diatom flora, if the water happened to be polluted. The growth forms like *Fragillaria. Gomphonema, Nitzshia* and *Pinnularia* could be regarded as the forms avoiding pollution conditions as sustained by Rathweshwar wetland.

In the present study, the diatoms reach their maximum during summer (May–June) when the water temperature ranges from 28-33° C. The growth decreased during rains followed by winter when the temperature is low. Seenayya (1972) similarly noted that the diatoms grow profusely when the temperature is high during summer (March–June). Pahwa and Mehrotra (1966) recorded two peaks of diatoms during summer and winter, but the summer peak constituted the higher numbers. Lund (1965, 1955) and Zafar (1964) attributed the responsibility of the production of diatoms to phosphate and silica. Therefore, nitrate, phosphate and silica are significantly important for the growth of Bacillariophyceae.

## Diversity Index

A species diversity index is a biological indicator which integrates information on the kinds and numbers of organisms present in aquatic community. In the present study, Shannon and Weavers index of diversity was calculated. High index values in Ratheshwar wetland revealed its low pollution level (Woodwell, 1970). Wilhm (1970b) concluded that in clear water bodies the Shannon–Weaver's index values of diversity ranges from three to four, while in heavily polluted waters it is usually or around one.

## Chlorophyll Content

The phytoplankton biomass can be measured by estimating the total chlorophyll content of the algae (Kopf, 1983). The growth, reproduction and chlorophyll content of aquatic microorganisms are affected by a great variety of physico-chemical and seasonal factors which in a multitude of ways, may act with or against one another (Bozniak and Kennedy, 1968). In the present study, the highest quantity of total chlorophyll was recorded in the month of June, while the lowest average of total chlorophyll was recorded in the months of July and February. The low value of chlorophyll in July can be explained probable due to high dilution of water by Rains. In winter (February) temperature affects the growth of plankton.

The significant increase in overall total chlorophyll contents during June can be attributed to higher temperature and nutrients, particularly nitrogen and phosphorus (Hutchinson, 1967). The characteristic seasonal variations in total chlorophyll content reported in the study are in agreement with earlier studies (Sondergaard and Janson, 1979 and Marshall and Robert, 1989.)

The good positive correlationship between chlorophyll and phosphate was in accordance with the past studies (Joseph and Harvey, 1982), which supports our findings. Chlorophyll content can be used as an ecological index of production (Billore and Mall, 1975) and also as an index of primary productivity of phytoplankton (Anwar *et al.*, 1987). In the present study, chlorophyll content has inverse relationship with species diversity and direct relationship with density of phytoplankton. These findings are supported by (Brenner *et al.*, 1989).

## The Macrophytes

Aquatic vesicular plants are important indicators of water quality (Seddon, 1972; Shimoda, 1984). Ratheshwar wetland harboured a luxuriant growth of aquatic macrophytes. The prolific growth of sub-merged macrophytes of Ratheshwar add high amount of dissolved oxygen, particularly in winters as observed by (Sahai and Sinha, 1969 and Sahai and Srivastva, 1976).

The important nutrients like phosphate, nitrate and sulphate which recorded in low quantity during winter (January–February) might be due to its consumption by these autotrophs during early winter (November–December) This is in agreement with the findings of Sahai and Sinha (1969). Water temperature is one of the vital factor which monitors the growth of vesicular plants (Pipe, 1989).

The composition and species diversity of hydrophytes were greater during winter (December–February), whereas poor vegetation was observed during summer (March–June) Similar observation have been made by (Venu and Seshavatharam, 1988).

Plants of Ratheshwar wetland at stations one and three differ from each other. The difference in the community of the two stations might be due to local pollution, as station one is used for washing, bathing and drinking water for human and cattle. According to Haslam (1978), growth and distribution of plants are determined by complex of chemical, physical and biotic factors (Table 13.8).

**Table 13.8: Monthly Variation of Macrophytes in Ratheshwar Pond (1988-89)**

| *Sl.No.* | *Name of the Macrophytes* | *1988* | | | | | | | *1989* | | | | |
|---|---|---|---|---|---|---|---|---|---|---|---|---|---|
| | | *June* | *July* | *Aug.* | *Sep.* | *Oct.* | *Nov.* | *Dec.* | *Jan.* | *Feb.* | *March* | *April* | *May* |
| 1. | *Typha angustata* | + | + | + | + | + | + | + | + | + | + | + | + |
| 2. | *Echinochloa colonum* | + | + | + | + | + | + | + | + | + | + | + | + |
| 3. | *Marsilea quadrifolia* | + | + | + | + | + | + | + | + | + | + | + | + |
| 4. | *Ipomoea aquatica* | + | + | + | + | + | + | + | + | + | + | + | + |
| 5. | *Fimbristylis ferruginea* | + | + | + | + | + | + | + | + | + | + | + | + |
| 6. | *Cyperus alopecuroides* | + | + | + | + | + | + | + | + | + | + | + | + |
| 7. | *Nymphoides cristata* | – | – | – | – | + | + | + | + | + | + | – | – |
| 8. | *Potomogeton pectinatus* | + | + | + | + | + | + | + | + | + | + | + | + |
| 9. | *P. nodosus* | – | – | – | + | + | + | + | + | + | + | + | – |
| 10. | *P. crispus* | – | – | – | – | – | + | + | + | + | + | – | – |
| 11. | *Hydrilla verticillata* | + | + | + | + | + | + | + | + | + | + | + | + |
| 12. | *Najas minor* | + | + | + | + | + | + | + | + | + | + | + | + |
| 13. | *Vallisneria spiralis* | – | – | – | – | – | + | + | + | + | + | – | – |
| 14. | *Ottelia alismoides* | – | – | – | – | + | + | + | + | + | – | – | – |
| 15. | *Chara* sp. | – | – | – | – | + | + | + | + | + | – | – | – |

+ = Present: – = absent.

**Table 13.9: The Correlation Co-efficient 't' of Physico-chemical and Biological Parameters of Ratheshwar Pond (Site-I)**

| Sl.No. | Parameters | Parameters | | | | | | | | | | | | | | | | | |
|---|---|---|---|---|---|---|---|---|---|---|---|---|---|---|---|---|---|---|---|
| | | 1 | 2 | 3 | 4 | 5 | 6 | 7 | 8 | 9 | 10 | 11 | 12 | 13 | 14 | 15 | 16 | 17 | 18 |
| 1. | Temperature | 0.0 | | | | | | | | | | | | | | | | | |
| 2. | pH | 8.5 | 0.0 | | | | | | | | | | | | | | | | |
| 3. | Total alkalinity | 6.9 | 6.2 | 0.0 | | | | | | | | | | | | | | | |
| 4. | Total solids | 6.8 | 7.1 | 8.4 | 0.0 | | | | | | | | | | | | | | |
| 5. | Carbonate | 5.1 | 11.0 | 4.0 | 4.9 | 0.0 | | | | | | | | | | | | | |
| 6. | Bi–carbonate | 0.1 | 0.6 | 0.1 | 0.4 | 1.0 | 0.0 | | | | | | | | | | | | |
| 7 | Free $CO_2$ | –4.5 | –3.7 | –2.2 | –2.2 | –3.5 | 0.6 | 0.0 | | | | | | | | | | | |
| 8. | Dissolved oxygen | –7.7 | –13.6 | –7.4 | –6.7 | –8.2 | 0.4 | 3.3 | 0.0 | | | | | | | | | | |
| 9. | Phosphate | 6.8 | 15.3 | 5.0 | 5.7 | 17.0 | 0.9 | –4.0 | –11.7 | 0.0 | | | | | | | | | |
| 10. | Nitrate | 7.9 | 21.3 | 7.4 | 8.2 | 10.6 | 0.5 | –3.2 | –20.0 | 14.7 | 0.0 | | | | | | | | |
| 11. | Silica | 3.4 | 5.4 | 2.9 | 3.8 | 5.5 | 0.7 | –2.6 | –5.4 | 5.7 | 5.3 | 0.0 | | | | | | | |
| 12. | Sulphate | 8.2 | 10.2 | 13.4 | 12.8 | 5.8 | 0.5 | –2.7 | –10.7 | 7.7 | 12.0 | 4.1 | 0.0 | | | | | | |
| 13. | Calcium | 10.2 | 8.6 | 12.8 | 14.2 | 5.2 | 0.3 | –2.8 | –10.1 | 6.7 | 10.3 | 3.9 | 21.8 | 0.0 | | | | | |
| 14. | Magnesium | 4.8 | 5.8 | 9.2 | 9.9 | 4.3 | 0.2 | –1.6 | –6.3 | 4.6 | 7.1 | 3.4 | 10.6 | 8.5 | 0.0 | | | | |
| 15. | Hardness | 8.6 | 16.8 | 8.7 | 7.5 | 8.6 | 0.5 | –3.2 | –16.5 | 11.9 | 24.3 | 4.3 | 13.1 | 10.6 | 6.8 | 0.0 | | | |
| 16. | Chloride | 10.5 | 11.1 | 12.1 | 12.1 | 6.0 | 0.3 | –3.0 | –13.5 | 7.9 | 14.1 | 4.3 | 26.3 | 31.5 | 9.1 | 13.9 | 0.0 | | |
| 17. | Phytoplankton chlorophyll | 1.2 | 0.8 | 1.0 | 1.8 | 0.7 | 0.7 | –0.5 | –0.7 | 1.0 | 0.9 | 0.3 | 1.1 | 1.4 | 1.0 | 0.8 | 1.2 | 0.0 | |
| 18. | Phytoplankton density | 1.6 | 1.2 | 1.6 | 2.5 | 1.0 | 0.6 | –0.5 | –1.1 | 1.2 | 1.3 | 0.7 | 1.6 | 1.9 | 1.6 | 1.1 | 1.6 | 12.6 | 0.0 |

The correlation between various physico-chemical factors and density of plankton indicates that the index has indirect relationship with these factors (Table 13.9). The decrease in the value of index with the increase in nutrients such as carbonates, total solids, phosphate, nitrate, sulphate etc. induces quantitative increase of certain population tolerant forms and disappearance of sensitive algal species.

Diversity index of Shannon–Weavers hold a good index for assessing and ranking the water quality of water bodies. According to Shannon's index Ratheshwar wetland can be treated as a freshwater or unpolluted (oligtrophic) water body

Hence, it can be concluded that chemical status of the water appeared to be the most vital factor, significantly influencing the general distribution of aquatic flora. The abiotic factors such as bottom soil, physical nature of the wetland and fluctuations of water level greatly affects the distribution of plants within the range of chemical tolerance (Kaul *et al.*, 1978).

## References

Adoni, A. D. (1985). Work Book in Limnology. Pratibha Publishers, C–10 Gour Nagar, Sagar, India.

Ali Arshad, Reddy, K. R. and Debusk, W. F. (1988). Seasonal changes in sediment and water chemistry of a sub-trophical shallow •eutrophic lake. Hydrobiologia, 159 (2): 159-165.

Anderson Robin, M. and Jacob Calf. (1988). Sub–merged macrophyte biomass in relation to sediment characteristics in ten temperate lakes. Freshwater Bio. 19: 115-112.

Anwar, H. O., Teru, I. and Yusho, A. (1987). The distribution of chlorophyll-a, as an index of primary productivity of phytoplankton in Khor EL Ramla of the high dam lake, Egypt. J. Tokyo Univ. Fish. 74 (2): 145-158.

A. P. H. A. (1976). Standard methods for the Examination of Water and Waste Water American Public Health. Association, New york.

Barnett, P. H. (1953). Relationship between alkalinity, absorption and regeneration of added phosphorus in fertilized trout lakes. Trans. Am. Fish Soc. 82: 79-80.

Beeton, A. M. (1965). Eutrophication of the St. Lawrence Great lakes. Limnol. Oceanogr. 10: 240–254.

Berendse, Frank, Poland Boffink and Gerrit Rouwenhorst. (1989). A comparative study on nutrient cycling in wet health land ecosystem. II. Litter decomposition and nutrient mineralization. Ecologia 78: 338-348.

Billore, S. K. and Mall, L. P. (1975). Chlorophyll content as an Ecological index of dry matter production. J. Ind. Bot. Soc. 54 (I, II): 75-77.

Blum, J. L. (1956) The ecology of river algae. Bot. Rev. 22: 291-341.

Bozniak, G. E. and Kennedy, L. L. (1968). Periodicity and ecology of the phytoplankton in an oligtrophic and eutrophic lake. Can. J. Bot. 46: 1259-1271.

Brenner, F. J., Wayne, S., John, F. S., James, P. and Charles, M. (1989). Relationship between plankton communities and productivity of three surface mine lakes in. Pennsylvania. J. PA-Acad. Sci. 63 (1): 13-19.

Claesson, A. and Ryding S. O. (1977). Nitrogen–a growth limiting nutrient in eutrophic lakes. Prog. Water Tech. 8: 291-299.

Dakshini, K. M. M. and Gupta, S. K. (1984). Physiography and limnology of three lakes in environs of the Union territory of Delhi, India. Proc. Indian Natn. Sci. Acad. B 50: 417-430.

Desikachary, T. V. (1959). Cyanophyta. ICAR, New Delhi.

Dhakar, M. L. (1979). Studies in some aspects of the Hydrobiology of Indrasagar tank (South Rajasthan) Ph. D. Thesis. University of Udaipur, Udaipur.

Fjerdingstad. E. (1965). Some remarks on a new saprobic system. Biological problems in water pollution, Third Seminar 1962. Public Health Service Publ. 25: 232–235.

Forsberg, C. (1977). Nitrogen, a growth factor in freshwater. Water Prog. Tech. 8: 275-290.

Fritch, F. E. (1945). The structure and reproduction of the algae. Vol. I Cambridge Univ. Press, London.

Ganapati S. V. (1960). Ecology of tropical waters. Proc. Sym. Algol. ICAR, New Delhi, 214-218.

Georing, J. J. (1972). The role of Nitrogen in Eutrophi process. In. MICHELL, . (Ed). Water Pollution Microbiology, Wiley Interscience, New Delhi.

Goldman, C. R. (1972). The role of minor nutrients in limiting the productivity of aquatic ecosystems. Nutrients and eutrophication. Spec. symp. Am. Soc. Limnolo. Oceanogr. 1: 21-33.

Haslam, S. M. (1978). 'River Plants' (Eds). S. M. Haslam. "Nutrients", Cambridge University Press, London. pp. 120-130.

Hegde, G. R. (1985). Comparison of Phytoplankton biomass in four water bodies of Dharwad, Karnataka State (India). Proc. Indian Acad. Sci. (Plant Sci). 94: 583-587.

Henry, R., Tundisi, J. E. and Curi, P. R. (1984). Effects of phosphorous and nitrogen enrichment on the phytoplankton in a tropical reservoir (Lobo reservoir, Brazil). Hydrobiologia, 118 (2): 177–186.

Hargenrader, G. L. (1980). Eutrophication of the salt valley reservoirs 1968-1973. II. Changes in physical and chemical parameters of Eutrophication. Hydrobiologia 74: 225-240.

Hutchinson, G. E. (1957). A treatise on limnology. Vol. I. John Wiley and Sons. Inc., New York.

Hutchinson, G. E. (1975). A Treatise on limnology. Vol. II. Limnological Botany, John Wiley and Sons, New York, London, Sydney, Toronto, pp. 660.

Iyengar, M. O. P. and Desikachary, T. V. (1981). Volvocales. ICAR, New Delhi.

Jose, L. (1990). Algal flora of Sabarmati River of Gujarat. A Ph. D. thesis, S. P. University, V. V. Nagar, Gujarat.

Joseph, A. K. and Harvey, W. H. (1982). The phosphorous–chlorophyll–a, relationship in periphytic communities in a controlled ecosystem. Hydrobiologia 94: 173-176.

Kaul, V., Trisel, C. L. and Handoo, J. K. (1978). Distribution and production of macrophytes in some aquatic bodies of Kashmir. Glimp. Ecol. (Ed). J. S. Singh and Gopal, B., Prakash Publishers, Jaipur, India. PP. 313-334

Khan, A. A. and Siddiqui, Q. A. (1971). Primary production in a tropical fish pond at Aligarh, India. Hydrobiologia 37 (3-4): 447-456.

Khan, M. A. and Zutshi, D. P. (1980a). Primary productivity and trophic status of Kashmir, Himalayan lake. Hydrobiologia 68: 3-8.

King, D. L. 1970. The role of carbon in Eutrophication. J. Water Poll. Control Fed. 42: 2035-2051.

Kopf, W. (1983). Investigations of the dynamic of chlorophyll concentration in natural populations of phytoplankton. Arch. Hydrobiol. 98 (2): 173-180.

Lakshminarayana, J. S. S. (1965). Studies in the phytoplankton of river Ganges, Varanasi, India. I. Physico–Chemical characteristics of river Ganges. Hydrobiologia. 25: 119-137.

Lannineer, J., Lea Kauppi and Yrjana, E. R. (1982). The role of nitrogen as a growth limiting factor in the eutrophic lake Vesijarvi, Southern Finland. Hydrobiologia. 86-87: 81-85.

Likens, G. E. (1975). Primary production of inland aquatic ecosystem. In. H. Lieth and R. H. Whiyyaket (Ed). Primary Productivity of the Biosphere. Freshwater Biol. 2: 309-320.

Lund, J. W. G. (1950). Studies on *Asterionella formosa* Hass. II. Nutrient depletion and the spring maximum. J. Ecol. 38: 1-35.

Marshall, C. T. and Robert, H. P. (1989). General patterns in the seasonal development of chlorophyll-a for temperate lakes. Limnol. Ocenogr. 34 (5): 856-867.

Mishra, G. P. and Yadav, A. K. (1978). A comparative study of physico–chemical characteristics of AUD lake water in Central India. Hydrobiologia 59 (3): 275-278.

Morgan, N. C. (1970). Changes in the fauna and flora of a nutrient enriched lake. Hydrobiologia 35: 545-553.

Moyle, J. B. (1949). Some indices of lake productivity. Trans. Amer. Fish Soc. 76: 322-324.

Munawar, M. (1970). Limnological studies on freshwater ponds of Hyderabad, India. I. The biotope. Hydrobiologia. 35 (1): 127-162.

Munawar, M. (1970). Limnological studies on the freshwater ponds of Hyderabad, India. II. The Bioceonose, Distribution of unicellular and colonial phytoplankton 36 (1): 105-128.

Munawar, M. (1974). Limnological studies on freshwater ponds of Hyderabad. India IV. The bioceonose. Seasonal abundance of unicellular and colonial phytoplankton in polluted and unpolluted environments. Hydrobiologia 44 (1): 13-27.

Munawar, M. and Zafar, A. R. (1967). A preliminary study of vertical movement of *Eudorina elegans* and *Trinema linerare* during a bloom caused by them. Hydrobiologia 29 (1-2): 140-148.

Nirmal Kumar (1990). Ecological Studies of Certain ponds with reference to Eutrophication and weed growth around Anand Gujarat. A. Ph. D. thesis, S. P. University V. V. Nagar Gujarat.

Nirmal Kumar, J. I. (1991). Seasonal primary productivity of phytoplankton of temple tank–Vadtal, Gujarat. J. Ind. Bot. Soc, 70: 427-428.

Nirmal Kumar, J. I. (1992). Trophic status of certain lentic waters in Kheda district Gujarat, India. In: Ecology and Pollution of lakes and reservoirs. Ed. R. K. Trivedy. Ashish Publishers, New Delhi. 203-222.

Nirmal Kumar, J. I. and Rana, B. C. (1989). Variation in primary productivity of phytoplankton of Tarapur pond, Gujarat, Ind. Bot. Cont. 6 (3): 95-97.

Nirmal Kumar, J. I. and Rana B. C (1993). A composite distribution of nutrients in sediments of two ponds around Anand, Gujarat, Indian. J. Ecobiology. 6: 47-51.

Nirmal Kumar, J. I., Geetica Sharma, Rita N Kumar and Shintu Joseph (2002). An assessment of Eutrophication and weed growth of certain wetlands of Gujarat. In: International J. Pollution Research. (Accepted).

Pahwa, D. V. and Mehrotra (1966). Observations on fluctuations in the abundance of plankton in relation to certain hydrological conditions of river Ganga. Proc. Nat. Acad. Sci. India 34 B: 157-189.

Paramasivam, M. and Sreenivasan, A. (1981). Changes in algal flora due to pollution in Cauvery river. Ind. J. Environ. Hlth. 23: 222-238.

Patel, R. J. (1989). Algae Novo ". Avichal Science Foundation, Shastri Marg, Vallabh Vidyanagar, Gujarat (India).

Philipose, M. T. (1967). Chlorococcales, ICAR, New Delhi.

Pipe, E. (1989). Water temperature and freshwater macrophyte distribution. Aquatic Bot. 34: 367-374.

Prescott, G. W. (1962). Algae of the Western Great Lakes Area, PP. 977, W. M. C. Brown Company Publishers, Dubuge, Iowa,

Qasim, S. Z., Wellershaus, S., Bhattathiri, P. M. A. and Abidi, S. A. H. (1969). Organic production in a tropical estuary. Proc. Indian Acad. Sci. 69 (6): 51-94.

Rana, B. C. (1977), A note on the phytoplankton of a small tank. Geobios. 4: 206-208.

Rana, B. C and Nirmal Kumar, J. I. (1993). A Composite rating of trophic status of certain ponds of Gujarat India. J. Environmental Biology. 14: 113-120.

Rana, B. C, Nirmal Kumar and Sreenivas, S. S. (1995). Phytoplankton ecology of certain water bodies of Gujarat. India. In Algal Ecology (Eds): Kargupta, A. N and Siddique, E. N. International Book distributors, Dehradun, India. Pp. 101-129.

Rao, V. S. (1971). An ecological study of three freshwater ponds of Hyderabad, India. I. The environment. Hydrobiologia 38: 213-223.

Rao, V. S. (1972). An ecological study of three freshwater ponds of Hyderabad, India I. The environment. Hydrobiologia 39: 315-372.

Rao, P. B., Sharma, A. P. and Singh, J. S. (1982). Limnology and phytoplankton production of a high altitude lake. Int. J. Ecol. Environ. Sci. 8: 39-51.

Ruttner, F. (1953). Fundamentals of limnology, Univ. of Toronto Press, Toronto, Canada, PP. 342.

Sahai, R, and Sinha, A. B. (1969). Investigations on bio-ecology of inland waters of Gorakhpur (U. P)., India I. Limnology of Ramgarh lake Hydrobiologia. 34: 433-447.

Sahai, R. and Srivastva, V. C. (1976). The Physico–chemical complexes and its relationship with the macrophytes of Chilwa lake. Geobios 3: 15-19.

Saran, H. M. and Adoni, A. D. (1984). Studies on seasonal variation in orthophosphates and nitrates in Sagar lake, Sagar. Acta Bota. Indica 12: 223-225.

Schelske, C. L. (1979). Role of phosphorus in Grate lake. Eutrophication. Is there a controversy ? J. Fish Res. Board, Cana. 36: 286-288.

Schindler, D. W., Armstrong, E. A. J. Holmgren, S. K. and Bronnskill, G. J. (1971). Eutrophication of lake 227, experimental lakes areas, North–Western Ontario by addition of phosphate and nitrate. J. Fish. Res. Bd. Can. 28: 1763-1782.

Seddon, B. (1972). Aquatic macrophytes as limnological indicators. Freshwat. Biol. 2: 107-130.

Seenayya, G. (1971). Ecological studies in the plankton of certain ponds (Freshwater) of Hyderabad, India. Part II. Phytoplankton I. Hydrobiologia. 37: 55-68.

Seenayya, G. (1972). Ecological studies in the plankton of certain freshwater ponds of Hyderabad. India. II. Phytoplankton–2. Hydrobiologia, 39 (2) 247–271.

Seshavatharam, V. and Venu, P. (1980). Dry matter production and energy content of two sub-merged aquatic macrophytes of Kondakar AVA lake. Third All–India Botanical Conf.: 59: 92.

Shaji (1989). Algal Flora of polluted waters of Gujarat. A Ph. D. thesis, S. P. University, V. V. Nagar, Gujarat.

Shimoda, Michiko (1984). Macrophytic communities and their significance as indicators of water quality in two ponds in the Saijo basin, Hiroshima prefecture, Japan. Hikobia 9: 1-14.

Singh, V. P. (1960). Phytoplankton Ecology of Inland waters of Uttar Pradesh. In. Proc. of the Sympo. On Algolo. P. 243-271.

Smith, V. H. (1983). Low Nitrogen to phosphorus ratios favour dominance by blue-green algae in lake Phytoplankton Science. 221: 669-671.

Sondergaard, M. and Jenson, K. S. (1979). Physico–chemical environmental, phytoplankton biomass and production in oligotrophic, soft water lake, Kalgaard, Denmark. Hydrobiologia 63: 241-253.

Sreenivasan (1991): Studies of water pollution in lentic waters in Gujarat. A Ph. D. thesis, S. P. University, V. V. Nagar–Gujarat.

Sreenivasan, A. (1964b). The limnology, Primary production and fish production in a tropical pond. Limnol. Oceanogr. 9: 391-396.

Sreenivasan, A. (1965). Limnology of tropical impoundments III. Limnology and productivity of Amaravati reservoir (Madras State), India. Hydrobiologia 26: 501-576.

Sreenivasan, A. (1965). Remarkable growth of fish in Saudoimedu Demonstration tank (North Arcot. Dist. Madras St). with a note on its ecology. J. Bombay Nat. Hist. Soc. 62 (1): 165

Sreenivasan, A. (1968). Limnology of tropical impoundments. V. Studies of two impoundments in Nilgiris, Madras State, India Phykos. 7 (1): 144-160.

Stumm, W. and Stumm, Zollinger, E. (1972). The role of phosphorus in eutrophication. In Michell, R. (Ed). Water Pollution Microbiology, Wiley Interscience, New York.

Swarup, K. and Singh, S. R. (1979). Limnological studies of Suraha lake (Ballia). J. Inland fish Soc. India 11: 22-33.

Trivedy, R. K. and Goel, P. K. (1984). Chemical and Biological methods for water pollution studies. Environmental Publications, Karad (India).

Thomas, E. A. (1973). Phosphorus and eutrophication. in. Griffith, E. J. *et al.* (eds). Environmental Phosphorus. Handbook, Wiley and Sons, New York, pp. 585-611.

Unni, K.S. (1984). Limnology of a sewage polluted tank in Central India. Int. revue. Ges. Hydrobiol. 69: 553-565.

Venkateshwarlu, V. (1969). An ecological study of the algae of the river Mosi, Hydy. (India), with special reference to water pollution. IV Periodicity of some common species of algae. Hydrobiologia 36: 45-65.

Venkateshwarlu, V. (1981). Algae as indicators of river water quality and pollution. In. WHO works on Biological indicators and indices of environmental pollution. pp. 93-100.

Venkateshwarlu, V. (1986). Ecological studies on the rivers of Andhra Pradesh, with special references to water quality and pollution. Proc. Ind. Sci. Acad. 96 (6): 495-508.

Verma, S. R. and Shukla, G. R. (1970). The Physico–chemical conditions of Kamala Nehru Tank, Muzaffarnagar (U. P). in relation to the biological productivity. Envi. Hlth. 12: 110-128.

Vijayvergia, R. P. (1988). Ecological studies of lake Udaisagar, with reference to water bloom. Ph. D. Thesis, Sukhadia Uni., Udaipur.

Vollenweider, R. A. (1968). Scientific fundamentals of the eutrophication of lakes and flowing waters, with particular reference to nitrogen and phosphorus factors in eutrophication. Rep. Org. Econ. Coop. Dev. DAS/CSI/68. 27, Paris.

Vollenweider, R. A. (1979). Das Nahrstoffbelastungskonzept als Grundlage Fur den externen Eingriff in den Eutrophierungaproze Stehender and Talsperren. Z. Wasser Abwasserforsch. 12 (2)

Welch, P. S. (1952). Limnology, Mc Graw Hill Co., New York, PP. 538.

Wetzel, R. G. (1975). Limnology. Saunders Compnay, Philadelphia, U. S. A.

Wetzel, R. G. and Hough, R. A. (1973). Productivity and role of aquatic macrophytes in lakes. An assessment. Pol. Arch. Hydrobiol. 20: 9-19.

Wilhm, J. L. (1970b). Effect of sample size on Shannon's formula. South western Naturalist. 14: 441-445.

Woodwell, G. M. (1970). Effects of pollution the structure and physiology of Ecosystems Science 168: 429-433.

Wuhrmann, K. (1976). Chemical impact on inland aquatic ecosystem. Pure Appl. Chem., 45: 193-198.

Zafar, A. R. (1964). On the ecology of algae in certain fish ponds of Hyderabad, India. II. Distribution of unicellular and colonial forms. Hydrobiologia 24 (4) 556-566.

Zafar, A. R. (1966). Limnology of Hussain Sagar Lake–Hyderabad, India. Phykos 5 (1 and 2): 115-126.

Zafar, A. R. (1968). Certain aspect of distribution pattern of phytoplankton in lakes of Hyderabad. Proc. Symp. Rec. Adv. Trop. Ecol., 368-375.

Zutshi, D. P. (1985). The Himalayan lake ecosystems. In: Singh, J. S. (ed). Environ. Regenration in Himalaya: Concepts and Strategies 325-338.

Zutshi, D. P. and Vass, K. K. (1984). Limnological studies on Dal lake, Kashmir, V. Impact of human activities and the evolution of lake environment. In national Seminar on Environment, Bhopal, Feb. 8-10.

**Aquatic Biodiversity in India: The Present Scenario, 2005** **285–296**

*Edited by*: D.R. Khanna, A.K. Chopra & G. Prasad

*Published by*: Daya Publishing House, New Delhi

# 14

# Status of Aquatic Biodiversity of Bhimtal Lake in Kumaon Region (Uttaranchal)

**Davendra S. Malik**

*Department of Zoology and Environmental Sciences, Gurukul Kangri University, Hardwar–249 404 (Uttaranchal)*

The Kumaon region lies in the lessor Himalayan zone at 28°43′55″ to 30°49′12″ N latitude and 78°44′30″ to 80°5′E longitude and provides an epitome of geological architecture of the whole Himalayan region.Hence, Kumaon region is the part of newly state Uttaranchal, blessed with enchanting beauty and varied natural water resources. The Nainital region is known as "West-moor land of Kumaon" has several lakes of large size and scenic beauty (Navill,1922). At present, there are only six perennial, five seasonal and many extinct lake depressions. Nainital, Bhimtal, Naukauchiatal, Sattal, Pannatal (Garudtal) and Khurpatal are the perennials, while Sariatal,Malwatal are in the form of flow through depressions and Sukhatal, and Khorital are seasonal.

Bhimtal lake is situated 22 km. away from Nainital at an altitude of 1332 m above msl consisting the surface area approximately 85.26 hectare. The ecosystem of Bhimtal lake has been showing a degrading trend. There has been very rapid settlement and encroachment upon the lake environment by human habitation beyond its carrying capacity. This has resulted in the disturbance of its ecology and bio-diversity studies conducted during the past fifteen years, the observed data revealed positive indication that the ecology and biotic life of Bhimtal lake has deteriorating due to release of domestic sewage and the sediments debris derived from the surrounding unstable mountain slopes, prone to land slides, rock fall deforestation, drainage complications, overgrazing and intensive agricultural practices. These natural and anthropogenic activities directly or indirectly affected the aquatic biodiversity in the lentic ecosystem of Bhimtal lake.

## Materials and Methods

The methods adopted in investigating the aquatic biodiversity (Phytoplankton, Zooplankton and Macrobenthos) are based on standards evolved in different research laboratories both in India and abroad. Wherever necessary, the methodology was modified to suit the local conditions. Surface plankton were collected from different stations in the Bhimtal lake using conical hand plankton net. Plankton were counted by micro transect methods (Lackey, 1938 and Edmondson,1974). The identification were performed using keys provided by (Edmondson,1959; Yeatman,1959; Fernando 1974 and Smirov,1974 and 1976).

The benthic biota in the lake was sampled by Ekman's dredge at each sampling station to collect the samples of soft sediments (deeper area), whereas a scoop type bottom sampler was used to collect the samples of sandy sediments (shallower area). The macro benthos retained through a brass sieve of a mesh size of 0.5/0.5 mm and identified by help of literature mentioned as Pennak,1953; Needham and Needham,1972).

The fish catch practice were adopted by operating deep water gill nets and cast net along the shore line having mesh ranging 75-120 mm and 20-30 mm knot to knot respectively. The fish caught were collected and preserved in 5 per cent formaline for identification.

Fishes were identified according to (Gunther *et al.*, 1959; Day,1879 and Jayaram,1987).

## Results and Discussion

An accurate bathymetric mapping is the primary requirement for limnological and biotic investigations of Bhimtal lake. The maximum depth of Bhimtal lake was 23.83 meter about 70 meter from island towards south east.The mean depth recorded was 12.10 meter. The length of the shoreline was 4025.0 meter with 1.67 shoreline index. In Bhimtal lake about 33.84 per cent area has a gentle slope *i.e.* between 0-10 m depth, 29.80 per cent of increasing trend of gentle slope between 10-20 m depth, 27.60 per cent with a very gentle slope between 20-24 m depth. The islands, forest nurseries and swamps in around the lake catchment basin has a total area about *i.e.* 10.40 per cent of the total basin area of the lake (Tables 14.1 and 14.2). There was evidence to show that the depth of Bhimtal lake reduced due to erosion of surrounding hills and water sheds. There was sufficient siltation at the bottom especially during monsoon seasons.

**Table 14.1: Physical Geographical and Morphometric Characteristics of Bhimtal Lake**

| | | |
|---|---|---|
| Altitude | (msl) | 1332 |
| Longitude | (E) | 79°34 |
| Latitude | (N) | 29°21 |
| Lake basin area | ($km^2$) | 11.70 |
| Lake surface area | (ha) | 85.26 |
| Maximum length | (m) | 1915.5 |
| Maximum effective length | (m) | 1468.5 |
| Maximum width | (m) | 486.5 |
| Maximum effective width | (m) | 537.0 |
| Mean width | (m) | 321.5 |
| Maximum depth | (m) | 23.83 |
| Mean depth | (m) | 12.10 |
| Length of shore line | (m) | 4025.0 |
| Development of shore line | (m) | 1.67 |
| Volume | ($V \times 10^3$ $m^3$) | 4075.7 |
| Development of volume | | 1.43 |

**Table 14.2: Percentage of Area (Lake Basin) in Relation of Depth Level of Bhimtal Lke**

| Depth Group (m) | Lake Basin Area (ha) | Percentage |
|---|---|---|
| 0–5 | 17.35 | 20:35 |
| 5–10 | 11.50 | 13.49 |
| 10–15 | 13.60 | 15.95 |
| 15–20 | 11.80 | 13.84 |
| 20–25 | 22.70 | 26.62 |
| Island | 0.91 | 1.07 |
| Swamps | 6.40 | 7.51 |
| Forest nursery | 1.00 | 1.17 |

Plankton plays a significant role in aquatic ecosystem as producers and consumers. Plankton maintains the equilibrium of food spectrum in natural food chain. The population abundance of phytoplankton in Bhimtal lake did not show very stable pattern due to a frequent changes in community structure. The annual average percentage of major phytoplankton were for Chlorophyceae 46.92 per cent, Bacillariophyceae 33.68 per cent, Dinophyceae 14.30 per cent, Chrysophyceae 3.45 per cent, and Cynophyceae 1.56 per cent.The chlorophyceae dominated during warmer months (May–October) contributing 50.0–72.40 per cent, while Bacillariophyceae during January-April (46.2-73.8 per cent) (Fig. 14.1). The richness and availability of phytoplankton in Bhimtal lake is exhibited by about 50 species. The species–wise dominance were in order of Chlorophyceae (28 species), Bacillariophyceae (12 species) Cynophyceae (7 species) Dinophyceae (2 species) and chryesophyceae (1 species). The dominant genera recorded were *Closterium, Staurastrum,Pediastrum, Scenedesmus, Ceratium, Peridinium, Meelosira, Fragilaria, Synedra and Cymbella*. Certain aspects of phytoplankton community of some Kumaon lakes have been studied by (Sharma *et al.*,1982 and Pant *et al.*, 1983).A large number of diatoms and green algae were studied in Sattal by Joshi *et al.* (1981). Therefore, Bhimtal lake with remarkable species richness can be considered as a temperate lake. The Bhimtal lake has moderate amount of nutrients and could be categorized as a mesotrophic lake. The seasonal changes in phytoplanktonic populations were mainly observed in the population of annual species or in the abundance of perennials

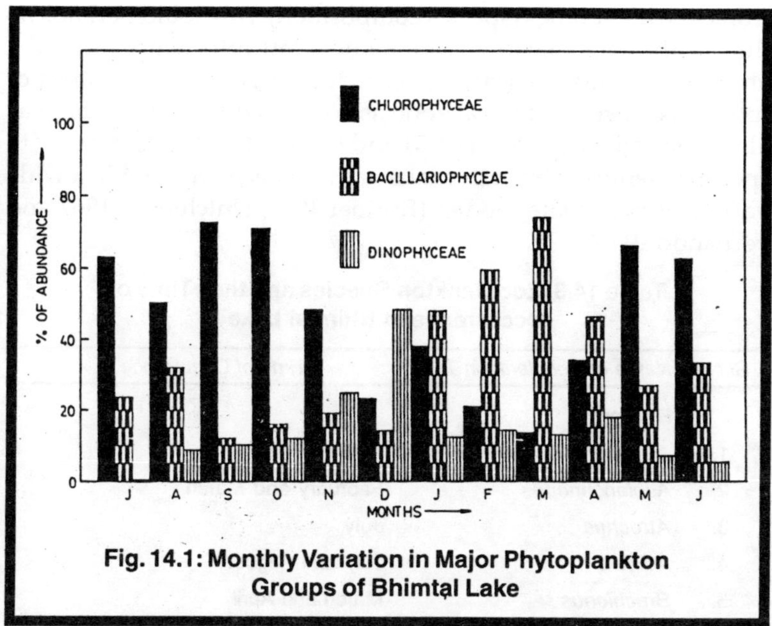


**Fig. 14.1: Monthly Variation in Major Phytoplankton Groups of Bhimtal Lake**

in different aquatic ecosystem (Neushal 1967; Dayton 1970 and Tenore, 1974).

The zooplanktons constitute a specific group of organism providing a link between producers and successive consumers in an aquatic ecosystem. The zooplanktonic population were characterized by a bimodal pattern of distribution during the study periods. Numerically, the rotifers were the most abundant followed by Copepoda and Cladocera. The highest densities of rotifers were recorded during March to July (51.50–68.25 per cent). However, minimum value (44.0 per cent) was recorded in January. Copepoda did not show any seasonal variation and Cladocera did not indicate any specific variation. The composition of zooplankton comprised 65 species belonging to rotifera (47), caldocera (12) and copepoda (6). Out of 65 species, 49 were seasonal, while 16 species (5 Rotifera, 7 Copepoda and 4 cladocera) were perennial (Table 14.3). Therefore, it stands to reason that food cycle specially predation plays an important role in determining the seasonal cycles of zooplankton population like the tropical lakes (Lewis, 1979). Bhimtal lake, thus shared the characteristics of both temperate (Species number) and tropical water bodies (species composition).This could perhaps be due to the fact that this water body fall into temperate area altitudinally and tropical one latitudinally. A greater numbers of rotifers occurred in littoral zone as compared to limnetic zone as also reported by Pennak (1957) and George (1966). Cladocera (12 species) identified from this lake has also been reported in a wide variety of fresh water bodies (Ruttner 1930, Hutchinson 1967 and Fernando 1980).

**Table 14.3: Zooplankton Species and their Time of Occurrence in Bhimtal Lake**

| *Sl.No.* | *Name of Zooplankton Taxa* | *Time of Occurrence* |
|---|---|---|
| | **Rotifera** | |
| 1. | *Asplanchna* | April and May |
| 2. | *Asplanchnopus* | February and March |
| 3. | *Atrochus* | July |
| 4. | *Adineta* | July and August |
| 5. | *Brachionus* | March and April |
| 6. | *Cephalodella* | April and May |
| 7. | *Cephaloda* | March |
| 8. | *Colurella* | Round the year |
| 9. | *Conochilus* | September |
| 10. | *Dicranophorus* | September and October |
| 11. | *Oidymodactylus* | May–September |
| 12. | *Epiphanes* | May–June |
| 13. | *Euchalinus* | April–June |
| 14. | *Entroplea* | May–June |
| 15. | *Eothinia* | March–September |
| 16. | *Harringia* | May–August |
| 17. | *Keratella* | August–October |
| 18. | *Lepadella* | Perennial |
| 19. | *Lacane* | May–June |
| 20. | *Lophocaris* | September–October |
| 21. | *Macrofrachela* | July, September and October |
| 22. | *Monoostyla* | March–May |

*Contd...*

**Table 14.3–Contd...**

| *Sl.No.* | *Name of Zooplankton Taxa* | *Time of Occurrence* |
|---|---|---|
| 23. | *Manfridium* | March–May |
| 24. | *Mytilina* | February–June |
| 25. | *Platylas* | March–May |
| 26. | *Polyarthra* | March–June |
| 27. | *Phyllodina* | Perennial |
| 28. | *Philodenaveous* | May–September |
| 29. | *Phinoglena* | March–October |
| 30. | *Rotaria* | May–December |
| 31. | *Trichocerca* | May, June, September and October |
| 32. | *Trichotria* | May–June |
| | **Cladocera** | |
| 1. | *Alona* | Perennial |
| 2. | *Alonella* | February–September |
| 3. | *C. hydorus* | Perennial |
| 4. | *Ceriodaphnia* | July–December |
| 5. | *Daphnia* | Perennial |
| 6. | *Hyocryptus* | March–July |
| 7. | *Moina* | July–October |
| 8. | *Macrothrix* | July–October |
| 9. | *Pleuroxus* | February–November |
| 10. | *Simocephalus* | January–April, July–December |
| | **Copedoda** | |
| 1. | *Phyllodiaptomus* | Perennial |
| 2. | *Mesocyclops* | Perennial |
| 3. | *Microcyclops* | Perennial |
| 4. | *Eucyclops* | Perennial |
| 5. | *Cyclops* | Perennial |
| 6. | *Ectocyclops* | July–December |

Benthos are important components of the bottom biocoenosis. Invertebrate macrobenthic animals mainly consisted of sessile or sedentary animals. Unlike planktonic organisms, benthic components form relatively stable community in the sediments and reflect characteristics of both sediments and the water column. The highest abundance of total macrobenthos were recorded in May, while lowest density occurred in September. Numerically, the oligochaetes (65.0 per cent) and diptera (30.0 per cent) were the chief components of macrozoobenthos and species *Tubifex-tubifex* alone contributed much (60.0 per cent) while *Chironomus plumosus* contributed (6.0 per cent) to the total macrozoobenthic standing community. The predominant macrobenthic taxa recorded have been listed in (Table 14.4).

**Table 14.4: Major Macrobenthic Taxa of Bhimtal Lake**

| *Sl.No.* | *Macrobenthos Groups* | *Abundant Species* |
|---|---|---|
| 1. | Turbillaria | Dugesia species |
| 2. | Oligochaeta | Tubifex tubifex, Limnodrilus sp. |
| 3. | Hirudinea | Glossiphonia weberi, Hemiclepsis marginata, Barbronia webri |
| 4. | Decapoda | Potaman koolooense |
| 5. | Ephemeroptera | Boetis sp., Ephemera varia, Heptagenis aphrodite, Caenis sp. |
| 6. | Odonata | Libellula quadrimaculata; Enallagma sp. |
| 7. | Trichoptera | Mystacides sepulchrallis, Polycentropus sp. |
| 8. | Lepideptera | Unidentified sp. |
| 9. | Coleoptera | Hyphoporus aper; Helochares lintus; Trophisternus sp., Bidessus sp. |
| 10. | Diptera | Chironomus plumosus, Forcipomyia sp., Procladius sp., Chaoborus sp., Polypedilum helterale, Xenochironomus sp., Stenochironomus sp., Glyptotendipes sp. |
| 11. | Gastropoda | Lymnaea accuminatae; Gyraulus convexiusculus |

In Bhimtal lake, benthic biota principally consist of oligochaetes, insect larvae, flatworms and molluscs. The degrative nature of lake ecosystem was supported by the abundance of oligochaetes and larvae of diptera. These organisms can be considered as biological indicators for organic pollution emerging in the lake. The abundance of *Tubifex* and *oligochaetes* indicates that the bottom is polluted with organic material. The abundance of insect larvae at the bottom was observed indicating their selective food habit. Lake Bhimtal, with its advanced mesotrophy supported a higher standing crop of the low oxygen tolerant macrozoobenthos since profundal zone never experienced anoxia. However, deficit in hypolimnetic oxygen has been reported by Gupta and pant. (1989). Species diversity can be considered as an effective tool in evaluating the community structure Wilhm and Dorris (1968) and Tenore (1974). Hence, the present observations have exhibited that Bhimtal lake possessed a stable and uniform benthic community structure. Benthic fauna have increased with eutrophication and productivity of lake.

**Table 14.5: Major Species of Fish Recorded from Bhimtal Lake**

**Order: Cypriniformes**

**Family: Cyprinidae**

**Sub-family: Cyprininae**

1. *Cyprinus carpio* var. *specularis* (Linnaeus)
2. *Cyprinus carpio* var. *communis* (Linnaeus)
3. *Cyprinus carpio* var. *nudus* (Linnaeus)
4. *Hypophthalmichthys molitrix* (Valenciennes)
5. *Ctenopharyngodon idella* (Valenciennes)
6. *Punitius conchonius* (Hamilton)
7. *Puntius ticto* (Hamilton)
8. *Tor-tor* (Hamilton)
9. *Tor putitora* (Hamilton)
10. *Cirrhinus mrigala* (Hamilton)
11. *Catla catla* (Hamilton)
12. *Labeo rohital* (Hamilton)

**Sub-family: Rasborinae**

13. *Barilius bendelisis* (Hamilton)
14. *Barilius bola* (Hamilton)
15. *Barilius vagra* (Hamilton)
16. *Schizothorax plagiostemus* (Grey)
17. *Schizothorax richardsonii* (Grey)

Sub-family: Nemacheilinae

18. *Namacheilus montanus* (McClelland)
19. *Nemacheilus rupicola* (McCelland)

**Family: Channidae**

20. *Channa gachua* (Hamilton)

In Kumaon region, the water bodies were used primarily for sport fishing. During past two decades, Bhimtal lake was unscientific exploited for increasing fish yield. Disregarding their ecology, exotic and endemic fish species were introduced leading to no benefit. In the present study, twenty fish species belonging to two families and twelve genera were recorded (Table 14.5). The perennial fish species available throughout the year are *Tor putitora* and *Cyprinus carpio*. These two species significantly contributed the bulk of the lake fishery. Snow-trouts (Schizothroacids) are principally available during winter season. Majority of the species composed small fishes *i.e. Puntius* spp., *Barillus* spp., and *Nemacheilus* species., realized as forage species. Analysis of fish catches made during the study periods showed that the catch per unit effort ranged between 30 to 425 grams/man/hour. The catches were maximum during spring followed by summer. Sehgal (1974) has recently warned that introduction of the exotic mirror carps have drastically reduced the endemic fishes *viz.* mahseer and snow trout. However, the population of endemic fishes have significantly dwindled in the past few years.

## Acknowledgement

The author is highly grateful to Dr. K.L. Sehgal (Rt. Director) N.R.C. Cold Water Fisheries (ICAR) Bhimtal, Nainital (Uttaranchal) for providing all research facilities and constant inspiration. Lastly, I gratefully acknowledge financial assistance from Indian Council of Agricultural Research, New Delhi.

## References

Day, F. (1979). The fishes of India, Burma and Ceylon, being a natural history of fishes known to inhibit sea and fresh water. William Dawson, London. 788pp.

Dayton, P. K. (1970). Competition, predation and community structure in tropical lake. Ph. D. Thesis, Univ. of Washington D. C.; 245pp.

Edmondson, W. T. (1974). A simplified method for counting phytoplankton in 'a manual on methods for measuring primary production in aquatic environment' Vollen Weider R. A. (Ed)., 14-16, Oxford, Black Well Sci. Publications.

Edmondson, W. T. (1959). Rotifera–fresh water biology, 2$^{nd}$ edn. Wily and Sons, New York,: 420-484.

Fernando, C. H. (1980). The fresh water zooplankton of Srilanka, with a discussion of tropical fresh water zooplankton composition. Int. Revue Ges. Hydrobiol., 65: 85-125

George, M. C. (1966). Comparative plankton ecology of five fish tanks in Delhi, India. Hydrobiologia, 27 (1-2): 81-108.

Gunther, S., Drer, N. and Habil (1959). Fresh water fishes of the World. Vol. 1 and 2 translated by Tucker, D. W. (1973). T. F. H. publication Inc. Ltd., London 750pp.

Gupta, P. K. and M. C. Pant (1989). Sediment chemistry of Lake Bhimtal U. P., India. Int. Revue Ges., Hydrobiol., 74 (6): 679-687.

Hutchinson, G. E. (1957). A treatise on Limnology, Vol. 2 and 3 John Willy and Sons, New York. 1115 and 660 pp.

Jayaram, K. C. (1987). The fresh water fishes of India, Pakistan, Bangladesh, Burma and Srilanka. Zoological survey of India, Calcutta 480pp.

Joshi, A., A. P. Sharma and M. C. Pant (1981). Limnological studies in a subtropical system, lake Sattal, India. Acta. Hydrochim. et. Hydrobiol., 9 (4): 407-425.

Lackey, J. B. (1938). U. S. public health reports. 53: 2080-2093.

Lewis, W. M. (1979). Zooplankton. A community analysis approach, New York, Springer Verlag. 163pp.

Needham, J. G. and P. R. Needham (1972) A guide to the study of fresh water biology. Holden-Day I. N. C. San Francisco. Calif., 94 (3): 1-108.

Neushal, M. (1967). Studies of subtidal vegetation in western Washington, Ecology, 48: 83-93.

Nevill, H. R. (1922). District gazetter of the United Province of Agra and Outh. Vol. 34, 450 pp.

Pant, M. C., A. P. Sharma and O. P. Chaturvedi (1983). Phytoplankton population and diel variation in a subtropical lake. J. Environ. Biol. 4 (1): 15-25.

Pennak, R. W. (1957). Species composition of limnetic zooplankton communities. Limnol. and Oceanogr., 2: 222-232.

Pennak, R. W. (1953). Comparative limnology of eight Colorado mountain lakes. Univ. Colo. Studies Ser. Biol., 2: 1-75.

Ruttner, F. (1930). Das plankton des lunzer untersees. Verteilung in Raum and Zeit Wahrend der Jahre, 1903-1913. Int. Revue. Ges. Hydrobiol. Hydrogr., 23: 1-138.

Sehgal, K. L. (1974). Fisheries survey of Himachal Pradesh and some adjacent areas with special reference to Trout, Mahseer and allied species. J. Bomb. Nat. Hist., 70 (3); 456-474.

Sharma, A. P., S. Jaiswal, V. Negi and M. C. Pant (1982) Phytoplankton community analysis in lakes of Kumaon Himalaya. Arch. Hydrobiol., 93: 173-193.

Smirnov, N. N. (1974). Chydoridae of the fauna of theUSSR crustacea academy of sciences of the USSR. Zoological Institute New series no 101 Vols. 1 no. 2 English translation Israel program for scientific translation: 644pp.

Smirnov, N. N. (1976). Macrothricidae and Moinidae fauna of the World, Fauna of the USSR academy of sciences of the USSR, Zoological Institute, New Series no. 102 Crustacea, Vol. 1. 236pp.

Tenore, K. R. (1974). Macrobenthos of the Palmlico river estuary, North Carolina, Ecol. Mono., 42; 51-69.

Wilhm, J. L. and T. C. Dorris (1968). The biological parameters for water quality criteria. Biosciences, 18; 477-481.

Yeatman, H. C. (1959). In: W. T. Edmondson (ed). Fresh water Biology 2nd edn. Wiley and sons, New york, Encyclopedia. 1530pp.

**Aquatic Biodiversity in India: The Present Scenario, 2005** **297–303**
*Edited by*: D.R. Khanna, A.K. Chopra & G. Prasad
*Published by*: Daya Publishing House, New Delhi

# 15

# Effect of Pulp and Paper Mill's Effluent on Microbial Diversity of River Gola

**A.K. Singh, Rajeev Rajput*, Rakesh Bhutiani*, Mukesh Ruhela** and V. Singhal*****

*Department of Environmental Sciences, G.B.Pant University of Agriculture and Technology, Pant Nagar*

** Department of Zoology and Environmental Sciences, Gurukul Kangri University, Hardwar*

*** Department of Environmental Sciences, Guru Jambeshwar University, Hissar*

****Department of Biotechnology, IIT, Roorkee – 247 667*

## Introduction

The progress of the country since independence has been phenomenal but rapid industrialization has also brought with it the problem of environmental pollution. Today almost everything around us *viz.* the air we breathe, the water we drink and even the soil we grow our food on, are heavily polluted.

Pollution of natural water bodies is increasing day by day due to rapid population growth, industrial proliferation, urbanization, increasing living standard and a wide range of other human induced development activities.

Industrial development has come to be regarded as a measure of progress of country. In any developed or developing country industrialization assumes a great importance for its economic growth. However, due to industrialization the scale of sophistication of industrial pollution has become alarming all over the world. The pollution caused to the aquatic resources of the country by the indiscriminate growth of industry, is greatly damaging.

Industrial washes consist of a variety of chemicals, out of which some are extremely toxic to living beings. Industries including pulp and paper release oils, greases, solvents, heavy metals suspended solids, inorganic and organic pollutants and non-biodegradable materials in to the soil.

Effluent of pulp and paper industry is known to contain toxic as well as mutagenic and carcinogenic components. Some of these compounds appear in effluent from chlorine bleaching plant and have been identified as chlorinated phenols, catechols and trihydroxybenzenes as well as dehydroabetic and unsaturated carboxylic acids. The partial degradation of lignin during production of pulp and paper results in the formation of both chlorinated monomeric and polymeric phenols (chlorolignins).

The effluent from Century Pulp and Paper Mill Lalkuan submerges a large track of arable land before joining the river Gola thus damaging the health (biodiversity) of soil, crop and surface water.

## Materials and Methods

In the present study various physico-chemical parameters of Century Pulp and Paper Mill has been studied. It is located at Lalkuan (Dist. Nainital Uttaranchal) and is 8 km. away from Pantnagar.The effluent after discharge from this industry falls to river Gola. The study has been done to evaluate the effect of effluent on diversity on river Gola.

The samples were collected from the following four sites during October 2000 to July 2001.

| | | |
|---|---|---|
| SITE No. I | : | 2 km upstream of river Gola (control) |
| SITE No. II | : | The paper mill effluent channel-joining river Gola. |
| SITE No. III | : | Confluence point of effluent with river Gola. |
| SITE No. IV | : | 2 km away from site 3 towards down stream of river. |

The physico-chemical and microbiological parameters were analyzed using standard methods (APHA,1998; Trivedy and Goel,1986). The soil samples were collected in sterilized plastic bags from effluent fed soil. These were then taken to laboratory and stored at 2° C, until they were analyzed.

The water and soil samples were also analyzed for quantitative and qualitative estimation of bacteria, fungi and coliform. Microbial population was estimated by pour plate technique using agar media for bacteria, potato dextrose agar media for fungi. Coliform count was done by most probable number (MPN) method using Mc Conky broth media and nitrifying bacteria were analyzed by nitrogen free growth medium (Jenkinson and Powlson, 1976).

## Results and Discussion

During the present study the characteristics of water samples of river Gola were analyzed. The results of various physico-chemical parameters and soil microbial population are shown in Tables 15.1 and 15.2. The microbial diversity of river Gola at different sites is shown in Table 15.3 whilst the names of various species of fungi and bacteria isolated from river Gola are given in Tables 15.4 and 15.5.

It was observed that most of the physico-chemical parameters are found within the range except DO, COD, BOD, alkalinity and lignin which were found maximum at site II and minimum at site I *i.e.* control site. The pollution load decreases as distance increases from site II to site IV. Similar observations are also reported by (Achari *et al.*,1999and Nemade and Shrivastava,1997).

The value of pH ranged between 7.20 to 7.96 (Abbassi,1985) reported that in combined wastewater pH ranged from 6.2 to 8.6, while (Singh *et al.*,1996) reported that the pH ranges from 6.9 to 8.6 in combined effluent of Shreyon paper mill ltd. The value of chlorides, and free $CO_2$ ranges between 22.2±0.62 mg/l to 213.3± 8.11 mg/l and 23.11±0.35 mg/l to 92.5± 2.10 mg/l. respectively.

**Table 15.1: Physico-chemical Characteristics of Water Samples Collected from Various Sites Along River Gola**

| *Parameters* | *Site I* | *Site II* | *Site III* | *Site IV* |
|---|---|---|---|---|
| Colour | 1129 | 5748 | 5132 | 4939 |
| Odour | No smell | Pungent | Slightly pungent | Mild pungent |
| Temperature ºC | 21.5 ±0.162 | 26.3± 0.82 | 24.2 ±0.61 | 23.19 ± 0.32 |
| pH | 7.20± 0.13 | 7.96± 0.33 | 7.52 ±0.31 | 7.32 ± 0.29 |
| Chloride (mg/l) | 22.2± 0.62 | 213.3± 8.11 | 199.3 ±17.21 | 153.1 ± 15.39 |
| Alkalinity (mg/l) | 342.6 ±20.0 | 822.1± 15.5 | 799.9 ±32.1 | 698.0 ± 30.19 |
| Free $CO_2$ (mg/l) | 23.11± 0.35 | 92.50± 2.10 | 87.9 ±0.92 | 77.9 ± 0.88 |
| Total alkalinity (mg/l) | 83.50± 2.58 | 455.0±12.3 | 395.5 ±11.9 | 295.10 ± 11.10 |
| Total solids (mg/l) | 449.0±5.90 | 4739.2 ±115.3 | 3929.1± 92.9 | 3119 ±83.5 |
| Total dissolved solids (mg/l) | 437.0±1.37 | 2618.0 ±93.2 | 2016.3± 76.2 | 1998.0 ± 68.3 |
| DO (mg/l) | 6.30 ±0.42 | Nil | Nil | 2.26± 0.31 |
| BOD (mg/l) | 312.3 ±0.54 | 1289.3 ±52.1 | 1209.3± 49.8 | 1111.4 ±38.7 |
| COD (mg/l) | 819.34± 0.77 | 3215.2 ±123.13 | 3039.1± 89.7 | 2981.0 ±77.7 |
| Lignin | Nil | 1600.1± 111.2 | 1513.2± 99.1 | 1232.2± 89.1 |
| Na | 0.72± 0.01 | 182.3 ±54.3 | 153.3 ±46.5 | 132.4 ±39.1 |
| K | 0.82± 0.21 | 21.40 ±0.43 | 18.1 ±0.32 | 16.9± 0.29 |
| Ca | 2.36± 1.01 | 84.3 ±2.32 | 77.3 ±2.10 | 73.4± 2.00 |
| Mg | 44.3± 12.3 | 68.63 ±4.12 | 57.32 ±3.39 | 54.31 ±2.99 |
| Zn | 39.31 ±11.9 | 65.25 ±3.11 | 58.11± 3.00 | 55.20 ±2.78 |
| $SO_4$ | 49.12 ±12.13 | 85.21 ±3.32 | 79.21 ±3.10 | 68.11 ±3.00 |
| $NO_3$ | 423.1 ± 49.2 | 87.53 ±3.00 | 80.13± 3.32 | 76.12±3.01 |

**Table 15.2: Soil Microbial Population as Influenced by Paper Mill's Effluent**

| *Month* | *Soil Microbial Population (× $10^5$/gm Dry Soil)* | | | |
|---|---|---|---|---|
| | *Bacteria* | *Fungi* | *Actinomycetes* | *Nitrifying Bacteria* |
| March | 163.0 | 12.00 | 135.00 | 70.00 |
| June | 158.00 | 11.00 | 132.00 | 69.00 |
| October | 172.00 | 13.00 | 138.00 | 74.00 |
| Average | 164.33 | 12.00 | 135.00 | 71.00 |

**Table 15.3: Microbial Diversity of River Gola at Different Sites**

| *Parameters* | *Site I* | *Site II* | *Site III* | *Site IV* |
|---|---|---|---|---|
| Fungi | 214.0 ±2.31 | 296.0±1.11 | 320.0 ±1.36 | 260.0±2.61 |
| Bacteria ($x10^3$/100ml) | 23.00± 1.11 | 36.0 ±2.34 | 49.0±1.24 | 42.0± 1.24 |
| Coliform group (x105/100ml) | 0.58±1.80 | 0.63± 2.12 | 0.78 ±2.50 | 0.69 ±1.60 |
| *Escherichia coli* ($x10^5$/100ml) | 0.46± 1.4 | 0.59± 2.20 | 0.71 ±1.50 | 0.64±2.9 |

**Table 15.4: Different Fungi Isolated from River Gola**

| *Sl.No.* | *Name of the Species* |
|---|---|
| 1. | *Aspergillus* sp. |
| 2. | *Penicillium* sp. |
| 3. | *Fusarium* sp. |
| 4. | *Trichoderma* sp. |
| 5. | *Gliocladium* sp. |
| 6. | *Alternaria* sp. |
| 7. | *Candida* sp. |
| 8. | *Chaetomium* sp. |
| 9. | *Cladosporium* sp. |
| 10. | *Curvularia* sp. |
| 11. | *Sclerotium* sp. |

**Table 15.5: Different Bacteria Isolated from River Gola**

| *Sl.No.* | *Name of the Species* |
|---|---|
| 1. | *Escherichia coli* |
| 2. | *Aerobacter* sp. |
| 3. | *Aeromonas* sp. |
| 4. | *Klebsiella pneumoniae* |
| 5. | *Pseudomonas* sp. |

The soil samples were collected in the month of March, June and October. Bacteria, Fungi Actinomycetes and nitrifying bacteria were found maximum in the month of October as 172.00 colonies/ml, 13.0 colonies/ml, 138.0 colonies/ml, and 74.0 colonies/ml respectively. Maximum number of colonies were reported in October because of maximum moisture content in this month and the minimum number of colonies are found in the month of June *i.e.* 158 colonies/ml, 11.0 colonies/ml, 132.0 colonies/ml, and 69.0 colonies/ml respectively. Similar observations have also been reported by (Ramaswami,1996; Gupta and Sharma,1975; Gupta *et al.*,1980 and A. Kumar, 2002). Occurrence of maximum species during winter seasons might be due to moderate temperature and slightly higher values of pH and organic matter.

The minimum microbial diversity was found at site I and maximum at site III due to regular mixing of effluent with the river water. Such differences and fluctuations in fungal numbers may be due to surface run-off, washing of clothes, mixing of effluents anf other such activities.

## Conclusion

It was concluded that due to mixing of paper mill's effluent in river Gola affects its aquatic biodiversity. As microbial biodiversity increases many problems like eutrophication, algal bloom etc. that badly affects the ecostatus of thisa river. So effluents must be discharged after proper treatment.

## References

Abbassi, A. (1985). Occurrence, toxicity and treatment of lignin in pulp and paper mill effluent. State of Art J. Env., 65: 1-8.

Achari, M. S., Dhakshinamurthy, M and Arunachalam, G. (1999). Study on the influence of paper effluent on the yield availability and uptake of nutrients in rice. J. Ind, Soc. Soil Sc., 47 (2). 276-280.

APHA, AWWA, WPCF (1998). Standard methods for examination of water and wastewater, 20$^{th}$ ed. American public health Association, Washington D. C.

Gupta, R. D., and Sharma, G. J. (1975). Microbiological properties of certain soils of Himachal pradesh. Soil Tech. 12(1): 37-41.

Gupta, R. D., Jha, K. K. and Sharma, P. K. (1980). Distribution of microorganisms in relation to physico-Chemical properties in soils of Jammu and Kashmir. J. Ind. Soc. Soil Sc. 28: 259-262.

Jenkinson, D. S. and Powlson, D. S. (1976). The effect of biocidal treatment on metabolism in soil-V. Soil Biol. Biochem. 8: 209-213.

Kumar, A. (2002). Impact of irrigation by pulp and paper mill effluent on soil and ground water quality around Lalkuan. Ph. D. thesis, Gurukula Kangri Vishwavidyalaya, Haridwar pp. 1-168.

Nemade, P. N. and Shrivastava, V. S. (1997). Detection of metals in pulp and paper mill effluent by ICP-AES and flame photometry and their impact on surrounding environment. J. Ind. Poll. Contrl. 13 (2). 143-149.

Ramaswami, P.P. (1996). Studies on the physico-chemical and biological properties of soil of Madras. Madr. Aric. J. 53: 388-397.

Sarkar, J. P. and Chaudhari. S. (1991). Forms and distribution of fungi in paper and Pull manufacture wastewater. J. Mycopathol. Res. 29 (1). 87-92.

Singh, R. S. Marwaha, S. S and Khanna, P. K. (1996). Characteristics of pulp and paper mill effluent. J. Ind. Poll. Cont., 12 (2). 163-172.

Trivedy, R. K. and Goel P. K. (1986). Chemical and biological methods for water pollution studies, Environmental publications, Karad.

**Aquatic Biodiversity in India: The Present Scenario, 2005** **304–327**
*Edited by*: D.R. Khanna, A.K. Chopra & G. Prasad
*Published by*: Daya Publishing House, New Delhi

# 16

# Latitudinal Variation in Phytoplankton Assemblage

**A. Wanganeo, S. Gagroo, A.R. Yousuf* and R. Wanganeo**
*Department of Limnology, Barkatullah University, Bhopal*
**Department of Zoology, Kashmir University, Kashmir*

The population density in water bodies varies considerably according to nutrient supply and body size. Changes in water quality exert a selective action on the flora and fauna, which constitute the living population of water, and the effects produced in them can be used to establish biological indices of water quality. Like other living organisms, the environment in their vicinity also largely affects phytoplankters. Therefore, it is quite significant to enumerate phytoplankton of a water body along with the quality of water.

The water-bodies studied in the present work are situated at various altitudes ranging from plains to high mountains (*i.e.*, from 494 m.a.s.l to–3000 m.a.s.l). Besides their variation in altitudes, these water bodies have different climatic conditions and are subject to varied degrees of human impacts (*i.e.*, both rural and urban) resulting in variation of their biotic populations.

The high altitude mountain lakes (alt. > 3000 m.a.s.l) are an ideal habitat for cold-water fisheries (Salmo *truta fario*). Water of these lakes is also used for potable purposes. Because, of their placement (location) these water bodies are best reservoirs for generating hydroelectric generation. The valley lakes (alt.>1500 m.a.s.l) besides being an attraction for tourists and naturalists are also the main sources of potable water. These lakes also serve as source of food and fodder in terms of macrophytes and fish. They are also important for transportation, irrigation and agriculture. Upper lake (Bhopal) situated at 494 m.a.s.l in Vindhyan range is the chief source of drinking water. About 30 mgd of potable water is supplied to the Bhopal City. Besides this, the water-body is also being used as a sanctuary for the visiting migratory birds, raising fish and for aquatic sport. The human activities are causing undesirable ecological changes because of which water quality deteriorates followed by major changes in their biota.

## Methodology

Sampling at different altitudes has been carried out following uniform sampling pattern. Both physical and chemical characteristics have been analyzed following the methodology as given in APHA (1985) and Welch (1948). Quantitative estimation of phytoplankton was carried out by passing a known volume of water through a nylobolt plankton net No. 130. Taking 1 ml. of the preserved sample in a Sedgwick rafter chamber and counting its entire contents did enumeration of algae. The results have been expressed as units per liter after (Wanganeo and Wanganeo,1991). The algae, where possible, was analysed to species, otherwise only to generic level. The works of the following were consulted for identification: Smith (1950); Hirano (1955, 1966; Desikacharya (1959); Edmondson (1959); Prowse (1962); Weber (1971); Wanganeo and Wanganeo (1991) and others. Similarity index was calculated by Jaccard measure Maugurran (1988); Shannon and Weavers (1964) index was used for the calculation of species diversity. (Nygaard's,1949) index was used for the computation of trophic state index.

## Results and Discussion

### Physico-chemical Characteristics

Present study has been carried out on seven water bodies. The morphometric data of various lakes are presented in Table 16.1.

**Table 16.1: Morphometric Values of the Lakes**

| *Lakes Characteristics* | *Units* | *Gangabal* | *Nundkol* | *Manasbal* | *Khushalsar* | *Malpursar* | *Anchar* | *Upper Lake* |
|---|---|---|---|---|---|---|---|---|
| Attitude | m | 3570 | 3507 | 1584 | 1584 | 1592 | 1584 | 494 |
| Latitude | | 34'51'N | 34'48'N | 34'15'N | 34.2'N | 34'10'N | 34'0'N | 23'20'N |
| Longitude | | 74'73'E | 74'78'E | 74'40'E | 74.60'E | 74.25'E | 74'12'E | 77'15'E |
| Area | ha | 172.5 | 42.5 | 281 | 123 | 0.28 | 680 | 2169 |
| Max. length | km | 3 | 1.25 | 3.5 | 2.25 | 0.1 | 6 | 10.6 |
| Max. breadth | km | 0.95 | 0.5 | 1.5 | 0.75 | 0.04 | 2.76 | 3.75 |
| Max. depth | m | 83.5 | 67 | 12.5 | 4 | 1.5 | | 7.2 |

Temperature, both atmospheric and water was recorded at the sampling sites and as such there is no similarity between the present data with that of meteorological records reported by the respective local observatories. Altitude plays a significant role in governing the thermal regimes of the water bodies. It has been seen that with the increase in the altitude there was a corresponding decline in the water as well as the air temperatures, which in turn control the biological activity of these water bodies.

The lakes situated at an average altitude of 1500 m.a.s.l. showed maximum difference in their water temperature range values compared to the tropical lakes situated at 494 m.a.s.l. (Wanganeo,1980) reported meteorological conditions to be responsible for thermal stratification in Manasbal Lake and other valley lakes.

| *Altitude* | *Mean Air ºC* | *Mean Water ºC* |
|---|---|---|
| >3000 m.a.s.l (Summer only) | 10.5 | 11.1 |
| >1500 m.a.s.l. | 15.7 | 13.9 |
| 494 m.a.s.l. | 27.8 | 23.8 |

Lakes above 3000 m.s.l. recorded a mean transparency value of 2.42 m. among the lakes situated above 1500 m.s.l., Manasbal Lake recorded the maximum transparency value (mean value of 3.34 m.). Lake situated at 494 m.s.l. recorded a mean Secchi value of 0.91 m. Thus the valley lakes (above 1500 m.s.l.) recorded a higher transparency value compared to the high altitude lakes (above 3000 m.s.l.), followed by Upper lake (494 m.a.s.l.) Table 16.2. The lake above 3000 m.s.l. recorded near neutral pH values while the lakes above 1500 m.s.l. and 494 m.a.s.l. recorded pH in an alkaline range. A direct correlation between the photosynthesis and pH has been encountered, that is why higher average pH values (8.5 units) have been recorded in the lakes situated at lower altitudes (*i.e.*, 494 m.a.s.l.). Further, a reduction in pH values with the increase in altitude has been recorded, thus confirming the importance of photosynthetic activity on pH values. On average basis pH values in the valley lakes (>1500 m.a.s.l.) have been found to be influenced by the presence of $CO_2$.

| *Lakes* | *pH* | $CO_2$ |
|---|---|---|
| Manasbal | 8.0 | 20 |
| Anchar | 8.0 | 24 |
| Malpursar | 7.8 | 27 |
| Khushalsar | 7.4 | 52 |

**Table 16.2: Secchi Variation in Different Lakes**

| *Lake* | *Minimum* | *Maximum* | *Mean* | *SD±* |
|---|---|---|---|---|
| | *Meters* | | | |
| Gangabal | | | | |
| St. 1 | 2.5 (IX) | 2.80 (V–VIII) | 2.42 | 1.45 |
| Manasbal | | | | |
| 1987–88 | | | | |
| St. 2 | 2.70 (V) | 4.10 (I) | 3.28 | 0.39 |
| 1988–89 | | | | |
| St. 2 | 2.60 (VI) | 4.10 (XII) | 3.34 | 0.39 |
| Khushalsar | | | | |
| St. 1 1987–88 | 0.10 (VIII–X) | 0.55 (V) | 0.33 0.14 | |
| 1988–89 | 0.10 (VIII) | 0.50 (I, VI–XI) | 0.35 | 0.14 |
| St. 2 1987–88 | 0.20 (VIII) | 0.70 (I, XI–XII) | 0.52 | 0.17 |
| 1988–89 | 0.02(VIII) | 0.70 (XI–XII) | 0.45 | 0.18 |
| Malpursar | | | | |
| St. 2 1987–88 | 0.30 (IV, VI, XII) | 0.50 (IX–X) | 0.38 | 0.04 |
| 1988–89 | 0.30 (IV–XII) | 0.50 (I, II, III, V, VI, VII, IX) | 0.45 | 0.07 |
| Anchar | | | | |
| St. 2 1988 | 0.80 (I, II, VIII–XII) | 0.95 (V, VII) | 0.85 | 0.06 |
| Upper Lake | | | | |
| St. 1 1992 | 0.46 (VII) | 1.73 (I) | 0.91 | 0.36 |

Thus fall in pH may be related to the presence of high concentration of $CO_2$ in water.

The lakes situated on an average altitude of above 1500 m.s.l. and at 494 m.a.s.l. recorded higher concentration of dissolved oxygen in surface water as compared to bottom waters. The high values can be the result of vigorous photosynthesis. The alkalinity at the three altitudes is mainly of bicarbonate type. Higher chloride values were recorded in valley lakes (av. Alt. 1500 m.a.s.l.) compared to the tropical lake (494 m.a.s.l.) and the mountain lakes (> 3000 m.a.s.l.) This might be due to the direct influence of human and cattle excreta entering into the systems from their respective catchment areas. The valley lakes, on an average value basis, showed the following sequence.

| **Khushalsar** | **>** | **Anchar** | **>** | **Manasbal** | **>** | **Malpursar** |
|---|---|---|---|---|---|---|
| **(39.80 mg/l)** | | **(35.91 (mg/l)** | | **(20.3 mg/l)** | | **(19.58 mg/l)** |

Comparing the mean $NO_3$-N values in the lakes situated at the three altitudes, the lakes above 3000 m.s.l., recorded low concentration of $NO_3$ -N (49.52 μg/l) compared to lake at 494 m.a.s.l. (86.26 μg/l) and lakes above, 1500 m.s.l. (537.81 μg/l). The high concentration of $NO_3$–N in the lakes situated at an average altitude of 1500 m.a.s.l. can be attributed to the extensive use of the catchment area for agricultural purposes. The lakes above 3000 m.s.l. recorded the least values of phosphorus on average basis (8.0 μg/l) followed by lakes at > 1500 m.a.s.l. (54.0 μg/l) and at 494 m.a.s.l. (121.0 μg/l). Based on the (Wetzel's,1983) classification, the lakes above 1500 m.s.l. and 494 m.s.l. have been put under eutrophic category. The mountain lakes (> 3000 m.s.l.) have been found to be oligotrophic, while the tropical lakes (494 m.a.s.l.) fall under the hypereutrophic category. The higher values of phosphorus in valley (1500 m.a.s.l.) and the Upper lake (494 m.a.s.l.) is on account of excessive input of sewage and domestic waste waters from their respective catchment areas.

Applying the classification proposed by Wetzel (1983) the four valley lakes can be categorized as under:

| *Lake* | *Average Value* | *Status* |
|---|---|---|
| Malpursar | 21.0 μg/l | Mesotrophic |
| Manasbal | 23.0 μg/l | Mesotrophic |
| Anchar | 31.0 μg/l | Mesotrophic |
| Khushalsar | 135.0 μg/l | Hypereutrophic |

## Biological Characteristics

The number of phytoplankton species was low in the tropical lakes situated at an altitude of 494-m.a.s.l (59) compared to the lakes situated above 3000 m.s.l (63) and the valley lakes situated at an altitude >1500 m.a.s.l (112). Compere (1983) also observed a greater number of taxa in the valley lakes of Kashmir. He related this to the presence of rich and varied macrophytic vegetation, which affords the algae the variety of ecological niches. A decline in the number of species towards the tropics has been observed. Lewis (1978) also suggested that the phytoplankton species abundance decline towards the tropics. Kalff and Watson (1986) could not furnish any evidence regarding the decline of phytoplankton species abundance with latitude temperate zone to the tropics.

Bacillariophyceae class recorded the maximum number of species (31) in the lakes above 3000 m.s.l. while the lakes above 1500 m.s.l. (55 spp.) and lakes situated at an altitude of 494 m.a.s.l. (27spp.) recorded their maximum number of species from the class chlorophyceae. Zutshi (1991) also recorded maximum number of diatom species from the high altitude lakes of Himalayan region. Raina (1981) also recorded maximum species from chlrophyceae in BodSar Lake, Kashmir while Wanganeo (1980) reported the maximum number of 55 chlorophyceaen species in Manasbal Lake.

Above 3000 m.s.l. both the lakes recorded their maximum population in September 1988 with the class Bacillariophyceae contributing the maximum. *Cymbella lanceolata* recorded the maximum number during the month in both the lakes. Olive *et al.* (1968) reported that the large, deep and relatively unproductive lakes in temperate region usually exhibit a rise in total phytoplankton in the spring and early summers, reaching a maximum in summer. Wanganeo and Wanganeo (1991) recorded the dominance of class Bacillariophyceae in the lakes above 3000 m.s.l., contributing approximately 80 per cent towards the phytoplankton while in the valley lakes (>1500 m.a.s.l.) the authors recorded the contribution of Bacillariophyceae to be only 40 per cent.

Among the valley lakes (>1500 m.a.s.l.) Manasbal Lake recorded a single peak in summers. Class Bacillariophyceae with maximum contribution from the species Synedra ulna dominated the peak. (Wanganeo and Zutshi, 1981) also recorded dominance of Bacillariophyceae in Manasbal Lake. While conducting the diurnal studies, they reported that the phytoplankton population of Manasbal lake to be non-motile. The abundance of diatoms in deeper regions has been attributed to their high sinking rate by the authors. Khushalsar Lake recorded two peaks in summer only and the class Chlorophyceae dominated both the lakes. *Zygnema pectinata* and *Scenedesmus quadricauda* species dominated the peak populations respectively. Malpursar Lake recorded a summer and winter peak. During summer, Bacillariophyceae recorded dominance while during winter Chlorophyceae was the dominant class.

Blooms of *Navicula radiosa* and *Spirogyra* sps. dominated the peak populations respectively. One smaller peak was observed in between the two peaks during early winter. Class Bacillariophyceae with *Cymbella lanceolata* contributing the maximum dominated the peak. Anchar Lake recorded two peaks, a winter and a summer peak. The class Bacillariophyceae dominated both the peaks. The dominant species contributing to the class were *Fragillaria crotenensis*, Melosira sps. and *Cymbella lanceolata*. Wanganeo and Wanganeo (1991) reported that the valley lakes depicted varied types of algal patterns, because of their different morphological and chemical characteristics. Kant and Kachroo (1977) reported maximum phytoplankton density in autumn in Dal and Nagin lakes of Kashmir. (Khan, 1978) obtained unimodal behaviour of phytoplankton in Naranbagh Lake, Kashmir. Raina (1981) observed two peaks, one in winter and another in summer in Bod–Sar Lake, Kashmir.

In the lake situated at 494 m.a.s.l. a large peak was recorded during monsoon period, which may be on account of the high nutrient influx from its catchment area. Class Bacillariophyceae dominated the peak. *Melosira* sps. was the main contributor to the class. Two smaller peaks were recorded in the late summer and post monsoon period. Class Cyanophyceae dominated the peaks. *Microcystis aeruginosa* contributed the maximum number towards the peak. Release of phosphate and nitrogen from soil colloids, decomposition of organic matter and conversion of insoluble salts contribute to the sudden rise of nutrients enriching the water body during monsoon season (Hutchinson, 1941). This might be the reason of obtaining the monsoon peak in the Upper Lake.

The similarity index for various classes between the lakes is shown in Table 16.3. The total phytoplankton species composition

**Table 16.3: Similarity Index in Different Lakes**

| | *N* | *M* | *K* | *MP* | *A* | *U* |
|---|---|---|---|---|---|---|
| Total Phytoplankton | 0.84 | 0.52 | 0.51 | 0.5 | 0.49 | 0.38 |
| | | 0.45 | 0.43 | 0.47 | 0.41 | 0.36 |
| | | | 0.57 | 0.56 | 0.79 | 0.56 |
| | | | | 0.61 | 0.44 | 0.48 |
| | | | | | 0.62 | 0.48 |
| | | | | | | 0.39 |
| Bacillariophyceae | 0.93 | 0.54 | 0.56 | 0.53 | 0.56 | 0.37 |
| | | 0.56 | 0.56 | 0.53 | 0.56 | 0.41 |
| | | | 0.82 | 0.86 | 0.82 | 0.62 |
| | | | | 0.75 | 0.79 | 0.54 |
| | | | | | 0.69 | 0.54 |
| | | | | | | 0.57 |
| Chlorophyceae | 0.69 | 0.55 | 0.47 | 0.54 | 0.52 | 0.43 |
| | | 0.39 | 0.38 | 0.36 | 0.36 | 0.36 |
| | | | 0.58 | 0.54 | 0.68 | 0.48 |
| | | | | 0.49 | 0.34 | 0.48 |
| | | | | | 0.54 | 0.42 |
| | | | | | | 0.3 |
| Cyanophyceae | 1 | 0.44 | 0.58 | 0.27 | 0.47 | 0.22 |
| | | 0.44 | 0.58 | 0.27 | 0.47 | 0.22 |
| | | | 0.75 | 0.56 | 0.7 | 0.55 |
| | | | | 0.5 | 0.68 | 0.41 |
| | | | | | 0.61 | 0.4 |
| | | | | | | 0.5 |
| Dinophyceae | 1 | 0.5 | 0.5 | 0.5 | 0.5 | 0.5 |
| | | 0.5 | 0.5 | 0.5 | 0.5 | 0.5 |
| | | | 1 | 1 | 1 | 1 |
| | | | | 1 | 1 | 1 |
| | | | | | 1 | 1 |
| | | | | | | 1 |
| Euglenophyceae | – | – | – | – | – | – |
| | | – | – | – | – | – |
| | | | 0.41 | 0.18 | 0.5 | 0.58 |
| | | | | 0.14 | 0.83 | 0.75 |
| | | | | | 0.16 | 0.25 |
| | | | | | | 0.6 |

G = Gangabal Lake; N = Nundkol Lake; M = Manasbal lake; MP = Maloursar lake; A = Anchar lake; U = Upper lake.

revealed maximum similarity between the two lakes situated above 3000 m.s.l. The lake situated above 1500 m.s.l. showed little similarity among them. This is indicative of the fact that the four lakes do not enjoy the same trophic status. Comparing the three altitudes, the high altitude lakes (>3000 m.a.s.l.) showed least similarity with the lake situated at 494 m.a.s.l. Among the valley lakes (>1500 m.a.s.l.) Mansbal Lake recorded the maximum similarity (0.56) with the tropical lake (494 m.a.s.l.). (Kalff and Watson, 1986) recorded only 13 per cent of species from tropical lakes, which were not recorded from temperate lakes. Both the authors were of the view that a similarity of environmental conditions is not essential to have the same species. Lewis (1978) also noted a large degree of similarity between tropical and temperate lake flora.

From the lake diversity index it has been observed that the lakes situated above 3000 m.s.l. recorded the highest diversity ($\xi = 4.82$) Compared to lakes situated above 1500 m.s.l. ($\xi = 3.32$) and lakes at 494 m.a.s.l. ($\xi = 2.07$). According to Elber and Schanz (1989), high diversity index is recorded in oligotrophic lakes while low is recorded in eutrophic lakes. They were also of the view that increase in diversity was as a result of nutrient availability and grazing pressure.

Among the valley lakes, Khushalsar recorded the maximum nutrient concentration and also the maximum species diversity, which goes against the above finding. This can be attributed to the fact that the less density of phytoplankton in this lake compared to the other valley lakes seems to have been caused by increase in grazing by the large zooplankton species. This encouraged the development of other phytoplankton species and thus lead to an increase in diversity. Elber and Schanz (1989) were also of the opinion that density reduction in the phytoplaknton community seems to have been caused by increase in grazing by the larger zooplankton species. This encourages the development of a large number of other phytoplankton species and in turn lead to an increase in the diversity.

The dominance of any algae in a lake may indicate its trophic level and as such can act as indicator species.The dominant species of Bacillariophyceae encountered in the present study are: *Cymbella* spp., *Fragillaria crotonesis, Melosira spp., Navicula radiosa* and *Synedra ulna* (Table 16.4). An increase in *Fragillaria crotonesis* suggests a

**Table 16.4: Trophic Status of Lakes Situated at Different Altitudes on the Basis of Indicator Species**

| | *Oligotrophic* | *Mesotrophic* | *Eutrophic* | *Remark* |
|---|---|---|---|---|
| Total Phosphorous X = mg/l | G:8.16 | MP:21.32 | K:135.31 | – |
| | N:8.25 | M:23.35 | B:121.54 | – |
| | A:31.35 | B:121.54 | – | |
| Species Diversity | G:4.63 | MP:1.99 | K:4.28 | – |
| | N:5.06 | M:3.38 | B:2.07 | – |
| | A:3.22 | | – | |
| **Taxa** | | | | |
| **Bacillariophyceae** | | | | |
| *Cocconeis placentula* | G,N | M,MP,A | K,B | Water rich in organic content (Kamat, 1981). Clean water (Palmer, 1980) |
| *Cyclotella* sp. | G,N | M,MP,A(D) | K,B | Oligotrophic and mesotrophic (Renold, 1980) |
| *Cymbella Cymbifonnis* | G,N | – | – | Clean water (Venkateshwarlu, 1981) |
| *Cymbella lanceotata* | G,N | M,MP,A | K(D) | – |
| *Cymbella ventricosa* | G,N(D),MP | M(Dom),A | K,B | Polluted waters (Dickman, 1975) |
| *Eunotia pectinalis* | G,N | M,MP,A | K,B | Acid water (Kamat, 1981) |
| *Fragillaria crotonensis* | G,N | M,MP,A | K,S | Oligotrophic, eutrophic (Huthchinson, 1967) |
| *Gornphosphaeria* sp. | G,N | – | – | Oligotrophic, mesotrophic water (Reynold, 1984) |

*Contd...*

**Table 16.4–Contd...**

| | *Oligotrophic* | *Mesotrophic* | *Eutrophic* | *Remark* |
|---|---|---|---|---|
| *Meiosira* sp. | – | M,MP,A | K,B(D) | Oligitrophic, eutrophic (Hutchinson, 1967) |
| *Navicula radiosa* | G,N | M,MP(D),A | K,B | Wide range of nutrient conc. (Round, 1959) |
| *Pinnularia stretoraphae* | G,N(D) | M,MP,A | K | |
| *Suirella ovata* | G,N | M,MP,A | K,B | Organically rich water (Palmer, 1980) |
| *Synedra ulna* | G,N | M,MP,A | K,B | Oligotrophic eutrophic (Hutchinson, 1967) |
| **Chlorophyceae** | | | | |
| *Ankistrodesmus falcatus* | G | M,MP,A | K,B(D) | |
| *Closterium eherenbergii* | G,N | M,MP,A | K,B | Mesotrophic (Coesal *et al.*, 1978) |
| *Closterium parvulum* | G,N | – | – | Oligotropnic (Coesal *et al.*, 1978) |
| *Coelastrum microporum* | – | M,A | K,B | Eutrophic (Palmer, 1980) |
| *Cosmarium contactum* | – | MP,A | – | Mesotrophic (Coesal *et al.*, 1978) |
| *Cosmarium granatum* | – | M,MP,A | K,B | Mesotrophic (Coesal *et al.*, 1978) |
| *Cosmarium margaritatum* | G,N | M,MP,A | K,B | Mesotrophic (Coesal *et al.*, 1978) |
| *Cosmarium moniliformes* | – | M,MP,A | K,B | Mesotrophic (Coesal *et al.*, 1978) |

*Contd...*

**Table 16.4–Contd...**

| | *Oligotrophic* | *Mesotrophic* | *Eutrophic* | *Remark* |
|---|---|---|---|---|
| *Cosmarium subcrenatum* | G(D),N(D) | M,MP,A | K,B | |
| *Cosmarium subtumidum* | – | M,A | K | Mesotrophic (Coesal *et al.*, 1978) |
| *Cosmarium reneiformis* | – | M,MP,A | K | Mesotrophic (Coesal *et al.*, 1978) |
| *Desmidium swartzii* | G,N | M,MP,A | – | Mesotrophic (Coesal *et al.*, 1978) |
| *Euastrum binale* | G,N | M,MP,A | K,B | Transitional species (Coesal *et al.*, 1978) |
| *Pandorina* sp. | – | M,MP,A | K(D),B | Polluted waters (Venkateshwarlu, 1981) |
| *Padiastrum simplex* | G,N | M,MP,A | K,B(D) | Eutrophic (Huchinson, 1967) Polluted water (Nordie, 1976) |
| *Scenedesmus bijugatus* | – | M,MP,A(D) | K,B | Eutrophic (Hutchinson, 1967) Polluted water (Nordie, 1976) |
| *S. quadricauda* | G,N | M,MP,A | K(D),B(D) | Eutrophic (Hutchinson, 1987) Polluted water (Nordie, 1976) |
| *Spirogyra* sp. | G,N | M,MP(D),A | K,B | Eutrophic (Mittal and Sengar, 1991); Eutrophic (Palmer, 1980) |
| *Straustrum deckii* | G | M,MP,A | K | Mesotrophic (Coesal *et al.*, 1978) |
| *S. furcatum* | G,N | M,MP,A | K,B | Oligotrophic (Coesal *et al.*, 1978) |
| *S. furcerigerum* | G | M,MP,A | K | Mesotrophic (Coesal *et al.*, 1978) |
| *S. orbiculare* | – | – | – | Mesotrophic (Coesal *et al.*, 1978) |

*Contd...*

**Table 16.4–Contd...**

| | *Oligotrophic* | *Mesotrophic* | *Eutrophic* | *Remark* |
|---|---|---|---|---|
| **Cyanophyceae** | | | | |
| *Lyngbya* sp. | – | M,A | K | Eutrophic (Palmer, 1980) |
| *Merismopedia glauca* | G,N | – | – | Oligotrophic (Hutchinson, 1967, Palmer, 1980) |
| *Merismopedia punctata* | G,N | M,MP,A | K | Oligotrophic (Hutchinson, 1967) |
| *Microcystis acruginosa* | – | M,A | K,B(D) | Eutrophic (Sreenivasan, 1963) |
| *Oscillatoria curviceps* | – | M,MP,A | K | – |
| *Oscillatoria limosa* | G,N | M(D)MP(D),A | K(D),B | Eutrophic (Palmer, 1980) |
| *Oscillatoria* sp. | G,(D)N(D) | M,MP,A | K | |
| **Dinophyceae** | | | | |
| *Ceratium hirudinella* | – | M,MP,A | K,B(D) | Mesotrophic (Rawson, 1956) |
| *Peridinium* sp. | G,N(D) | M(D),MP(D),A | K(D),B | Oligotrophic and Eutrophic (Hutchinson, 1987) |
| **Euglenophyceae** | | | | |
| *Euglenaacus* | – | M(D),MP,A(D) | K,B | Eutrophic (Palmer, 1980); Polluted (Venkatesh) |
| *Euglena eherenbergii* | – | MP(D),A(D) | K | Clean water (Palmer, 1980) |
| *Euglena gracilis* | – | A | K(D),B(D) | |
| *Phacus longicauda* | – | M,A | K(D),B | Clean water (Palmer, 1980) |
| *Phacus pleuronectes* | – | M(D),MP,A | K | Polluted waters (Palmer, 1980) |

G = Gangbal lake; N = Nundkol lake; M = Manasbal lake; K = Khushalsar lake; MP = Malpursar lake; A = Anchar lake; B = Upper lake; D = Dominant; A = Absent response to nutrient levels and was associated with the eutrophication of lake Washington. Stockner and Benson (1967) and Rawson (1956) also considered *F. crotonesis* to indicate eutropy. *Synedra ulna* has also been observed to prefer eutrophic water Lowe (1972). *Cymbella ventricosa, Synedra ulna* and *Fragillaria capucina* have been recorded as B-mesosaprobic indicator species which are commonly found in organically polluted water Dickman (1975). Palmer (1980) reported oligotrophic lakes to be accompanied by some genera of Diatomaceae (*Tabellaria spp.* and *Cyclotella spp.*). Butcher (1940) reported *Cocconies placentula* to be indifferent to mild organic pollution. Cholnoky (1968) and Palmar (1980) attributed it to a pH optimum of around 8.0 units.

Main dominant forms like *Cosmarium* spp., *Scenedesmus* spp., *Pediastrum* spp., *Straustrum* spp., *Pandorina* spp. and *Spirogyra* spp. (Table 16.4) represented Chlorophyceae.

Coesel *et al.* (1978) were of the view that the desmids showed great diversity of species in mesotrophic waters and any change in the nutrient condition in these waters can lead to deterioration of the desmid flora. Many desmid species are most frequent in oligotrophic waters, while a few are more frequent in eutrophic water bodies (Brook, 1965).

Rawson (1956) and Hickel (1973) considered Demidiaceae (*e.g., Stauastrum*) to indicate oligotrophy. Some *Staurastrum* spp. and other desmids have been indicated to form blooms Palmer and Maloney (1955). Fritsch and Rich (1990) noted a domination of Desmids from June to August correlated with high light intensity. Moss (1973) suggested that many desmids selected again hard water habitats because of typically higher pH, which limit the availability of free $CO_2$. Gough and Woelkering (1976) found *Staurastrum* to be the most prevalent genus in the plankton and the only one recorded from hard water lakes and calcareous spring ponds, which they studied. (Brook, 1965) found a high number of *Staurastrum* in eutrophic waters. (Morgan,1970) reported that as a result of eutrophication there is a decline in desmids followed by prolonged dense algal blooms in Loch Leven.

Amongst Chlorococcales, *Ankistrodesmus* spp., *Pediastrum* spp., *Scendesmus* spp. are most likely to be found in plankton of ponds or shallow fertile lakes Hutchinson (1967). Sommerfield *et al.* (1975) considered Canyon lake to be eutrophic on the basis of significant population of organism such as *Ankistrodesmus* sp., *Scenedesmus sp.*, and *Tetraedon* sp.. Nordile (1976) found *Ankistrodesmus falcatus, Pediastrum simplex* and *Scenedesmus* sp. in two polluted lakes of central Florida. Rodhe (1948) and Hutchinson (1967) suggested thermal optimum to be one of the important factors for the abundance of Scenedesmus population. Scenedesmus has been seen to need high concentration of phosphate phosphorus by Vollenweider (1968). Palmer (1980) reported *Anacsystis* sp., *Mougetia* sp., and *Spirogyra* sp., as pollution indicator algae. Mittal and Sengar (1991) reported that *Scenedesmus quadricauda* and *Spirogyra* sp. attain maximum growth with high amount of hardness while *Cosmarium* sp. and *Pediastrum simplex* record enhancement in their growth with appreciable amount of hardness.

Amongst Cyanophyceae dominant species were *Oscillatoria* sp., *Merismopedia* Sp., and *Microcystis* sp. (Table 16.4). Many blue green algae occur in nutrient poor waters while others are tolerant to high pollution Brook (1965). According to Rawson (1956) eutrophic lakes are characterized by blue green algae such as *Anabaena* sp. and *Microcystis* sp. Munro (1966) reported that the presence of very large numbers of *Anabaena* sp. and *Microcystis* sp. indicates a fairly high level of eutrophication. Gerloff and Skoog (1957) found that nitrogen was the primary limiting factor for the growth of *Microcystis aeruginosa* in several Wisconsin lakes. Okino (1973) related dense population of *M. aeruginosa* in lake Sowa an oligotrophic Japanese lake to the continuous supply of nutrients by inflow of domestic wastes and rapid turnover of nutrients. *Oscillatoria rubens* for a long time has been considered as a typical algal species for eutrophic and artificially eutrophicated lakes and has become among other Cyanophyceae, a biological indicator of changing tropic conditions (Ravera and Vollenweider, 1968).

Dinophyceae in the present study recorded two genera only, *Peridinium* sp. and *Ceratium hirudinella* (Table 16.4). Rawson (1956) classified *Ceratium* sp. as a mesotrophic form while Reynolds (1973) reported *Ceratium hirudinella* from nutrient rich water. Among Euglenophyceae the genera recorded during the study period are *Euglena* sp. And *Phacus* sp. (Table 16.4).

Palmer (1962) was of the view that phytoplankton dominated by Euglenophyceae is charecteristic of hard water lakes without an outlet. Forsyth and McColl (1975) reported that the presence of euglenophyceae in the water suggest an abundance of nitrogen rich organic matter in the lake sediments. (Wetzel,1975) also found euglenoids in shallow waters rich in organic matter. Seenaya (1972) showed that actively decaying detritus, which simultaneously release iron and consume oxygen, which in turn induce the growth of euglenoids. The majority of the species of *Euglena* sp. and *Phacus* sp. are found in small, and sometimes in very minute bodies of water, which often have high organic content (Hutchinson, 1967). Palmer (1980) reported *Euglena gracillis., E. acus* and *Phacus pleuronectes* as indicators of eutrophy.

Though no single index has yet proved adequate in characterizing the tropic levels of all lakes, but many provide an additional clue in determining the lake types. The relative number of species of Chlorococcales and Desmidiaceae present in particular lakes has been used to recognize trophic types. These are called the simple quotients c.f. Rawson (1956) and are obtained by dividing number of species of Chlorococcales by the number of species of Desmidiaceae. The Upper lake (494 m.a.s.l.) recorded the highest quotient of 1.2 followed by the valley lakes (>1500 m.a.s.l.), Khushalsar lake recorded a quotient of 0.9 followed by Manasbal lake (0.85) which in turn was followed by Anchar lake (0.59) and Malpursar lake (0.52). The high altitude lakes (>3000 m.a.s.l.) recorded the lowest quotient (0.50).

**Rawson's Simple Quotient Index**

| | | |
|---|---|---|
| Gangabal lake | 7/14 | 0.50 |
| Nundkol lake | 5/10 | 0.50 |
| Mansbal lake | 17/20 | 0.85 |
| Khushalsar lake | 16/17 | 0.90 |
| Malpursar lake | 11/21 | 0.52 |
| Anchar lake | 13/22 | 0.59 |
| Upper lake | 12/10 | 1.20 |

It seems rather odd to determine the trophic status from Desmid and Chlorococcales alone when other groups form more important components of the phytoplankton assemblages. Compound index as proposed by (Nygaard,1949) provides a better index of the trophic type

**Nygaards Compound Trophic State Index**

| *Lakes* | *Chloroco-ccales* | *Myxo-phyceae* | *Centri-ales* | *Eugle-niaceae* | *Desmids* | *Compoud Index* |
|---|---|---|---|---|---|---|
| Gangabal | 7 | 5 | 1 | - | 14 | 13/13=0.9 |
| Nundkol | 5 | 5 | 1 | - | 10 | 11/10=1.1 |
| Manasbal | 17 | 8 | 2 | 3 | 20 | 30/20=1.5 |
| Kaushalsar | 16 | 7 | 2 | 6 | 17 | 31/17=1.8 |
| Malpursar | 11 | 5 | 2 | 3 | 21 | 21/21=1.0 |
| Anchar | 13 | 9 | 2 | 5 | 22 | 29/22=1.3 |
| Upper lake | 12 | 7 | 2 | 3 | 10 | 24/10=2.4 |

Applying this index to the water bodies under present consideration, the lakes are categorised as:

| | |
|---|---|
| Gangabal lake | Oligotrophic |
| Nundkol lake | Mesotrophic |
| Manasbal lake | Mesotrophic |
| Khushalsar lake | Slightly eutrophic |
| Malpursar lake | Mesotrophic |
| Anchar lake | Mesotrophic |
| Upper lake | Slightly eutrophic |

Based on the species distribution in various water bodies, it has been observed that lakes above 3000 m.s.l. recorded high species diversity, low total phosphorus content besides the presence of species preferring nutrient poor water (Table 16.4). The presence of *Merismopedia glauca, M. punctata, Closterium parvulum, Cymbella cymbiformis* and *Gomphosphaeria* sps. further confirms that the lakes *viz.*, Gangabal and Nundkol are oligotrophic in nature. The presence of following species *Colosterium eherenbergii., Cosmarium contractum., C. margaritum, Desmidium swartzii* and *Staurastrum furcerigerum* gives an idea about mesotrophic nature of water bodies present at an altitude above 1500 m.a.s.l. except Khushalsar lake. Malpursar lake depicting very low species diversity (Table 16.4) would have been categorized among the eutrophic water bodies but on account of relative less nutrient status compared to Upper lake and the presence of species not preferring eutrophic water bodies suggest the inclusion of this water body among mesotrophic ones. The low concentration of nutrients especially 'P' may be on account of the luxuriant uptake of 'P' by the standing crop of algae especially (*Spirogyra* sp.). Thus emphasizing the importance of study of plankton population without which it would have been difficult to assign true trophic status to the water body of suh a type. Khushalsar lake (>1500 m.a.s.l.) recorded the presence of eutrophic species like *Actinastrum hantzschii.*, *Hydrodictyon* sp., *Lyngbya* sp. and *Euglena gracilis*. The dominance of *Pandorina* sp. and Scenedesmus quadricauda further confirm its trophic status. Upper lake (494 m.a.s.l.) reported the presence of eutrophic species like *Ankistrodesmus falcatus*, *Euglena gracilis* and more over, the dominance of species like *Melosira* sp. And *Microcystis aeruginosa* which prefer highly enriched water bodies confirms its eutrophic status. The high total phosphorus (Table 16.4) in Khushalsar lake and Upper lake further support the fact that these aquatic systems have undergone a trophic evolution. while considering the various indices *i.e.*, Nygaard's trophic status index, species diversity index, besides the nutrient status of various water bodies a general consensus has been arrived at, in assigning the present trophic status to the various aquatic ecosystems under consideration.

| *Altitude* | *Lakes* | *Trophic Status* |
|---|---|---|
| >3000 | Gangabal lake | Oligotrophic |
| | Nundkol lake | Mesotrophic |
| >1500 | Manasbal lake | Mesotrophic |
| | Khushalsar lake | Slightly eutrophic |
| | Malpursar lake | Mesotrophic |
| | Anchar lake | Mesotrophic |
| 494 | Upper lake | Slightly eutrophic |

The trophic status assigned is in conformity with the findings of Wanganeo (1984). The author classified the lakes on the basis of their nutrient status and primary production values. The lakes above 3000 m.s.l., have been categorized as low productive (Oligotrophic), while the lakes with high carbon uptake values and low species diversity have been categorised among higher productive lakes (Eutrophic). The lakes recording a transitional stage between the low and the high productive stage have been categorised as moderate (Mesotrophic lakes).

It can safely be concluded that the interference in the catchment area of various water bodies present at different latitudinal planes gets reflected in the biological populations especially the autrotrophs they support. Thus any change or shift in the biological population can safely be related to its trophic status.

## References

APHA (1985). Standard methods for the examination of water and wastewater. 16 th ed., APWA, AWWA, WWCF, U. S. A.

Brook, A. G. (1965). Planktonic algae as indicators of lake Types, with special reference to the Desmidiaceae. Limnol. Oceanogr., 10: 403-411.

Butcher, R. W. (1940). Studies in the ecology of rivers. IV. Observations on the growth and the distribution of sessile algae in the river Holl, Yorkshire. J. Ecol., 28: 210-233.

Cholnoky, B. J. (1968). Die okologie der diatomcen in Binnengewassen. 699 pp. Cramer Publisher.

Coesel, T. M. P., Kwakkastein, R. and Verschoor, A. (1978). Oligotrophication and Eutrophication tendencies in some Dutch Moorland pools, as reflected in their desmid flora. Hydrobiologia, 61 (1): 21-31.

Cole, G. A. (1975). Textbook of Limnology. C. V. Mosby company. Saint louis.

Compere, P. (1983). Some algae from Kashmir and Ladakh, W. Himalayas. Bulletin Society Royal Botany, Belgium., 116: 141-160.

Desikacharya, T. V. (1959). Cyanophyta. Indian Council of Agricultural Research, New Delhi.

Dickman, M. (1975). A comparison between the effect of sporadic and continuous loading from sulfite paper mills on attached algae in the Ottawa river near Ottawa, Canada. Proc. 10th., Canadian. Symposium on Water. Pollution Research. Canada, 65-72.

Edmondson, W. T. (1959). Fresh water biology. 2 nd ed., John Wiley and Sons Inc., New York, London.

Elber, F. and Schanz, F. (1989). The causes of change in the diversity and stability of phytoplankton communities in small lakes. Fresh water biology, 21: 237-251.

Forsyth, D. J. and Mc Coll, R. H. S. (1975). Limnology of lake Nagahewa, north island Newzealand. N. Z. J. Mar. Fresh wat. Resi. 9 (3): 311-332.

Fritsch, F. E. and Rich, F. (1990). Studies on the occurrences and reproduction of British fresh water algae in nature. II A five year observation of the fishpond, Abbot's Leigh, near Bristol. Proc. Bristol. Nat. Soc. Ser., 4 (2): 27-54.

Gerloff, G. C. and Skoog, F. (1957). Nitrogen as a limiting factor for the growth of Microcystis aeruginosa in southern Wisconsin lakes. Ecology, 38: 556-561.

Gough, S. B. and Woelkerling, W. J. (1976). Wisconsin desmids. II. Aufwuchs and plankton communities of selected soft water lakes, hard water lakes and calcareous spring ponds. Hydrobiol., 49 (1): 3-25.

Hickel, B. (1973). Limnological investigation in lakes of the Pokhara valley Nepal. Int. Rev. Ges. Hydrobiol., 58: 659-672.

Hirana, M. (1955). Fresh water algae of Bhutan. II. Actr. Phytotax. Geobot., 22 (1–2): 43–48.

Hirano, M. (1966). Fresh water algae of Kora–Koram and South west Himalaya. The Kyota Univ. Scientific Expedition to the Kora–Koram and Hindukush, 8.

Hutchinson, G. E. (1941). Limnological studies in Connecticut. IV. Mechanism of intermediary metabolism in stratified lakes. Ecol. Monogr., 11; 21-60.

Hutchinson, G. E. (1967). A treatise on Limnology. II. Introduction to the lake biology and the limnoplankton. John Wiley and Sons, Inc., New York, London.

Kalff. J. and Watson, S. (1986). Phytoplankton and its dynamics in two tropical lakes: A tropical and temperate zone comparison. Hydrobiologia, 138: 161-167.

Kamat, N. D. (1981). Diatoms and diatom populations indicating water quality and pollution. W. H. O.

Kant, S. and Kachroo, P. (1977). Limnological studies in Kashmir lakes. I. Hydrobiological features, composition and periodicity of phytoplankton in the Dal and Nagin lakes. Phykos, 16 (1-2): 77-97.

Khan, M. A. (1986). Hydrobiology and organic production in a marl lake of Kashmir Himalayan Valley. Hydrobiologia. 135: 233–242.

Lewis, W. M. Jr. (1978). Analysis of succession in a tropical phytoplankton community and a new measure of a succession rate. American Naturalist, 112: 401-414.

Lowe, R. L. (1972). Diatom Population dynamics in a Central Iowa drainage ditch. Iowa State J. Res., 47 (1): 7-59.

Magurran, A. E. (1988). Ecological diversity and its measurements, Croom Helm Limited, 11, New Felter Lane, London paper 1-178.

Mittal, S. and Sengar, R. H. S. (1991). Studies on the distribution of algal flora in polluted region of Karnan river at Agra (India) 1: 221–230. Current Trends in Limnology Editor, Nalin K. Shastree

Morgan, N. C. (1970). Changes in the fauna and flora of a nutrient enriched lake. Hydrobiologia, 35: 544-553.

Moss, A. (1973). The influence of environmental factors on the distribution of fresh water algae. An experimental study. II. The role of pH and the carbon dioxide bicarbonate system. J. Ecol., 61: 157-177.

Munro, J. L. (1966). A Limnological survey of lake Mc Iiwaine, Rhodesia. Hydrobiol., 28: 281-308.

Nordlie, F. G. (1976). Plankton communities of three Central Florida lakes. Hydrobiol., 48 (1): 65-78.

Nygaard, G. (1949). Hydrobiological studies on some Danish ponds and lakes. II. The quotient hypothesis and some new or little known phytoplankton organisms.

Okino, T. (1973). Studies on the blooming of Microcystis aeruginosa. Jap. J. Bot. 20 (6): 381-402. Palmer, C. H. (1962). Algae in water supplies. Pub. Health Ser. Pub. No. 657, U. S. Govt. Print Office, Washington, D. C. 88.

Palmer, C. H. (1980). Algae and water pollution, Castle House Publication Ltd.

Palmer, C. M. and Maloney, T. E. (1955). Preliminary screening for potential algicides. Ohio J. Sci., 55: 1-8.

Prowse, G. A. (1962). Diatoms of Malayan Freshwaters. Gard. Bull. Singapore. 19: 1.

Raina (1981). Plankton dynamics and Hydrobiology of Bod-sar lake, Kashmir. Ph. D. Thesis of Kashmir University (unpublished).

Ravera, O. and Vollenweider, R. A. (1968). Oscillatoria rubescens, D. C. as an indicator of lago Maggiore eutrophication. Schwezerische zeitschrift fur Hydrologie, 30: 347-380.

Rawson, D. S. (1956). Algal indicators of trophic lake types of Sasketchwan (Canada). Limnol. Oceanogr., 1: 18-25.

Reynolds, C. S. (1973). Phytoplankton periodicity of some North Shropshiremeres. British Phycological journal, 8; 301-320.

Reynolds, C. S. (1984). The ecology of freshwater Phytoplankton. Cambridge University Press.

Shannon C. E. and Weaver, W. (1964). The mathematical theory of communication. Univ. Illinois Press, Urbana 3: 125.

Smith, G. M. (1950). The Freshwater algae of the United States. Mc Graw Hill Book Co. Inc., New York.

Sreenivasan, A. (1963). Primary production in three upland lakes of Madras state, India. Curr. Sci., 32: 130-131.

Stockner, J. G. and Benson, W. W. (1967). The succession of diatom assemblages in the recent. sediments of lake Washington. Limnol. Oceanogr., 12: 513-532

Venkateswarlu, V. (1981). Algae as indicators of river water quality and pollution. Cent. Bd. Prev. Cont. Water Poll. Osm. Univ. Hyderabad, India. WHO workshop on biological indicators and indices of environmental pollution, 93-100.

Vollenweider, R. A. (1968). The scientific basis of lake and stream eutrophication, with particular reference to phosphorus and nitrogen and eutrophication factors. Tech. Rep. OECD Paris DAS/CSI/68, 27: 1-182.

Wanganeo. A. (1980). Phytoplankton photosynthesis, nutrient dynamics and trophic status of Manasbal lake, Kashmir. Ph. D thesis of Kashmir University. (Unpublished).

Wanganeo, A. (1984). Primary production characteristics of a Himalayan lake in Kashmir. Int. Revue. Ges. Hydrobiol., 69 (1): 79-90.

Wanganeo, A and Zutshi, D. P. (1981). Diurnal rhythm of plankton and associated environmental factors of a high altitude monomictic lake. Kashmir Univ. Res. J., 1: 1-9.

Wanganeo, A and Wanganeo, R. (1991). Algal population in valley lakes of Kashmir Himalayas. Arach. Hydrobiol., 121 (2): 219-233.

Weber, C. I. (1971). A guide to the common diatoms of water pollutions surveillance system station. U. S. Environmental protection Agency. Natl. Eviron. Res. Center. Analytical quality control laboratory Cincinnaohic.

Welch, P. S. (1948). Limnological methods. Mc Graw Hill Book Co. (2nd ed). New York and London.

Wetzel, R. G. (1975). Limnology. W. B. Saunders Co., Philadelphia, 743pp.

Wetzel, R. G. (1983). Limnology. W. B. Saunders Co., Philadelphia, 2nd. Edition.

Zutshi, D. P. (1991). Limnology of high altitudes lakes of Himalayan region. Verh. Internat. Verein., 24: 1077-1080.

**Aquatic Biodiversity in India: The Present Scenario, 2005** **328–331**
*Edited by*: D.R. Khanna, A.K. Chopra & G. Prasad
*Published by*: Daya Publishing House, New Delhi

# 17

# Studies on Biotic Diversity of Macro Invertebrates in Bihar River

**U. Awasthi and AshokAwasthi**
*Governmentt Girls P.G. College, Rewa, M.P.*

The river Bihar has its stretch of about 70 km. between Bartona, Amarpatan to Rewa (M.P.). It is most vital life of the Vindhya region and one of the key factor in geographical importance in the history of this region. Under Bihar river project a large number of barrages are to be constructed in the northern Zone of Bihar River.

The invertebrates, which live in water, are highly sensitive to environmental changes than any other organisms. Even a slight change in their optimum, environmental conditions shows its effect on them. They may die or show less growth or less density or they become infected with diseases.

In Rewa district 6227 hectares of water area is available in the form of village from riverine irrigation pond, tank and reservoirs. This water used for water supply, irrigation and fishery development, the invertebrate fauna of Bihar is very rich, beautiful and vast. They included various species of phylum protozoa, porifera, coelentrata, minor phyla, annelida, arthropoda and mollusca.

## Materials and Methods

The macro invertebrates were collected from five study sites of Bihar *viz. (i)* Raj ghat *(ii)* Bihar Bridge ghat *(iii)* Karahia ghat *(iv)* Boda bagh ghat and *(v)* Ajagarha ghat. The collected material was fixed in formaline and Lugols fixative. Microscopic preparations were made and identified. Monthly collections of this material were made to a period one-year from 2000-2001 to find out the seasonal fluctuations.

## Observation

As the environment approaches saturation, it becomes progressively more and more difficult for a new species to establish, species number and diversity thus become a constant. 54 species of macro invertebrates were collected. These include 6 species of oligochaeta, 1 species of Hirudenaria, 05 species of crustacea, 31 species of Insecta, 08 species of mollusca and 3 species of Nematoda.

## Discussion

In the present work, a number of physico-chemical parameters were studied because these also interfere their physiological process which reduce living organism ability to compete with the other population with in the environment. In station Rajghat aquatic crustacean were recorded where the bottom substrate is made up of pebbles stones. In Bihar bridge ghat site of the Bihar river the soil was rocky and silty most dominant crustacea were recorded Karahia ghat area of Bihar River was made up of large stone and sand, the most dominant macro invertebrates were crustaceans.

In Bodabagh ghat site, where the benthos was mostly stony and sandy, the large numbers of molluscan (Gastropods and Bivalves) dominant were observed.

In the Ajagarha ghat where the bottom was made up of sand and silt dominant oligochaeta population over crustacean highest population were insect and their larvae.

*Tubifex tubifex, Limnodrilus hoffmeisteri, Chironomus tentanus* and *Nematodeswee* the most resistant forms, *Branchiura sowerbyi,*

*Chaetogaster limnaei, Nais cummunis, Antocha* sp., *Psychoda* sp., *Hydrophilus* sp., *Enochrus* sp. were tolerant.

**Table 17.1: Species Diversity Index in Bihar River at "A" Site for 2000-2001**

| Sl.No. | Phylum/Class | Total No. of Individuals | No. of Individuals in the Species | Diversity Index |
|---|---|---|---|---|
| 1. | Nematoda | 3 | 1 | 1.58 |
| 2. | Oligochaeta | 4 | 1 | 1.99 |
| 3. | Hirudenaria | 1 | 1 | 0.00 |
| 4. | Crustacea | 12 | 1 | 3.49 |
| 5. | Insecta | 9 | 1 | 3.16 |
| 6. | Mollusca | 8 | 2 | 3.12 |

**Table 17.2: Species Diversity Index in Bihar River at "B" Site for 2000-2001**

| Sl.No. | Phylum/Class | Total No. of Individuals | No. of Individuals in the Species | Diversity Index |
|---|---|---|---|---|
| 1. | Nematoda | 3 | 1 | 1.58 |
| 2. | Oligochaeta | 3 | 1 | 1.58 |
| 3. | Hirudenaria | 1 | 1 | 0.00 |
| 4. | Crustacea | 9 | 1 | 3.16 |
| 5. | Insecta | 11 | 1 | 3.43 |
| 6. | Mollusca | 6 | 1 | 2.57 |

**Table 17.3: Species Diversity Index in Bihar river at "C" Site for 2000-2001**

| Sl.No. | Phylum/Class | Total No. of Individuals | No. of Individuals in the Species | Diversity Index |
|---|---|---|---|---|
| 1. | Nematoda | 1 | 1 | 0.06 |
| 2. | Oligochaeta | 2 | 1 | 0.99 |
| 3. | Hirudenaria | 1 | 1 | 0.00 |
| 4. | Crustacea | 1 | 1 | 0.00 |
| 5. | Insecta | 6 | 1 | 2.57 |
| 6. | Mollusca | 3 | 1 | 1.58 |

The species diversity index was higher (above 3.0) at station A and B in cleanwater. The highest densities of *T. tubifex* (582/sq.ft.) were recorded at station C where the species diversity index was the least.

The species diversity index of site D is similar to site A and the site E is similar to site C in species index.

## References

Bhatnagar, G. P., A. Purushothaman and G. S. Vijailakshmi (1978): Distribution of Plankton Biomass and chlorophyll–A in Vellar Estuary Bioresearch 2 (1 and 2) 83-87.

Dad, N. (1978): Population dynamics of benthic macro invertebrate's communities with reference to pollution in River Khan (Indore, M. P) M. Phil. Dissertation, Vikram Univ. Ujjain, INDIA.

Joshi, H. C. (1978): Observation on natural tabilisation of city sewage in the river Ganga near Allahabad, IAWPC Tech. Annual (V): 1157-1159.

Menon, K. G. and Murthy, S. S (1977): A summary of applied research and development of Sewage reclamation for industrial use at Madras IAWPC Conv. Vol. IV, 66-74.

**Aquatic Biodiversity in India: The Present Scenario, 2005** **332–338**
*Edited by*: D.R. Khanna, A.K. Chopra & G. Prasad
*Published by*: Daya Publishing House, New Delhi

# 18

# Coastal Biodiversity of Maharashtra: An Insight

**G.N. Kulkarni**
*College of Fisheries, Ratnagiri*

The concept of biodiversity and its conservation has grown significantly of late, in view of the human impact on the systems supporting life on the earth. In order to prevent the ecological crisis, stress therefore, is given to maintain the diversity of life, which includes floral, faunal and the habitat components. This requires balanced management of human use of the resources offered by the biosphere. Maharashtra is recognized as an important state contributing to the wealth of west coast and thus the nation. An attempt therefore is made here to provide an insight to the state's coastal diversity of the life. This would help bringing about future strategy towards sustainability of healthy and sound ecosystem to the benefit of generation to come.

## Coastal Habitat Features

The state of Maharashtra is endowed with a coastline extending almost 720 km, which is traversed, by a large number of estuaries and creeks. Varied ecosystems including mangroves, salt marshes, lagoons, estuaries, sea grass beds; coral reefs, small or inaccessible islands and beaches are encountered along the coast, supporting plethora of floral and faunal attributes. The coastal belt forms an important component of the Konkan region–an official revenue division of the state. The region comprises of five districts namely Sindhudurg,Ratnagiri, Raigad,Thane and Mumbai. The first two form the South Konkan while the rest, North Konkan.

The Konkan region geographically is trapped between Sahyadri mountain ranges in the east and Arabian Sea in the west. The terrain, here, is hilly in general with laterite soil dominating especially south and the red loamy alluvial occurring northwards. The region on average experiences rainfall of over 2500 mm annually and the air temperature ranges of 20-35°C. The climate besides being warm is also humid (humidity ranges–60-90 per cent)..

Table 18.1 shows district wise data pertaining to the coast.

**Table 18.1: Coastal Features of Maharashtra**

| *Feature* | *District* | | | | | *Total (State)* |
|---|---|---|---|---|---|---|
| | *Sindhu-durg* | *Ratnagiri* | *Raigad* | *Thane* | *Mumbai* | |
| Coastline (km) | 120 | 168 | 240 | 112 | 80 | 720 |
| Continental shelf (Sq km) | 16,000 | 36,000 | 21,000 | 10,512 | 28,000 | 1,11,512 |
| Area up to 17 m depth (sq km) | 1440 | 1060 | 2000 | 1000 | 700 | 6200 |
| Estuaries and creek (no.) | 14 | 18 | 15 | 18 | 5 | 70 |
| Kharland area (ha) | 8,000 | 12,000 | 15,000 | 35,000 | 10,000 | 80,000 |
| Distance from the shore line to the extent of Cont. shelf (km) | 88 (Malvan) | 128 | 248 | – | 272 | – |

From the Table 18.1–it is discernible that

1. Continental shelf area mainly belongs to Ratnagiri and Mumbai (60 per cent).
2. Continental shelf is wider off north Konkan and narrows down southwards.

3. Sindhudurg and Ratnagiri contribute nearly 40 per cent of the shallow sea area (*i.e.* up to 17 m depth).
4. Significant area is under tidal influence (80,000 ha) tidal force decreasing from north to south. Of this (Kharland area) about 18 per cent is reported to be suitable for aquaculture purposes, mainly in north Konkan.

It is also pointed out that numerous creeks and estuaries exist along the Maharashtra coast, 8 of them being major found in the north Konkan. The other (minor) creeks and estuarine regions are distributed all over the coastal belt of the state. The south Konkan is characteristic with small estuaries surrounded by hills or hillocks.

## Floral Components

The important flora of the coast mainly belong to 2 major categories–marine algae and mangroves.

### Marine Algae

According to a survey made by NIO (Goa), 94 species belonging to 51 genera are encountered along Maharashtra with total annual yield of 20000 tons. Malvan and Ratnagiri coast are found to have luxuriant growth, producing 450 and 400 tons respectively.

### Mangrove

Based on a satellite survey it is seen that mangrove vegetation covers significant area of the coast occupying 210 km$^2$, mostly in Raigad and Thane districts. Of these 15 and 33 km$^2$ belong to Sindhudurg and Ratnagiri respectively.

It has been possible to record at least 17 species of mangroves belonging to 11 genera and 8 families.

## Faunal Composition

### Fish

Konkan region is the prime fish producer of the state, contributing 80-90 per cent of the fish production. The coastal economy depends on this oceanic resource to a large extent. The finfish and shellfish of Konkan belong to several hundred species (mainly inshore and brackish water in habitat, of which about 90 are of commercial importance.

Fish composition shows variation between northern and southern parts of the coast according to nature of bottom and the oceanic current pattern.

## Avifauna

Mangrove ecosystem as well as coastal shallow seas support good bird population. A study made in and around Ratnagiri has revealed 121 species of birds belonging to 82 genera of 39 families. About 20 per cent of these are reported to be true migrants.

## Turtles

Along the Konkan belt, 4 species of turtle are known to occur. Of these Olive ridley (*Lepidochelys olivaceae*) is the most common. However, near Malvan, green turtle (*Chelonia mydas*) is commonly noticed. The turtle population need to be thoroughly assessed in area specially in Bagmandala (Raigad), Mithbav and Kash (Sindhudurg). In addition to the above, the biotic components include, microbial population, numerous invertebrate and vertebrate forms and other plants of terrestrial, aerial and aquatic ecosystem, occurring in conjunction and composing the overall biodiversity.

## Human Role

The coast of Maharashtra is very much influenced by various developmental activities. The northern part (Thane and Mumbai) is thickly populated, 3 districts down south, showing comparatively less population density.

The prime activities of the region include fishing, trade and industrial processes, and agriculture and horticulture to a smaller extent.

Most of the industries are concentrated in Mumbai and Thane part and to some degree Raigad and Ratnagiri districts. As a result, the ecosystems, in the north Konkan are subjected to high degree pollution. The southern part namely Ratnagiri and Sindhudurg are much cleaner. The major industry of Ratnagiri include Dabhol power project (currently not in operation) and a resin plant.Besides, some fish processing units also are also functional.

Fishing clearly is the major business activity of the entire coast, with an involvement of over 15000 mechanized and non mechanized vessels producing more than 4.0 lakh tons of sea food resources for the purpose of domestic and international trade. The coast has 184 small and 3 major ports.

The brackish water farming has shown phenomenal growth during last 7-8 years. As per the available record Raigad, Thane and Sindhudurg districts are leading in this sector with prawn culture units 48, 42, and 35 respectively.

The state Government has recently declared separate status for Ratnagiri and Sindhudurg, the former having been identified as the district for horticulture while the latter for tourism. Accordingly the development plans are afoot.

In view of the developmental activities, the coastal biodiversity of the state is influenced. The following observations (Table 18.2) point out at the incidences of ecological significance having occurred in different parts of the coast during last decade.

**Table 18.2: Coastal Incidences of Ecological Importance**

| Year | Incident/Fact | Effect/Crisis |
|---|---|---|
| 1993 | Oil spill from Bombay high | 5 km stretch near Raigad affected |
| 1996 | Green tide along south Konkan | Discoloration and mortality of several varieties of finfish and shellfish |
| 1990-2000 | Contamination of Dabhol creek north Ratnagiri | Fish kills as a result of industrial discharge from Lote complex |
| 2001-2002 | Red tide | Reddish tinge due to dinoflagellates |

Literature points out the extinct status of 2 mangrove plant genera *viz., Xylocarpus* and *Lumnitzera*.along the Ratnagiri stretch. Further it is observed that habitat alteration,due to fishing jetty and bundh construction as well as indiscriminate discharge of organic wastes, appear to have led to reduced abundance (by 70–85 per cent) of shore bottom dwellers in about last 10-15 years at Ratnagiri. Prominent among these are the star fish (*Astropecten*), lamp shell (*Lingula*), burrowing sea anemones, tusk shell (*Dentalium*), the wedge clam *Donax* and the mole crab (*Hippa*).

## Conservation Measures

Efforts have already been initiated towards the objective of long-term protection of the state's biodiversity. The NIO (Goa) has identified following areas of the coast for this purpose.

Ratnagiri (open coast including Shirgaon creek), Vikroli (Thane creek), Mumbra-Diva, Achara stretch near Malvan, Purnagad (near Ratnagiri), Colaba (south Mumbai) and Malvan.

Of these Malvan coast is considered as a favourable site for establishing a bio-reserve. The characteristics of this site are as under.

1. *Location*–16° 15′ and 16° 50′ N Latitude

   73° 27′ and 73° 31′ E Longitude
2. *Geomorphology*–indented with bays and creeks–Kolamb, Kalavati and Karli
3. *Life*–Rich in mangroves, sea grasses and sand dune vegetation especially 94 species of marine algae, 198 species of fauna including sponges, sea fans, corals (hard and small), sea anemones etc.

The state Government has already declared its intent, in 1995, to set up a marine park at the above site (near Malvan) for which an area of 30 sq. miles is considered suitable. However the proposal has met opposition from the resident people mainly fishermen folk who expect threat to the fisheries due to the said park. In Malvan area fishing no doubt is main activity with more than 500 trawlers operating. But owing to high diversity and pollution free environment it would be advisable to go ahead with the plan of setting up the marine park. The state Government as a part of management, has also established 20 artificial reefs in 10 ft. depth in the Malvan sea.

Towards the conservation of biodiversity, following measures need attention:

1. Collection of primary data on biodiversity, it's uses and susceptibility to developmental processes, with immediate and top priority to Ratnagiri and Malvan regions of the South Konkan coast. Early survey and inventory of the coastal biodiversity is needed.
2. Revision of fishery related rules and mechanism of implementation with regard to carrying capacity of the region/state is to be encouraged.
3. Control of erosion by ecofriendly measures such as replanting sand dune vegetation, beach nourishment need be strengthened. It may be noted that the state

Government,under social forestry scheme, has done commendable work by planting *Casurina* and mangrove plants like *Rhizophora,* in and around Ratnagiri. More concerted efforts are needed in this line, in other areas as well.

4. Evaluation of environmental impact due to industrial discharge, agriculture and aquaculture farms and mining activities along the coast is required to be undertaken. Further in order to control pollution mandatory pretreatment of effluents, pollutant dispersal models, and sitespecific carrying capacity models would be essential.
5. Planning and promotion of ecotourism with proper EIA protocol, based on sitespecific carrying capacity.
6. Awareness of the value of habitat, biodiversity and need of it's conservation involving women,school children, NGO's. It is praiseworthy to note here that voluntary organisations like Sahyadri friends club,Chiplun of the district of Ratnagiri, has initiated turtle conservation measures.
7. Creation of biosafety units in view of the impact of GMO, both domestic and introduced.
8. Formation of separate biodiversity authority at state/ national level for regular monitoring of the programs as well as changes occurring.

## Concluding Remarks

Realizing the social attitude of the local people participatory approach involving them would be most ideal and beneficial. There should be sound co-ordination between different department *viz.* forest, fisheries, industries, environment, tourism and those concerned at administrative level, as well as NGO's of the region like Indian Society of Environmental Science and Technology. Consequently every sincere and constructive effort would ensure the conservation of the coastal biodiversity of this important state.

**Aquatic Biodiversity in India: The Present Scenario, 2005** **339–342**

*Edited by*: D.R. Khanna, A.K. Chopra & G. Prasad

*Published by*: Daya Publishing House, New Delhi

# 19

# Biodiversity: A Present Scenario

**Pramod Kumar Joshi**

*Government P.G. College, Mhow, M.P.*

If we look back to our ancient heritage, we will find that ever since the period of Ramayana and Mahabharat or the time of great Indian physician Dhanvantri, they were quite about the natural wealth of India in terms of biodiversity. For curing Lakshman, Hanumanji; from extreme south went up to north for picking up a particular plant. There is also a myth about Dhanvantri that every plant used to speak before him revealing it's utility for human beings. According to Hindu philosophy which believes in rebirth every soul has to pass 84 millions animal types before birth as a human being. This itself shows the importance of different animal life or importance of biodiversity.

India was probably the first nation on earth to practice land conservation and wildlife management. The late E.P.Gee, one of the India's few noted conservationist found that in the year 300 B.C. government decreed to set aside certain areas where the extraction of timber, burning of charcoal, collection of grass, fuel and leaves, the cutting of cane and bamboo, trapping fur skins and tooth and bone were totally prohibited. Since that time sense of regulations has been forgotten and large part of India have been lost to both men and nature through indiscriminate lumbering and overgrazing leading in turn to desiccation of remaining vegetation.

In present stage of the world when there is general agreement on the conservation of the global atmosphere which should be free from chloro floro compound which is vastly reducing ozone layer, or smog from automobiles and fine carbon particles or hazardous chemicals from factory outlets or poisonous pesticides from fields into our aquatic reservoir. This has drawn the attention of our Scientists towards earth's natural heritage-'biodiversity'.

To study the biodiversity means to study the sum of total life forms at all levels of organization. According to Denny (1997) "Biodiversity includes assemblages of plants and microorganisms their genetic variability expressed and populations, their habits ecosystem and natural areas, the mosaic of which constitutes the landscapes which gives the richness to the natural environment.

Loss of plant and animal variety or their extinction was earlier due to the natural evolution process in that the rate of loss was not much more but now a days the human activities is the main cause for this fast depletion of biodiversity. Increasing human populations and their multifarious need or 'greed' is the main cause of biodepletion.

It is evident that human race of the 20$^{th}$ century will be known as biggest destroyer of the natural environment and biodiversity. Man with his strong desire and sharptools, hazardous chemicals has allowed only species of plants and animals which is directly beneficial to him in terms of many food or medicines. Thus he has destroyed much other variety of animals and plants, which were indirectly useful.

For our planet and its healthy atmosphere biodiversity or the totality of life in forms of animals and plants is of great importance.It is essential for the welfare of all the living beings in various aspects. Its different aspects may be summarized as:

## Ecological Aspects

Biodiversity provides interactive dynamics of the ecosystem of the plant it regulates the gaseous mixture of the atmosphere controlling soils in fertilized state, control pests and maintains bio geo chemical cycles. It maintains proper flourishing atmosphere in the aquatic ecosystem for plants and animals.

## Food and Nutritional Aspects

Biodiversity provides a peaceful coexistence among plants and animals. This enables us to get different variety of food with nutrition value thus serving the human society since thousands of years. It covers from mushrooms to sea algae and variety of aquatic animals amphibians, fishes, reptiles, birds and mammals.

## Economic Aspects

If we think about the most important contribution of biodiversity on earth then definitely it is food. To feed the global population there is a wide range of plants and animals from lower animals to fishes and from amphibians to mammals.

## Medicinal Aspects

Man since his knowledge of medical science has never conquered upon the diseases; what ever the remedial approaches and the drugs have been invented by him is either based on medicinal plants or from animals. In the process of human civilization man has discovered many plants and animals directly or indirectly useful for mankind.

## Aesthetic Aspects

After getting food, medicine, shelter and clothes man has always been fascinated by natural beauty, which is also contributed by our natural biodiversity.

In fresh water aquatic environment there flourishes many microorganisms-phytoplankton and Zooplankton along with different types of plants like *Hydrilla verticillata, Cerato phylum, Demersum cyperus, Marsilea minuta, Vallisneria spiralis, Eichhornia crassipes* and *Polygonum glabrum* among animals are protozoan's, porifers, coelentrates, arthopods, molluscans are the main phyla. These plants and animals constitute natural food chain, thus maintaining ecosystem of biodiversitic discipline.

Today man with his vested interest and heavy tools exploiting animals and plants, polluting our aquatic bodies with hazardous chemical and factory outlets and pesticides runoff from fields causing gradual depletion of plants and animal varieties. This is the sole cause of the present awakening from rescue operation and maintaining our natural wealth.

Fresh water fishes are specialty of the specific area whether they are from tropical or temperate region or the arctic zone. In fishes there are always possibility of natural hybridization hence they need special taxonomic attendance. At present there are 19000 known species of fishes out of which 10 per cent are still unknown. India due to it's diversified geographical position we have a big range of fishes.

In a survey 1974 there was 85 lacs hectare water bodies including 30 lacs hectare ponds, lakes and reservoirs 39 lacs hectares rivers and canals 12 lacs hectares estuarine and 14 lacs hectares of mixed area of wetland. In 1975 in India there were total 2479270 water resources for fish cultivation out of which 51940 in U.P., 32470 in Himachal Pradesh, 272500 in Bihar and 252000 in Madhya Pradesh. Due to increasing civilization there is a gradual reduction in their number and area.

The aquatic biodiversities of fresh water fisheries in Madhya Pradesh and it's adjacent provinces, there are mainly 24 families of 60 genus and approximately 115 species of fish varieties. Some important genera and species of fish varieties are as follows:

*Catla catla, Labeo rohita, L.calbasu, L.fimriatus, Cirrhinus mrigale, Tor tor, Mystus seenghala, M. cavasius, M. vittatus, Wallago attu, Ompok bimaculatus, Clarius mangur, C. batrachus, Heteroprieustes fossilis, Notopterus chitala, N. notopterus, Mastacambelus armatus, Cirrhinus reba, Labeo bata, Oxygaster bucaila, Ophiocephalus punctatus, Chanda nama, C. ranga, Nandus nandus, Anabas testudineus, Rasbora daniconius, Rita rita, Badis badis, Xenodon cancila, Puntius ticto, P. sarna, Amphinous cuchia and Mugil carsula.*

Among exotic fishes-*Ctenopharyngodon idellus, Tilapia mossambica* and *Cyprinus carpio.*

At present it is very important to maintain the fishes of pure strain and there is a worldwide organization to identify each country to maintain and conserve certain fishes. The worldwide awakening and joint approach is the only remedy to conserve our aquatic biodiversity.

**Aquatic Biodiversity in India: The Present Scenario, 2005** **343–349**
*Edited by*: D.R. Khanna, A.K. Chopra & G. Prasad
*Published by*: Daya Publishing House, New Delhi

# 20

# Biodiversity and Conservation of Algae

**Narendra Mohan and Jitendra Mohan**
*Department of Botany, DAV College, Kanpur*

It was estimated that there are about 1,25,000 species of all organism in India and about 4,00,000 would probably discovered in the future (Gadgil,1996). Of the 1,25,000 species, 85,000 are animals. There are about 18,000 species of angiosperms in the country. To this number, we have to add the species of algae, fungi, bryophytes, pteridophytes and gymnosperms, even if we exclude the bacteria. Obviously, the estimates of species number in India are on the far lower side. Thus, we do not have dependable estimates of the number of plant species in India. In addition, we are not certain as to what to conserve. Till all issues related to conservation are resolved, no meaningful conservation action is possible. Khoshoo, 1996 revived the problems and prospects of conservation of biodiversity in India and proposed sitting up of a National Biodiversity Conservation Board, which reflects the fluidity of the situation.

Phycology is the branch of science dealing with the study of algae. The history of phycology is as old as man's interest in botany. References for algae were found in old Chinese literature where there economical value have been maintained. Certain references were also reported in old Roman and Greek literatures. Phycos is a Greek word for algae, Fucus for Roman and Tsao for Chinese.

In the earlier centuries progress was made in the scientific study of algae. In 12$^{th}$ century algae were being used for manufacturing purposes in the north east of France and then from here it spread to Great Britain.

The study of algae, their structure and reproduction is closely connected with the evolution and development of the microscope. Antoni Leeuvenhock reported a number of unicellular algae and flagellates on September 7, 1674 and this is known as date of birth of microbiology.

In India the much work has been done from 1918 onward. Iyengar, M.O.P., 1920 onwards published a series of papers of algae of South India. He is known as "father of modern algology in India".

Over a three and a half billion year history algae have evolved and adopted as primary plant colonizers of almost every known habitat in terrestrial and aquatic environments through out the Biosphere. Microalgae are major contributors to primary production throughout the world, especially when considering marine phytoplankton. The role of microalgae as primary substrate colonizers is scientifically recognized aquatic systems, and, although not well documented in terrestrial ecosystems. Finally, microalgae are also colonizers of harsh ecosystems, such as thermal springs, underground caves, snoe and sea-ice.

It is well known that the Algae have been in use for one purpose or the other in some parts of the word from time immemorial.

Algae are an informative group to monitor as indicator for environmental degradation. Algae are suspended freely in water called phytoplankton. All phytoplankton species are sensitive to environmental changes, but some are more sensitive than others. Changes in the diversity of phytoplankton (algae) species and their relative members can be used to sewage water quality changes. The greater the diversity of phytoplankton, in general, the better in the health of the water body. If your monitoring program record a drop in phytoplankton diversity, their may be indicate water quality is declaring. Some times natural seasonal changes may also cause fluctuations in the members and diversity of algae.

In addition to the prominent algal component, sea–ice biota also include other life forms such as bacteria, colourless flagellates, foraminiferans, ciliates, nematodes, copepods, amphipods, krill and fish (Horner, 1990 a, Horner *et.al.*, 1992). Ice microalgae are mainly dominated by the bacilariophyta (Diatoms), while other groups are also represented such as the chlorophyta, chrysophyta, craspedophyta, cryptophyta, cyanophyta, dictyochophyta, pyrrhophyta, euglenophyta, hectophyta (Prymnesiophyceae) and prasinophyta. At different northern conditions geographical locations, diatoms are the predominating microalgae in bottom ice communities, with a fairly high representation of pinnates forms (fragelariophyceae, bacillariophyceae), of which *Navicula, Nitzschia* and *Pinnularia* represent the main genera in terms of species richness (Poulin, 1990 b).

The monitoring of the biological diversity of sea-ice algal communities can not be done randomly, but must take in to account the fluctuations in the species composition based on seasonality and geography. The various kinds of communities also add to the difficulties in comparing them. The protocol describe below will serve the monitoring of the biodiversity of sea–ice algae, *viz.* the interstitial communities that develop in the bottom 10 cm. of congelation ice.

Knowledge about the types and number of algae can provide valuable insights into water quality and aquatic food web. Like most plants, algae produce oxygen and take up nutrients while they are growing, but when they die and decay they can release nutrients. They therefore have a major influence on the nutrient status of a water body. Further move, the decay process requires oxygen, so if large numbers of algae are decaying they use up oxygen in water and can cause fish and macro invertebrate to die.

Algae are extremely important in an ecological since in any aquatic environment for two reasons. Firstly they are the primary producers of all aquatic food chainse, and therefore all aquatic animals and decomposers are dependent on them. Secondly, in the process of food production through photosynthesis, they use up the dissolved carbondioxide in the water and release oxygen. The photosynthetic process is important for monitoring sufficient oxygen levels in the water to meet the respiration needs of aquatic animals.

The many kinds of phytoplankton are classified into many different classes and families. For our purposes we will group them into four groups, the blue-green algae, the green algae, the diatoms, and the flagellates.

A number of environmental factors influence the growth of phytoplankton in an aquatic environment. Some of the more important ones are nutrient levels, light, water temperature, pH, salinity and turbidity. Nutrient perhaps the most significant and of the nutrients, phosphorus and nitrogen greatly influence algal growth. Many studies have reported a high correlation between algal growth and nutrient availability in the water.

While algal blooms may occur under natural conditions, in the recent years they have increased in frequency and intensity. The reasons appear to include pollution of water ways with nutrients especially through sewage and industrial influents, the reduction in water flow through irrigation, industrial and domestic uses, and degradation of rivers and lake ecosystems.

Blue-green algae in particular are important to monitor because of their potential impact on human and animal health and because high numbers often indicate high nutrient levels in the water. These blue-green algae also called cynobacteria reduce water quality when they are present in large numbers. Some produce toxcins, odours or thick scum on the water surface called water blooms.

The toxin can cause death and illness in stock and gastroenteritis, liver damage and skin and eye erritation in humans. The blue-green algae to particularly watch for are *Microcystis, Anabaena, Oscillatoria, Spirullina, Aphanizomenon* and *Nodularia* species.

Algae, along with other marine organisms, produce an incredible diversity of secondry metabolites. An individual species contents more than 1,000 unique chemical entities (or the enzymatic machinery for producing compounds given the appropriate stimulus). One explanation for the waste chemical diversity to be found in the biological diversity of marine species, including algae, is that marine organism need to develop and survive in an environment of fierce competition for resource and nutrients making it necessary for them to have developed biochemical physiological mechanism that enable them to produce bio active compounds for many different purposes, such as protecting them selves against viral disease, pathogenic fungi and predators and for other functions like reproduction and communication. The biodiversity of marine algae species along with the chemical diversity found in each species, represents a practically unlimited resource that can be used beneficially, with the help of biotechnology, for developing products for agriculture, pharmaceuticals compounds, medical research materials, industrial enzymes, etc.

Nowhere on earth can be found the same wealth of biological diversity as in the oceans. In the biosphere, the most important classes of organism originated in fundamentally marine environments, so the sea is a unique source of genetic information. The oceans, therefore, offer abundant resources for research and development, but this potential remains practically unexplored as a base for new biotechnologies.

The role of marine algae natural products in the discovery of drugs that could rich the pharmaceutical market has increased notably in recent years, due to a substantial improvement in biological screening methods. Thus, natural marine products have mainly been investigated for their anti–micro, cytotoxic, anti–tumour and anti–inflammatory properties. This has now spread to include the search for marine metabolites with immunosupressor activity, and emerging therapeutic area.

Studies of red algae of the plocamium genus collected of the cost of chile and the Antarctic, around the Spanis anarctic base "Juan Carlos I" have led this group to isolate a series of poly halogenated substance (high degree of chlorine and bromine uptake in the molecule), some with anti–microbe properties that are comparable with those of the commercially used antidote erictomachine, used as a refernce in the analysis of biological activities. From a red algae of the Laurencia genus collected from Ester Island, they also obtained halogenated products with insecticides properties.

In culture conditions sewage effluent influenced in the growth and mineral compositions of algae. Increase in sewage effluent increased the bio volume of *Phormidium* and *Anabeana* upto 5 percent, of *Lyngbya* upto 1 per cent and of *Oscillatoria* upto 10 per cent concentration of sewage. Beyond these levels, further increase in sewage concentration decreased the biovolume of respective alga Dubey (2003). Similar results that at low concentration of tannery effluent the growth and biovolume of *Anabana sp.* increased is also observed by Mohan *et al.* (1989). Similarly low concentration of tannery effluent increased the bio volume of *Phormedium* Mohan *et al.* (2002). Mohan *et al.* (2002) also observed that influence of fertilizer effluent increased the bio volume of *Oscillatoria senuis*.

Upto 5 per cent fresh and dry matter yield of *Phormidium* and dry matter yield of *Lyngbya*, upto 10 per cent fresh matter yield of *Anabaena* and fresh matter yield of *Lyngbya* upto 15 per cent fresh matter yield and upto 20 per cent dry matter yield of *Oscillatoria* were observed with the increase in sewage effluent. Beyond these lavel further increase in effluent showed decrease in yield of respective alga Dubey (2003). Mohan *et al.* (1989) also observed that dry matter of yield of *Anabaena* of increased with low concentration of tennery effluent. The use in fresh and dry matter of *Oscillatoria* tenuis with lower concentration of fertilizer effluent was also observed by Mohan *et al.* (2002). Similarly Mohan *et al.* (2002) repeated increase in fresh and dry matter yield of *Phormedium* sp. by tannery effluent.

The mineral composition also increased with the increase in sewage concentration upto 1 per cent for phosphorus, sulphur, nitrogen and upto 5 per cent of calcium, potassium, iron and manganese, control of alga *Phormidium*. Beyond these revels further increase in concentration of sewage decreased the respective mineral elements in alga *Phormidium*. Mineral composition of *Oscillatoria* tennis shown some what similar results with lower concentration of fertilizer effluent was observed by Mohan *et al.* (2002).

Research in the XXI century for marine organism can provide the foundation for developing biosensors, bioindicators and diagnosis devices for medicine, aquiculture and environmental monitoring. One promising monitor is the gene probe, to be used to identify organism that can affect health, or that can be useful for research. Specific gene probes will be useful, for instance, for detecting human pathogens in food produce that has come from the sea, rational waters, pathogenic fish in farming systems, etc. More studies should be carried out in to marine ecological systems, in order to determine what would be the base line of normal function, and also to monitor ecosystem to be able to predict potential changes and disturbance due to physical, chemical or biological impacts.

**Aquatic Biodiversity in India: The Present Scenario, 2005** **350–353**
*Edited by*: D.R. Khanna, A.K. Chopra & G. Prasad
*Published by*: Daya Publishing House, New Delhi

# 21

# Similarity Index of Phytoplankton Species of River Kunda at Khargone, Madhya Pradesh

**S.K. Mahajan**
*Botany Department, Government Post Graduate College, Khargone – 451 001, M.P.*

## Introduction

For the sake of comparison of species diversities between different ecosystems in various climatic conditions, a useful and convenient method is to calculate an index for diversity and dominance. Czechanovski (1913) had developed an index which is popularly called similarity index and used to compare the species diversity of two aquatic ecosystems. Besides it is also convenient to determine the species diversity of an unpolluted site with that of polluted one. Earlier, Kar *et al.* (1987) studied the species diversity of the two sites of Ib river situated in Sambalpur district of Orissa. So far no such work has been done in Madhya Pradesh and therefore an attempt was made to study the species diversity of upstream and downstream waters of river Kunda, a tributary of Narmada at Khargone, Madhya Pradesh.

## Materials and Methods

Water samples were collected from upstream (near Dargah) and downstream (neart Meldeleshwar) of river Kunda during 2001-2002 and algal taxa identified following standard literature (Desikachary, 1959; Randhawa, 1959 and Philipose, 1967). Similarity index was calculated by applying Czechanovski's (1913) formula as follows:

$$S = 100 \times 2c/a + b$$

Where a is the number of species one site, b is the number of species in another site, c is number of species common to both sites and S is Czechanovski's similarity index.

The results are shown in Table 21.1. Enumeration of algal taxa is done classwise. Euglenoids and few microscopic invertebrates were also reported but they have not been considered in the present study. Physico-chemical parameters of water samples collected from upstream and downstream River Kunda at Khargone are shown in Table 21.2.

## Results and Discussion

In all 32 algal taxa were observed, out of which 14 taxa were from upstream and 24 taxa from downstream samples. The similarity indices for chlorophyceae, Cyanophyceae and Bacillariophyceae were as 0.33, 0.02 and 0.14 respectively. It is noted that one or more species can be considered as water pollution indicator. If the river had not been polluted the overall similarity indices would have been close to one. Pollution sensitive algae like *Schizomeris* and *Staurastrum* were not found in polluted downstream samples, while pollution indicator forms like *Oscillatoria* abounded the site.

## Acknowledgements

The author is thankful to the principal. Govt. P.G. College, Khargone for kind help.

**Table 21.1: Similarity Index of Phytoplankton Species of River Kunda at Khargone, M.P.**

| *Sl.No.* | *Name of Taxa* | *Upstream (Unpolluted)* | *Downstream (Unpolluted)* |
|---|---|---|---|
| | **Chlorophyceae (0.33)** | | |
| 1. | *Chlorella* sp. | + | + |
| 2. | *Closterium cyclicum* | – | + |
| 3. | *C. cynthea* | – | + |
| 4. | *Cosmarium* sp. | + | + |
| 5. | *Mougeotia abnormis* | + | + |
| 6. | *Scenedesmus armatus* | + | + |
| 7. | *S. brasiliensis* | – | + |
| 8. | *S. denticulatus* | – | + |
| 9. | *Schizomeris* sp. | + | – |
| 10. | *Spirogyra condensate* | + | + |
| 11. | *S. sinensis* | + | + |
| 12. | *Staurastrum* sp. | + | – |
| 13. | *Zygnema himalayansis* | + | + |
| | **Cyanophyceae (0.02)** | | |
| 1. | *Anacystis nidulens* | + | – |
| 2. | *Aphenocapsa muscicola* | – | + |
| 3. | *Gomphosphaeria* sp. | + | – |
| 4. | *Johanbaptista* sp. | – | + |
| 5. | *Nostoc* sp. | – | + |
| 6. | *Pscillatoria princeps* | + | + |
| 7. | *O. sancta* | + | – |
| 8. | *Spirulina* (=*Arthrospira*) sp. | – | + |
| | **Bacillariophyceae (0.14)** | | |
| 1. | *Cymbella affinis* | – | + |
| 2. | *Fragillaria rumpens* | + | + |
| 3. | *F. rumpens var. femiliaris* | – | + |
| 4. | *Gomphonema lanceolatum* | – | + |
| 5. | *G. sphaeroperum* | + | + |
| 6. | *Navicula cuspidate* | + | – |
| 7. | *N. pupula* | + | + |
| 8. | *N. reinhardii* | – | + |
| 9. | *Nitzschia* sp. | + | – |
| 10. | *Pinnularia brevicostata* | – | + |

+ = Absent; – = Present; Figures in paranthesis are similarity index.

**Table 21.2. Physico-chemical Parameters of Water Samples Collected from Upstream and Downstream of River Kunda at Khargone, Madhya Pradesh**

| *Sl.No.* | *Parameters* | *Upstream* | *Downstream* |
|---|---|---|---|
| 1. | pH | 7.8 | 8.63 |
| 2. | Total Hardness (mg/L) | 251 | 290 |
| 3. | Dissolved $O_2$ (mg/L) | 4.1 | 4.5 |
| 4. | Total dissolved solids (mg/L) | 55 | 70 |
| 5. | Chlorides (mg/L) | 126.5 | 139.9 |
| 6. | Phosphates (mg/L) | 25.2 | 0.15 |
| 7. | Calcium (mg/L) | 25.2 | 40.08 |
| 8. | Magnesium (mg/L) | 51.2 | 46.02 |

## References

Czechanovski (1913). (Quoted by Kar G. K., P. C. Mishra, M. C. Dash and R. C. Das, 1987) in "Pollution studies in River Ib III. Plankton population and primary productivity". *Indian J. Environ. Health*, 29 (4), 322-329.

Desikachary, T. V. (1959). Cyanophyta, I. C. A. R., New Delhi. P. 700.

Philipose, M. T. (1967). Chlorococcales, pub. ICAR New Delhi.

Randhawa, M. S. (1959). Zygnemaceae, Pub. ICAR, New Delhi.

**Aquatic Biodiversity in India: The Present Scenario, 2005** **354–358**
*Edited by*: D.R. Khanna, A.K. Chopra & G. Prasad
*Published by*: Daya Publishing House, New Delhi

# 22

# Limnological Assessment of Phytoplankton in a Reservoir of Khargone, Madhya Pradesh

**S.K. Mahajan**
*Botany Department, Government Post Graduate College, Khargone – 451 001, M.P.*

## Introduction

Water is a major element of all the components of biosphere and one of the most leading factors in existence of living organisms. Water is used in many ways like direct consumption by human beings, cattle etc. for domestic purposes, agriculture, industry, energy generation etc. The diverse uses of fresh water are based on its unique physico-chemical and biological properties which also render it unfit for one or several uses even after a minor change. The pollution of water bodies invariably affects the biota, fauna and flora, particularly the planktonic forms. A large number of studies were carried out in recent years to emphasize the importance of planktonic algae especially diatoms in the assessment of water quality (Palmer, 1969; Kamat, 1981; Somashekar and Ramaswamy, 1984) but so far no aquatic species have been found which can be regarded as parallel to the best known indicator plants, the terrestrial species like the 'copper mosses ' (Brooks, 1971) which are mainly restricted to such environments where a particular metal is present at high level. In the present investigation an attempt has been made to assess the feasibility of application of Nygaard 's phytoplanktonic quotient to evaluate and monitor the water quality and pollution level of Virla reservoir of Khargone, M.P.

## Description of the Study Area

Virla reservoir (21°50'N Lat., 75°23' E Long.,309 m above MSL) of is 25 km away from Khargone city on Julwania road. This reservoir was constructed during state time before 1947. Its catchment area is 50 sq.km and nearly 445 hactares of land is irrigated with the help of this reservoir.

## Materials and Methods

Water samples and algal material were collected in the month of May, 2001. Standard methods were followed during the collection and analysis of water samples (Golterman and Clymo, 1969 and APHA, 1980). The enumeration of algal taxa was done after careful identification with the help of standard literature (Desikachary, 1959 Randhawa, 1959 and Philipose, 1967). In order to assess the water quality Nygaard's (1949) Trophic State of Indices and Nygaard's (1976) modified compound quotient were applied. Results are shown in Tables 22.1–22.3.

**Table 22.1: Physico-chemical Data of Water Sample Collected from Virla Reservoir in May, 2001**

| *Sl.No.* | *Characteristic* | *Results* |
|---|---|---|
| 1. | Temperature | 36ºC |
| 2. | pH | 8.6 |
| 3. | Conductivity | 6.46 µmhos |
| 4. | Total hardness | 185 mg/L |
| 5. | Total alkalinity | 21 mg/L |
| 6. | Calcium | 31.86 mg/L |
| 7. | D.O. | 5.1 mg/L |
| 8. | B.O.D. | 3.3 mg/L |
| 9. | Chloride | 38.83 mg/L |

**Table 22.2: Algal Taxa Reported from Virla Reservoir of Khargone**

| *Sl.No.* | *Algal Group* | *Sl.No.* | *Algal Taxa Reported* |
|---|---|---|---|
| 1. | Myxophyceae | 1. | *Anabaena fertilissima* |
| | | 2. | Aphanocapsa sp. |
| | | 3. | *Aphanothece sp.* |
| | | 4. | *Aulosira fertilissima* |
| | | 5. | *Chamaesiphon rostaffinskii* |
| | | 6. | *Chroococcus minor* |
| | | 7. | *Cylindrospermum* sp. |
| | | 8. | *Dichothrix fusca* |
| | | 9. | *Gomphosphaeria* |
| | | 10. | *Merismopedia convoluta* |
| | | 11. | *Nostoc commune* |
| | | 12. | *Oscillatoria princeps* |
| | | 13. | *O. vizagapatensis* |
| | | 14. | *Phormidium* sp. |
| 2. | Bacillariophyceae | 1. | *Cymbella affinis* |
| | | 2. | *Fragilaria rumpens var. familiaris* |
| | | 3. | *Gomphonema sphaerophorum* |
| | | 4. | *Navicula mutica* |
| | | 5. | *N. rediosa* |
| | | 6. | *Nitzschia* sp. |
| | | 7. | *Pinnularia* sp. |
| | | 8. | *Pleurosigma spenceri* |
| | | 9. | *Stauroneis* sp. |
| 3. | Chlorophyceae | 1. | *Ankistrodesmus falcatus Var. acicularis* |
| | | 2. | *Chlorella* sp. |
| | | 3. | *Mougeotia* sp. |
| | | 4. | *Scenedesmus armatus var. exaculeatus* |
| | | 5. | *S. denticulatus* |
| | | 6. | *S. quadrifida* |
| | | 7. | *Spirogyra condensata* |
| | | 8. | *S. sinesis* |
| | | 9. | *Tetraedron trilobulatum* |
| 4. | Euglenophyceae | 1. | *Euglena* sp. |
| | | 2. | *Phacus* sp. |
| 5. | Chrysophyceae | 1. | *Synura* sp. |

**Table 22.3: Percentage Composition of Algal Groups at Virla Reservoir of Khargone (M.P.)**

| *Algal Groups* | *Percentage Composition* |
|---|---|
| Myxophyceae | 42.05 |
| Bacillariophyceae | 25.74 |
| Chlorophyceae | 25.71 |
| Euglenophyceae | 05.71 |
| Chrysophyceae | 02.85 |

## Results and Discussion

Table 22.1 revealed that the physico-chenical parameters are below the pollution level as per International standards. Table 22.2 revealed that out of total of 35 algal taxa, 14 members belong to blue greens, 9 to diatoms,9 to green algae, 2 to euglenoids and 1 member to Chrysophyceae. Besides, 2 microscopic invertebrates *i.e.* Cyclops and Rhabditis have also been reported. Table 22.3 revealed highest percentage of Myxophyceae (42.05) and lowest percentage of Chrysophyceae (2.85). Nygaard's (1949) Trophic State of Indices for Myxophyceae, chlorophyceae, Diatoms, Euglenophyceae were found to be 14.0, 6.0, 0.0, 0.10 respectively. The value of compound quotient was found to be 22.0 while the value of Nygaard's (1976) modified compound quotient was calculated as 31.0 indicating that Virla reservoir dam is under the process of high eutrophication. This is due to the abundance of green algae, blue greens and diatoms in comparison to negligible number of desmids. Aquatic animals have less preference to use green algae as food. Pollution of water due to various agencies is one of the important causes to increase the number of blue greens. These conditions promote eutrophication of water at a rapid rate due to which water is not suitable for drinking purposes but it helps in fish production to a certain extent.

## Acknowledgements

Author is thankful to the Principal, Govt. P.G. College, Khargone for facilities.

## References

APHA, AWWA and WPCF (1980). Standard methods for the examination of water and waste water, 15th ed. *Amer. Pub. Health. Assoc. Washington* DC, 113 pp.

Brooks, R. R.(1971). Bryophytes as a guide to mineralisation. Newzealand *J. Bot*. 9, 674-977.

Desikachary, T. V. (1959). Cyanophyta, Pub I. C. A. R., New Delhi.

Golterman, H. H. and R. S. Clymo (1969). Methods for the chemical analysis of fresh water B. P. Handbook no. 8, Blackwell, Sci. Pub. Oxford, 166 pp.

Kamat, N. D. (1981). Diatoms and diatom populations indicating water quality and pollution. WHO workshop on biological indicators and indices of environmental pollution. Cent. Bp. Prev. cont. Water Poll. And Osmania University, Hyderabad, 77-83.

Nygaard, G. (1949). Hydrobiological studies on some Danish ponds and lakes, Part-II. The quotient hypothesis and some new or little known phytoplankton organisms. K. denske. Vidensk, Selk. Skr. Biol 1: 1-293.

Palmer, C. M. (1969). A composite rating of algae tolerating organic pollution. J. Phycol. 5: 78-82.

Philipose, M. T. (1967). Chlorococcales: ICAR, New Delhi.

Philipose, M. T. (1967). Chlorococcales, ICAR, New-Delhi.

Randhawa, M. S. (1959). Zygnemaceae, ICAR, New Delhi.

Somashekar, R. K. and S. N. Ramaswamy (1984). Biological assessment of water pollution: a study of the river Kapila. Intern. J. Environ. Studies, 23: 261-267.

# Index


**A**

*Acanthus ilicifolius*, 105
*A. marina*, 105
*A. officinalis*, 105
*Acrostichum aureim*, 105
*Aegiceras comiculatum*, 105
Agarwal, N.K., 81
Algae, 343-349
Algal communities, 107-118
    bacillariophyceae, 117
    chlorophyceae, 116-117
    euglenophyceae, 117
    fresh water habitats, 107
    lake Kinneret, 108
    limnological studies, 109
    myxophyceae, 111-116
    periodicity, 113-115
    physico-chemical parameters, 112
    phytoplankton species, 108
    rainfall data of Bidar, 110
Algal taxa, 111
American endangered fishes, 8
*Anabas testiduents*, 76
Angadi, S.B., 107
Ansari, S.A., 184
*Apolochilus punchay*, 76
Aquaculture plan, 98
Aquarium, 39
Aquatic fungi, 194
– weeds, 154-174
*Astronotus ocellatus*, 57
*Avicennia alba* (Blume), 105
Awasthi, Ashok, 328

Awasthi, U., 328
*Azotobacter*, 22

**B**

*B. parviflora* (Raxb), 104
Badola, S.P., 146
Badola, Smita, 136
*Barilius*, 76, 90
*Betta splendens*, 57
Bhatt, Ira, 243
Bhimtal lake in Kumaon, 285-294
  benthic biota, 286, 292
  characteristics, 287
  chrysphyceae, 288
  fish catch practice, 286
  macrobenthic taxa, 292-
  macrobenthos, 286
  materials and methods, 286-287
  percentage of area, 288
  physical geographical and morphometric, 287
  phytoplankton, 286, 289
  species of fish recorded, 293
  zooplankton species, 290-291
  zooplankton, 286, 289
Bhutiani, Rakesh, 184, 297
Biodiversity, 339-342
  aesthetic aspects, 341-342
  ecological aspects, 340-341
  economic aspects, 341
  food and butritional aspects, 341
  medicinal aspects, 341
Biotic communities, 184-200
  acidophiles and alkalophiles, 199
  atmospheric water, 186
  barophiles, 199
  benthos, 193
  biodiversity in fresh water communities, 190-191
  – – marine communities, 195-197
  – of microorganisms in marine habitat, 197-200
  biota of limnetic zone, 193-194
  – – littoral zone, 193
  biotic communities of continental shelf, 196-197
  – – – coral reefs, 197
  – – – oceanic region, 195-196
  carbon dioxide, 189
  distribution of microorganisms, 187
  hydrogen ion concentration (pH), 190
  lentic ecosystems, 190-191
  light, 189
  nekton, 193
  neuston, 192
  oxygen, 189
  plankton, 192
  pressure, density and buoyancy, 188
  prokaryotic genera, 194
  psychrophiles, 198-199
  rapidly flowing water, 195
  salinity, 189-190
  slowly flowing water, 195
  stored water, 187

structure of aquatic ecosystems, 187-189
surface water, 186-187
temperature, 188-189
thermophiles, 200
types of water and distribution of microbes, 185-186
water resources in India, 185
Bohra, Chandan, 202
*Botia*, 90
*Bruguiera gymmnorrhiza* (L) Lank, 104
*Busycon canaliculatum*, 141
*Busycon*, 141

C

*Carassius auratus* (gold fish), 11
*Catla*, 76
Central Pollution Control Board (CPCB, 2001), 168
*Ceratophyllum demersum*, 167
*Cerethrium* species, 128
*Ceriops tugal*, 104
*Channa punctatus*, 76
Chauhan, S., 184
Chelonian fauna of Mizoram, 68, 69
*Chlorella* carpio nudus (leather carp), 11
*Chlorella fusca*, 116
*C. tumidium*, 116
*C. vulgaris*, 116
*Cirrhinus*, 76
*Closterium acerosum*, 116
CMFRI, Cochin, 4
Coastal biodiversity of Maharashtra, 332-338
Code of Conduct for responsible Fisheries (CCRF), 9
Convention on Biological Diversity (CBD), 3, 5
*Criceps decandra* (Griff), 104
*Crossochelius*, 90
Ctenopharyngodon ideall (grass carp), 11
*Cyanophylctis*, 77
*Cylindrocapsa geminella*, 116
*Cymbella lanceolata*, 310
Cyprinus carpio var. specularis (Miror carp), 11

D

*Dactylococcopsis*, 116
*Daneo rario*, 76
Dayal, R., 1
*Derris indica*, 105
*D. scandens*, 105
*D. trifoliate* (Lour), 105
Dexter Endangered Fish Farm, 8
Dissolved Inorganic Nitrogen (DIN), 20
Dissolved Organic Nitrogen (DON), 22
Diversity of fisheries resources of India, 3
*Domax*, 142

E

Ecological genomics, 18
Effluent of pulp and paper industry, 298
*Elodea canadensis*, 167
Endangered stocks, 8
*Ensis*, 142
Eukaryotic phytoplankton, 20
Eutrophication Gujarat, 243-275

Anand district, 245
bacillariophyceae, 261, 270-271
calcium, 267
carbonate and bi-carbonate, 264-265
chloride, 267
chlorophyceae, 256-257, 261, 269
chlorophyll content, 247, 261, 271-272
cyanophyceae, 254-260, 268-269
dissolved oxygen, 265
diversity index, 261-263, 271
euglenophyceae, 261, 270
eutrophic water, 244
eutrophication, 243, 244
free-carbon dioxide, 265
freshwater ecosystems, 244
hardness, 267
macrophytes, 247-248, 272-275
magnesium, 267
material and methods, 246-247
mesotrophic freshwater ecosystems, 244
nitrate, 265
pH, 264
phosphate, 265-266
physico-chemical characterization of water, 263
phytoplankton population, 254, 258-259, 268
pigment content of plankters, 263
Rateshwar pond, 249, 250, 251, 252
Ratheshwar wetland, 245, 346
sediment composition, 255, 268
silica, 266-267
sulphate, 266
temperature, 263-264
total alkalinity, 264
– solids, 264
wetlands, 243, 244

*Exoecaria agallocha*, 105

**F**

Finfishes, 4
Fish Biodiversity of India, 4
Fish fauna, 70, 146-151
material and methods, 147
physiogeographical features, 147
pisciculture practices, 148
recommendations for improvement of fisheries, 147-151
Fish genetic, 1-13
aquatic habitat potential, 3
cataloguing of fish genetic resources, 3-6
conservation of aquatic biodiversity, 12-13
conservation status, 6-7
creating fish sanctuaries and protected area, 10
habitat assessment and life history studies, 7
management of water resources on community basis, 9-10

potential species for breeding, 11-12
regulation of exotic species introduction, 10-11
restoration of endangered fish species, 7-8
sustainable fisheries resources management, 8-9

Fishes
species diversity of Indian fishes, 6

*Fragilaria*, 117
Francis day in 1889, 3-4
*Frustulia rhomboidies*, 117

G

Gagroo, S, 304
Gangori hatchery, 91
Garhwal Himalayan region, 81-95
capture fisheries, 88-91
fish biodiversity, 87-88
– seed resources, 91-92
fishery status, 88-90
Gangori hatchery, 91
Garhwal Himalayan lakes, 83-84
hatcheries, 91
Kaldyani hatchery, 91
lacustrine resources, 86-87
Mahseer hatchery, 92
miscellaneous fishery, 90
ponds, 87
potential of hill fisheries, 92
reservoir resources, 87
riverine resources, 86
snow trout fishery, 89
sport fishery, 91
strategies for development of hill fisheries, 93-95
Talwari hatchery, 91-92
trout fishery, 90
valley hydro-electric projects, 85
Varangana hatchery, 92

Gene Pool, 102
Genetical, 8
Genetically modified organisms (GMO), 12
Geographical information system (GIS), 7
*Gleocapsa*, 111
*Glyptothorax* sp., 58
*Glyptothorax*, 90
Gold fish, 11
Groundwater, 187

H

Halophiles, 199
Hatcheries, 91
Kaldyani, 91
Mahseer, 92
Talwari, 91-92
Varangana, 92

*Hedophyllum sessile*, 142
Herbivores, 20
*Hetropneustus fossilis*, 76
Hydrological cycle, 186

I

Indian Fishers Act of 1879, 2
– fishs, 3
Industrial development, 298
– washes, 298
International and Biological Program (IBP), 100

Intertidal region and biodiversity, 123-143
adaptation of cover all parameters, 129
– – strong wave action, 129-130
adaptations of intertidal organisms, 127-143
biology of intertidal region, 125-126
chiton trochids, 138
chitons and gastropods, 137
echinoderms, 136
fluctuations in salinity, 126-127
important of world ocean-conditions, 125
insulation, temperature and dessication, 127
littoral system, 135
marine farming, 124
molluscs, 137
oxygen stress, 127-128
patella, 138
physical factor influencing the interdial region, 126-127
sediments, 124
source of energy, 124
stephenson term, 135
tides, 130-143
tolerance to fluctuation in salinity, 129
waves, 126
Into-Tibbetan, 81

J

Jharkhand, 202-230
aquatic insects, 207
benthic animals, 203
– communities, 204
– macroinvertebrates, 205, 217
benthos of reservoir, 203
endobenthos, 204
epifauna, 203
gastropoda, 210, 223
haptobenthos, 204
infauna, 203
insecta, 210, 222-223
littoral zone, 203
macrobenthic fauna, 224-230
macrozoobenthic invertebrates, 210
macrozoobenthos, 210-215
oligochaeta, 210, 218-221
pelecypoda, 210, 223-224
psammon, 204
rhizobenthos, 204
riverine macroinvertebrates, 207
seasonal fluctuations, 217
shore fauna, 203
Joshi, Pramod Kumar, 339

K

*Kandelia candel*, 104
Kapoor, D., 1
Khanna, D.R., 81, 146
*Kirchnerilla lunaris*, 116
Konkan, 98-105
functions of wetlands, 101-102
future development and suggestions for wetlands in 21st century, 102-103

growth and yield of major mangroves, 104
hypersaline condition, 104
natural regeneration status, 104
nursery development techniques, 104
pathological investigation, 104
problems of South-Konkan wetlands, 103
qualitative status of mangroves in South Konkan, 104-105
regeneration of dominant species, 104
reproduction techniques, 103
scope's wetland types, 100-101
wetlands contribute, 102
wetlands ecosystem, 99-100
Kulkarni, K.N., 332
Kumaon hills, 146
– region, 285
Kumar, A., 154
Kumar, Anil, 297
Kumar, Arvind, 202
Kumar, N., 243
Kumar, Nirmal, 243

**L**

*Labeo*, 76, 90
*L. dero*, 11
*L. dyocheilus*, 11
*L. dyochelius*, 11
Lake diversity index, 313
Lakes of Kashmir, 310
*Laminaria setchelli*, 142
*Lemna minor*, 167
Lentic ecosystems, 191-193
*Limnocharis*, 77
*Limnodrilus*, 218
Limnological assessment, 354-357
*Limnonectes*, 77
Lotic ecosystems, 195
*Lumnitzera racmosa*, 105

**M**

Macro invertebrates in Bihar river, 328-331
discussion, 329-331
materials and methods, 329
observation, 329
Macroinvertebrate texa, 67
Mahajan, S.K., 350, 354
Maharashtra
avifauna, 335
coastal habitat features, 332-334
concluding remarks, 338
conservation measures, 336-338
faunal composition, 334
fish, 334-335
floral components, 334-338
human roles, 335-336
mangrove, 334
marine algae, 334
turtles, 335
Mahseer, 89
Malik, Davendra S., 285
Manasbal lake, 307
Mangrove forest, 98-99, 103
Mangrove timbers, 102
Marine bacteria, 18
– coral reserves, 10

Marine Product Export Development Agency (MPEDA), 39

Marine protected areas (MPAs), 10

Mixed tide, 132

Mizoram, 61-79

amphibian fauna of Mizoram, 78

Aquatic fauna of, 61-79

biodiversity of amphibians, 76-77

-- chelonian fauna, 67-69

-- fishes, 69-76

-- lower invertebrates, 66-67

conservation measures, 77-78

ecosystem diversity, 62-66

fish fauna in rivers of, 70

-- of Karnafuli (Tlabung) river, 75-76

--- Kolodyne river, 74

--- Mat river, 73

--- River Tuirini (Seruli) river, 72

--- Tlawng, 72

--- Tuirial river, 71

--- Tuivai river, 72

genetic diversity, 62

species diversity, 62

threat to biodiversity, 78-79

World Commission on Environment and Development (WCED), 61

*Modulus demises*, 128

*Mogeotia scalaris*, 116

Mohan, Jitendra, 343

Mohan, Narendra, 343

Molluscs of nudribranchi, 127

*Mugil*, 101

Munshi pond, 228, 229

Mushi pond, 226

*Mytilus*, 142

N

National Agricultural Technology Project (NATP), 8

National Bureau of Fish Genetic Resources (NBFGR), Lucknow, 2-8

National Water Resource Council (NWRC), 13

*Navicula rhomboidies*, 117

NGOs, 13

Nitrogen cycling in the marine environment, 18-30

*Nitrosococcus oceani*, 24

*N. europaea*, 28

*Noemacheilus*, 90

North-east India, 64

O

*O. irrigua*, 111

*Occnebra erinacea*, 137

*Oedogonium desikarcharyii*, 116

*O. hatei*, 116

Origin of tides, 131

Ornamental fish, 28-59

breeding of ornamental fishes, 54-58

common egg layer fishes, 42

- livebearer fishes, 40-43

- marine ornamental fishes, 51-53

culture of ornamental fishes, 53-54
egg buriers, 43
- depositors, 42-43
- scatters, 42
export of ornamental fishes, 58-59
fresh water ornamental fishes, 40, 44-50
guppy, 40-41
history of ornamental fish culture, 39
Indian wild caught fish, 43-50
indigenous colorful, 38
marine ornamental fishes, 51-53
marketing of accessories, 54
mollies, 43
mouth brooders, 43
nest builders, 43
ornamental fishes of India, 40
platy, 42
scope for entrepreneurs in ornamental fish trade, 53-59
swordtail, 41-42
trade, 38
world trade of ornamental fishes, 39
Oscar, 57
*Oscillatoria limosa*, 111
Oxidation reduction reactions, 19

**P**

Pandey, P.K., 18
Paper mill's effluent, 302
Papnash Pond Bidar Karnataka, 107
*Patella aspera*, 138
Patil, P.C., 107
Penaeid species, 99
Phytobenthos, 204
Phytoplankton assemblage, 304-32
Anchar, 309
bacillariophyceae, 310
biological characteristics, 310-323
chlorococcales, 320
cyanophyceae, 317
Khushalsar, 309, 313
Malpursar, 309
Manasbal, 309
pH, 308
physico-chemical characteristics, 305-309
rawson's simple quotient index, 320
secchi variation in different lakes, 308
similarity index in different lakes, 312
valley lakes, 309
water-bodies, 304
Phytoplankton species, 246, 310
of river Kunda, 350-353
Pilot Analysis of Global Ecosystems (PAGE), 2
*Pinnularia viridis*, 117
*Planctomyces*, 28
*Pseudoecheneis*, 90
*Pterorana khare*, 77
Pulp and paper mill's effluent, 297-302
bacteria, 301

effluent, 300
fungi, 301
microbial diversity, 301
pollution of natural water bodies, 298
soil microbial population, 300
value of pH, 299
*Puntius*, 76, 90
Purushothaman, C.S., 18

**R**

Rai, N., 184
Rainbow trout, 11
Rajput, Rajeev, 297
Ramanujam, S.N., 61
*Rasbora daniconius*, 76
Rateshwar pond, 249, 250, 251, 252
Rawat, U.S., 81
Remediation of soil, 167
*Rhizophora apiculate*, 104
*R. muconata*, 104
*Rhizosolenia*, 21
Rita, J.I., 243
River Bihar, 328
– Gola, 297, 298
– Kolodyn, 74
– Kosi, 7
– Ladhiya, 7
– Mahi, 245
– Mat, 73
– Tuirial, 71
– Tuirini (Seruli), 72
Ruhela, Mukesh, 184, 297

**S**

*Salmo truto fario*, 11
*Salvadora apetala*, 105
*S. casedaris*, 105
*S. persica*, 105
Sarkar, U.K., 1
Saxena, Amita, 123
*Scendesums aequinoctialis*, 116
*S. incrassatulus*, 116
*S. quadricauda*, 311
*Schizothorax* sp., 11
*S. gairdnery gairdneri*, 11
*Scylla serrate*, 101
Sea water, 123
Shiddamallayya, N., 107
Siamese fighter, 57
*Siliqua*, 142
Singhal, V., 154, 297
Sinha, Archana, 38
*Sonneratia alba*, 105
*Sphagnum* species, 100
*Spirogyra condensata*, 116
*Spisula*, 142
*Stauroneies anceps*, 117
Stream of Canada, 208
Sucker fish, 58
Sustainable development, 99
Sustainable Livelihood Approach Main and Coutts, 9
*Synedra ulna*, 117

**T**

*Thais lapillus*, 142
*Thais*, 142
Thapliyal, B.L., 81
Threatened fishes of India, 7
Tiger barb (*Puntius tetrazona*), 58
*Tivela*, 142
*Tor khudree*, 11

*T. putitora*, 11, 89

*Trichodesmium*, 22, 23

**U**

*Uca Pugnax*, 128

Udham Singh Nagar, 146

*Ulothrix cylindricum*, 116

*U. zonata*, 116

*Urosalpinx*, 137

Uttaranchal, 82

**V**

Valley lakes, 313

*Volvox aureus*, 116

**W**

Wanganeo, A, 304

Wanganeo, R., 304

Water bodies of Mizoram, 67

Water hyacinth aquatic weeds, 154-174

aerated aerobic water hyacinth systems, 161

duckweed systems, 161

factors responsible for phytoremediation of contaminants, 162

facultative anaerobic water hyacinth systems, 161

floating aquatic plant treatment systems, 160

mechanisms responsible for phytoremediation, 162

metal sequestration studies performed by various workers, 159

need for effluent treatment, 155-156

non-aerated aerobic water hyacinth systems, 161

on-site treatment, 168

phytodegradation, 157

phytoextraction, 157

phytoremediated plant biomass a source of energy, 170-174

phytoremediation, 156-157, 168

phytostabilization, 158

phytovolatilization, 158

research work on other aquatic weeds, 166-170

rhizofiltration, 157

selection of plant species, 158-160

use of plants to remove pollutants from air, 158

water hyacinth as phytoremediator, 162-166

water hyacinth systems, 161

Water pollution, 154

Watts Bar Reservoir, 206

Wetland of Indian continent, 99

Wetland taxonomy, 103

Wetlands, 103

World Resource Institute, USA, 2

**Y**

Yeragi, S.G., 98

Yeragi, S.S., 98

Yousuf, A.R., 304

**Z**

Zoobenthos, 202, 204

Zooplankton, 67

Zooplanktonic population, 289